Selected Titles in This Series

43 **Panos Pardalos, Sanguthevar Rajasekaran, and José Rolim, Editors,** Randomization Methods in Algorithm Design

42 **Ding-Zhu Du and Frank K. Hwang, Editors,** Advances in Switching Networks

41 **David Aldous and James Propp, Editors,** Microsurveys in Discrete Probability

40 **Panos M. Pardalos and Dingzhu Du, Editors,** Network Design: Connectivity and Facilities Location

39 **Paul W. Beame and Samuel R Buss, Editors,** Proof Complexity and Feasible Arithmetics

38 **Rebecca N. Wright and Peter G. Neumann, Editors,** Network Threats

37 **Boris Mirkin, F. R. McMorris, Fred S. Roberts, and Andrey Rzhetsky, Editors,** Mathematical Hierarchies and Biology

36 **Joseph G. Rosenstein, Deborah S. Franzblau, and Fred S. Roberts, Editors,** Discrete Mathematics in the Schools

35 **Dingzhu Du, Jun Gu, and Panos M. Pardalos, Editors,** Satisfiability Problem: Theory and Applications

34 **Nathaniel Dean, Editor,** African Americans in Mathematics

33 **Ravi B. Boppana and James F. Lynch, Editors,** Logic and Random Structures

32 **Jean-Charles Grégoire, Gerard J. Holzmann, and Doron A. Peled, Editors,** The SPIN Verification System

31 **Neil Immerman and Phokion G. Kolaitis, Editors,** Descriptive Complexity and Finite Models

30 **Sandeep N. Bhatt, Editor,** Parallel Algorithms: Third DIMACS Implementation Challenge

29 **Doron A. Peled, Vaughan R. Pratt, and Gerard J. Holzmann, Editors,** Partial Order Methods in Verification

28 **Larry Finkelstein and William M. Kantor, Editors,** Groups and Computation II

27 **Richard J. Lipton and Eric B. Baum, Editors,** DNA Based Computers

26 **David S. Johnson and Michael A. Trick, Editors,** Cliques, Coloring, and Satisfiability: Second DIMACS Implementation Challenge

25 **Gilbert Baumslag, David Epstein, Robert Gilman, Hamish Short, and Charles Sims, Editors,** Geometric and Computational Perspectives on Infinite Groups

24 **Louis J. Billera, Curtis Greene, Rodica Simion, and Richard P. Stanley, Editors,** Formal Power Series and Algebraic Combinatorics/Séries Formelles et Combinatoire Algébrique, 1994

23 **Panos M. Pardalos, David I. Shalloway, and Guoliang Xue, Editors,** Global Minimization of Nonconvex Energy Functions: Molecular Conformation and Protein Folding

22 **Panos M. Pardalos, Mauricio G. C. Resende, and K. G. Ramakrishnan, Editors,** Parallel Processing of Discrete Optimization Problems

21 **D. Frank Hsu, Arnold L. Rosenberg, and Dominique Sotteau, Editors,** Interconnection Networks and Mapping and Scheduling Parallel Computations

20 **William Cook, László Lovász, and Paul Seymour, Editors,** Combinatorial Optimization

19 **Ingemar J. Cox, Pierre Hansen, and Bela Julesz, Editors,** Partitioning Data Sets

18 **Guy E. Blelloch, K. Mani Chandy, and Suresh Jagannathan, Editors,** Specification of Parallel Algorithms

17 **Eric Sven Ristad, Editor,** Language Computations

16 **Panos M. Pardalos and Henry Wolkowicz, Editors,** Quadratic Assignment and Related Problems

15 **Nathaniel Dean and Gregory E. Shannon, Editors,** Computational Support for Discrete Mathematics

(Continued in the back of this publication)

Randomization Methods in Algorithm Design

DIMACS

Series in Discrete Mathematics
and Theoretical Computer Science

Volume 43

Randomization Methods in Algorithm Design

DIMACS Workshop
December 12–14, 1997

Panos Pardalos
Sanguthevar Rajasekaran
José Rolim
Editors

NSF Science and Technology Center
in Discrete Mathematics and Theoretical Computer Science
A consortium of Rutgers University, Princeton University,
AT&T Labs, Bell Labs, and Bellcore

American Mathematical Society

This DIMACS volume contains papers from the DIMACS Workshop on Randomization Methods in Algorithm Design, which was part of the DIMACS Special Year on Discrete Probability. The Workshop was held on December 12–14, 1997, at Princeton University, Princeton, NJ.

1991 *Mathematics Subject Classification.* Primary 03B05, 90A05, 68T15, 68Q42, 90C27, 90C30, 90B40, 68T01, 68Q15;
Secondary 68Q22, 68Q25, 68P10.

Library of Congress Cataloging-in-Publication Data

Randomization methods in algorithm design : DIMACS workshop, December 12–14, 1997 / Panos Pardalos, Sanguthevar Rajasekaran, Jose Rolim, editors.
p. cm. — (DIMACS series in discrete mathematics and theoretical computer science, ISSN 1052-1798 ; v. 43)
Workshop held at Princeton University.
Includes bibliographical references.
ISBN 0-8218-0916-4 (alk. paper)
1. Stochastic processes—Congresses. 2. Algorithms—Congresses. 3. Mathematical optimization—Congresses. I. Pardalos, P. M. (Panos M.), 1954– . II. Rajasekaran, Sanguthevar. III. Rolim, José D. P. IV. Series.
QA274.A1R36 1998
519.2–dc21 98-34537
CIP

∞ The paper used in this book is acid-free and falls within the guidelines
established to ensure permanence and durability.
Visit the AMS home page at URL: `http://www.ams.org/`

10 9 8 7 6 5 4 3 2 1 04 03 02 01 00 99

"And the science which knows to what end each thing must be done is the most authoritative of the sciences, and more authoritative than any ancillary science; and the end is the good of that thing, and in general the supreme good in the whole of nature."

- Aristotle (384-322 B.C.) (Metaphysics, Book 1, 2)

Contents

Foreword

The workshop on Randomization Methods in Algorithm Design was held in December 1997 at Princeton University, NJ. We would like to express our appreciation to Panos Pardalos, Sanguthevar Rajasekaran and Jose Rolim for their efforts to organize and plan this successful workshop. The workshop was part of the broader Special Year on Discrete Probability. We extend our thanks to Enrico Bombieri, Jeffrey Kahn and Peter Winkler for their work over many months as special year organizers. The workshop was a forum for bringing together researchers working in the theory and implementation aspects of algorithms involving randomization. DIMACS gratefully acknowledges the generous support that makes these programs possible. The National Science Foundation, through its Science and Technology Center program, the New Jersey Commission on Science and Technology, DIMACS' partners at Rutgers, Princeton, AT&T Labs Research, Bell Labs and Bellcore generously supported the special year.

Fred S. Roberts
Director

Bernard Chazelle
Co-Director for Princeton

Preface

In the context of the 1996-1998 DIMACS special year with focus on **Discrete Probability**, a three-day workshop entitled "Randomization Methods in Algorithm Design" was held on Dec. 12-14, 1997 at Princeton University. The workshop was organized by Panos Pardalos, Sanguthevar Rajasekaran, and José Rolim. More than 60 researchers from universities, institutes, governmental agencies, and industrial companies attended the workshop. Participants from Asia, Australia, and Europe gave the workshop an important international component.

Randomization has played an important role in the design of both sequential and parallel algorithms. This workshop was a forum for bringing together researchers working in the theory and implementation aspects of algorithms involving randomization.

The last decade has witnessed a tremendous growth in the area of randomized algorithms. During this period, randomized algorithms went from being a tool in computational number theory to finding widespread applications in many problem domains. Major topics covered in the workshop included randomization techniques for global and integer optimization problems, randomization in the design of approximate algorithms for combinatorial problems, randomization in parallel and distributed algorithms, practical implementation of randomized algorithms, de-randomization issues, and pseudo-random generators. Oded Goldreich delivered a distinguished lecture on "Combinatorial Property Testing (a survey)." A total of 33 invited talks were presented. Overall, the workshop achieved a great success. This volume contains a collection of refereed papers from the workshop.

The workshop was sponsored by DIMACS, through grants from the National Science Foundation, the New Jersey Commission on Science and Technology, and other sources. We would like to take this opportunity to thank the sponsors, Peter Winkler for his support, the DIMACS staff (and in particular Sandy Barbu), the participants, the authors, the anonymous referees, and the American Mathematical Society, for making the workshop successful and the publication of this volume possible.

Panos M. Pardalos, Sanguthevar Rajasekaran, and José Rolim
Gainesville and Geneva
March 1998

DIMACS Series in Discrete Mathematics
and Theoretical Computer Science
Volume **43**, 1999

Simple Randomized Mergesort on Parallel Disks*

Rakesh D. Barve†, Edward F. Grove‡, and Jeffrey Scott Vitter§

Abstract

We consider the problem of sorting a file of N records on the D-disk model of parallel I/O in which there are two sources of parallelism. Records are transferred to and from disk concurrently in blocks of B contiguous records. In each I/O operation, up to one block can be transferred to or from each of the D disks in parallel. We propose a simple, efficient, randomized mergesort algorithm called SRM that uses a forecast-and-flush approach to overcome the inherent difficulties of simple merging on parallel disks. SRM exhibits a limited use of randomization and also has a useful deterministic version. Generalizing the technique of forecasting, our algorithm is able to read in, at any time, the "right" block from any disk, and using the technique of flushing, our algorithm evicts, without any I/O overhead, just the "right" blocks from memory to make space for new ones to be read in. The disk layout of SRM is such that it enjoys perfect write parallelism, avoiding fundamental inefficiencies of previous mergesort algorithms. By analysis of generalized maximum occupancy problems we are able to derive an analytical upper bound on SRM's expected overhead valid for arbitrary inputs.

The upper bound derived on expected I/O performance of SRM indicates that SRM is provably better than disk-striped mergesort (DSM) for realistic parameter values D, M, and B. Average-case simulations show further improvement on the analytical upper bound. Unlike previously proposed optimal sorting algorithms, SRM outperforms DSM even when the number D of parallel disks is small.

1991 *Mathematics Subject Classification.* Primary 68P10, 68Q25, 68Q22, 68R05.

*An earlier version of the paper appeared in [BGV97].

†Support was provided in part by an IBM Graduate Fellowship.

‡Support was provided in part by the U.S. Army Research Office under grant DAAH04–93–G–0076 while the author was at Duke University.

§Support was provided in part by the National Science Foundation under grant CCR–9522047 and by the U.S. Army Research Office under grant DAAH04–96–1–0013.

1 Introduction

The classical problem of sorting and related processing consumes roughly 20 percent of computing resources in large-scale installations [Knu98b, LV85]. In the light of the rapidly increasing gap between processor speeds and disk memory accesses, popularly referred to as the *I/O bottleneck*, the specific problem of *external memory sorting* assumes particular importance. In external sorting, the records to be sorted are simply too many to fit in internal memory and so have to be stored on disk, thus necessitating I/O as a fundamental, frequently used operation during sorting.

One approach to alleviate the effects of the I/O bottleneck is to use *I/O algorithms* on parallel disk systems [HGK$^+$94, PGK88, Uni89, GS84, Mag87]. Aggarwal and Vitter [AV88], generalizing initial work done by Floyd [Flo72] and Hong and Kung [HK81], laid the foundation for I/O algorithms by studying the I/O complexity of sorting and related problems. Subsequently, Vitter and Shriver considered a realistic *D-disk* two level memory model in which secondary memory is partitioned into D physically distinct and independent disk drives or read-write heads [VS94]. Vitter and Shriver gave a randomized version of distribution sort that takes $\Theta(\frac{N}{DB}\frac{\ln(N/B)}{\ln(M/B)})$ I/O operations, which is optimal. Nodine and Vitter presented an optimal deterministic sorting algorithm *Greed Sort* that is a variation on the mergesort paradigm [NV90]. Nodine and Vitter also gave an optimal and more practical deterministic distribution sort called *Balance Sort* in [NV93].

In the D-disk two level memory hierarchy [VS94], there are two sources of parallelism. Firstly, as in traditional I/O systems, records are transferred concurrently in blocks of B contiguous records. Secondly, in a single I/O operation, each of the D disks can simultaneously transfer *at most* one block of B records, so that each operation can potentially transfer D blocks in parallel.

All the above mentioned optimal sorting algorithms for the D-disk model have relatively large constant factors and are complicated. They are not very attractive for implementation on existing parallel disk systems. As a result, the popularly known, asymptotically sub-optimal (by a multiplicative factor of $\Theta(\ln(M/B))$ technique of *mergesort with disk striping* (DSM) is commonly used to implement external sorting [NV90, NV93]. In DSM, the disks are coordinated so that in each parallel read and write, the locations of the blocks accessed on each disk are the same. DSM acts like external mergesort with $D' = 1$ disk and block size $B' = DB$. In each merge pass, $R = \Theta(M/DB)$ runs are merged together at a time, except possibly in the last pass. The number of passes thus required is $\ln(N/M)/\ln(M/DB)$, with an additional pass for initial run formation. The resulting I/O complexity of DSM is $\Theta(\frac{N}{DB}(1 + \frac{\ln(N/M)}{\ln(M/DB)}))$. The advantage of DSM is its simplicity, which results in very small constant factors, thus making it attractive for efficient implementation on existing parallel disk systems.

In this paper, we consider theoretical issues pertaining to *simple merging* on the D-disk two level memory hierarchy. By simple merging, we refer to the classical technique in which R sorted runs of records are repeatedly merged together until the file is sorted, preferably with R as close to M/B as possible. Vitter and Shriver [VS94] give some intuition regarding the difficulty of mergesorting on parallel disks. In Greed Sort [NV90], the trick of "approximate merging" is used to circumvent this difficulty. Aggarwal and Plaxton [AP94] use the Sharesort technique that does repeated merging with accompanying overhead.

Recently, Pai et al [PSV94] considered the average performance of a simple merging scheme for $R = D$ sorted runs, one run on each disk. Apart from their questionable block random depletion model and their relatively inefficient bounds for merging, their requirement that each run reside entirely on a single disk is fundamentally inefficient in the context of external sorting. A mergesort based on this merge scheme will require a transposition between merge passes so that striped output runs of one pass may be rearranged with each run on a disk.

1.1 Our Main Contributions

In this paper we present an efficient, practical, randomized algorithm for external sorting called *Simple Randomized Mergesort* (SRM). Our algorithm uses a generalization of the forecasting technique of [Knu98b] in order to carry out parallel prefetching of appropriate disk blocks during merging. Randomization is used only while choosing, for each input run, a disk on which to place the first block of that run so that the use of randomization is restricted to a bare minimum. The merging itself is deterministic. There are two aspects to our analysis of the algorithm.

The first (and main) part of our analysis is bounding the expected number of I/Os of SRM, where the expectation is with respect to the randomization internal to SRM and the bound on the expectation holds for any input. Using novel techniques, we are able to relate SRM's performance to a combinatorial problem we call the *dependent maximum occupancy* problem, which we further relate to the classical maximum occupancy problem [KSC78, VF90].

If internal memory available is of size M, with block size B and D parallel disks, SRM merges R runs together at a time in each merge pass, where R is the largest natural number satisfying $M/B = 2R+4D+RD/B$. We note that M/B is the number of internal memory blocks available. Hence the number of runs SRM merges together at a time is determined by the amount of memory at its disposal. As a function of M, B and D, R is given by $(M/B - 4D)/(2 + D/B)$. Under the assumption that $D = O(B)$ and $M = \Omega(DB)$, which is extremely realistic, we have that *SRM merges an optimal number, $R = \Theta(M/B)$, of runs together at a time.*

Given N records to be sorted, our SRM sorting algorithm first creates N/M sorted runs of length M during the run formation stage and then merges R (defined above) runs at a time using our merging technique. Any run written to disk at any step of the merge is written out in *forecasting format* with its first block on a random disk. The runs are cycled across disks, so I/O writes are fully parallel.

Our main theorem below gives expressions for the I/O performance of SRM for three patterns of growth rate between the number M/B of blocks in internal memory and the number D of disks in the parallel disk system. The different sizes of internal memory considered correspond to certain values of the merge order R as defined by the formula $R = (M/B - 4D)/(2 + D/B)$ given above. The general formulation in terms of occupancy statistics is given later.

Theorem 1 *Under the assumption that $D = O(B)$, the merge order $R = \Theta(M/B)$ of SRM is optimal. SRM mergesort uses $\frac{N}{DB}(1 + \frac{\ln(N/M)}{\ln R})$ I/O write operations*[1]*, which is optimal. The expected number $Reads_{SRM}$ of I/O read operations done by SRM can be bounded from above as follows:*

1. *If $R = kD$ for constant k, as R, $D \to \infty$, we have*

$$Reads_{SRM} \leq \frac{N}{DB} + \frac{N}{DB}\frac{\ln(N/M)}{\ln kD}\frac{\ln D}{k \ln\ln D}\alpha(k)$$

where $\alpha(k)$ is a constant determined by k, such that

$$\alpha(k) \leq \left(1 + \frac{\ln\ln\ln D}{\ln\ln D} + \frac{1 + \ln k}{\ln\ln D} + O\left(\frac{(\log\log\log D)^2}{(\log\log D)^2}\right)\right).$$

2. *If $R = rD\ln D$ for constant r, as R, $D \to \infty$, we have*

$$Reads_{SRM} \leq \frac{N}{DB} + c\frac{N}{DB}\frac{\ln(N/M)}{\ln(rD\ln D)} + O(1)$$

which is optimal within a constant factor, namely c. The magnitude of c depends on r.

3. *If $R = rD\ln D$ where $r = \Omega(1)$, we have*

$$Reads_{SRM} \leq \frac{N}{DB} + \frac{N}{DB}\frac{\ln(N/M)}{\ln(rD\ln D)}\beta(r)$$

where $\beta(r)$ is a constant determined by r, such that

$$\beta(r) \leq \left(1 + \sqrt{\frac{2}{r}} + \frac{\ln r}{\sqrt{2r}\ln D} + O\left(\frac{1}{r} + \frac{\log r}{r\log D} + \frac{1}{D\sqrt{r}}\right)\right).$$

The number of parallel reads is asymptotically optimal in this case; that is, the factor of proportionality multiplying the $\frac{N}{DB}\frac{\ln(N/M)}{\ln(rD\ln D)}$ term is 1.

Our algorithm can be made deterministic if the input file is initially in random order.

The mergesort algorithm by Pai et al [PSV94] uses significantly more I/Os and internal memory. In their scheme, the internal memory size needs to be $\Omega(D^2B)$ to attain a linear merging algorithm.

The other aspect of our analysis deals with the practical merit of our SRM mergesort algorithm on existing parallel disk systems. In Sections 8 and 9, we consider implementations of disk striping (DSM) and SRM mergesort algorithms for the interesting case $R = kD$, in which the number of blocks in internal memory scales linearly with the number of disks D being used. We demonstrate that

[1]Strictly speaking, the expressions for the exact number of "passes" and hence the number of I/O operations required to sort N records using a mergesort involves applying the ceiling ($\lceil\ \rceil$) function to certain logarithmic terms. Employing these exact expressions would mean dealing with a complicated dependence on N, without having any significant impact on our results for large N. Thus in this exposition we choose to overlook this issue and not apply the ceiling function, thus obtaining simplified expressions for I/O complexity.

SRM's good I/O performance is not merely asymptotic, but extends to practical situations as well. Like DSM, SRM also overlaps I/O operations and internal computation, which is important in practice.

In the next several sections we describe and analyze the merging procedure used by SRM. The next section discusses the basic idea of SRM and the way that records are distributed on the disks, which is important to attain efficient read and write parallelism. Section 3 discusses the forecasting data structure needed for parallel reads. Section 4 describes the merge process in detail. We analyze the merge process in Sections 5, 6, and 6.1 and prove our main result Theorem 1. In Section 7, we briefly discuss the potential benefits of a deterministic version of our algorithm. Practical comparisons with currently-used disk striping techniques are given in Section 8.

2 Basic Idea and Data Layout

Our primary goal in this paper is to take advantage of mergesort's simplicity and potential for high efficiency in practice when used with parallel disk systems. We use striped input runs in our merging procedure so that the output runs of one merge pass written with full write parallelism can participate as input runs in the next pass without any reordering or transposition. In this scheme, if the 0th block of a run r is on disk d_r, then the ith block resides on disk $(i + d_r) \bmod D$. Striping alone is not enough to ensure good performance since if the disk d_r for each run r is chosen deterministically, and a large number of runs is being merged, the merging algorithm can have abysmal worst case performance: throughout the entire duration of the merge, the R leading blocks of the R runs always lie on the same disk, thus causing I/O throughput to be only a factor of $1/D$ of optimal.

A natural alternative that we pursue in this paper is to consider a randomized approach to mergesort. We randomly assign the starting disk d_r for each input run r. Then, on average, it is not necessary to read many blocks from the same disk at one time. The analysis involves an interesting reduction to a maximum bucket occupancy problem.

In order to facilitate analysis *we choose the disk d_r on which any run r's initial block will reside, independently and identically with uniform probability from the set $\{0, 1, \ldots, D-1\}$.* The location of every subsequent block of that run is determined deterministically by cycling through the disks. We will measure our algorithm's performance by bounding the *expected* number of I/Os required to merge R runs for any *arbitrary (worst-case) input* runs, where the expectation is with respect to this randomization in the placement of the first block of runs. The actual records that constitute the different runs can be arbitrary and their relative order does not effect the bounds we derive.

If we consider the problem of mergesorting with random inputs, we believe that the same analysis technique can yield similar bounds for a version of our algorithm in which the starting disk d_r for run r is chosen to be uniformly staggered, so that $d_r = 0$ for $r = 0, 1, \ldots, R/D - 1$, $d_r = 1$ for $r = R/D, R/D + 1, \ldots, 2R/D - 1$ and so on. Moreover, as we indicate in Section 7, by taking into consideration the lengths of the runs being merged in different stages of the mergesort, it may be possible to improve the bound on overall average I/O performance. In this paper, however, we concentrate on our SRM method, since a very limited amount of explicit randomization in SRM (used to determine the random starting disks d_r) enables us to get around the average-case assumption of the deterministic version, thus facilitating the elegant and straightforward analysis presented here.

3 Forecasting Format and Data Structure

As observed earlier, each of the R runs that participate in a merge is output onto disk either during the run formation stage or during some earlier merge. This enables us to begin each run on a uniformly random disk. We can also format each block of every run so as to implant some *future* information as explained below. This is a generalization of the *forecasting* technique of [Knu98b]. We denote the ith block of a run r as $b_{r,i}$ and the *smallest (first) key* in block $b_{r,i}$ as $k_{r,i}$. Blocks of a run r are formatted as follows:

- Implanted in the initial block $b_{r,0}$ of run r are the key values $k_{r,j}$ for $1 \leq j \leq D$.
- Implanted in the ith block $b_{r,i}$, where $i > 0$, of run r is the single key value $k_{r,(i+D)}$.

The extra information for forecasting in each block is only one key value, not one record. So the extra space taken up by the forecasting format of runs is negligible. For simplicity, we assume that all key values are distinct.

Definition 1 Consider the unprocessed portions of the R runs at any point of time t during the merge. Among these records, the record with the smallest key is called the *next* record of the merge at time t. A block belonging to any run is said to begin *participating* in the merge process at the time t when its first record becomes the next record of the merge. A block ceases to participate in the merge as soon as all its records get depleted by the merge. At any point of time, consider all the blocks not in memory that reside on disk i. The *smallest block of run j on disk i* is the block that will be the earliest participating block among all the blocks of run j on disk i. The *smallest block on disk i* is the the earliest participating block on disk i at that time. At any time, the *leading* block of a run is the block (possibly partly consumed) that contains the smallest key value in that run at that time. Note that a block can become a leading block much before beginning to participate in the merge.

On an I/O read, the block read in by SRM from disk i is always the smallest block on disk i at that time. To be able to do this, SRM maintains a forecasting data structure.

Definition 2 The *forecasting data structure* (FDS) consists of D arrays $\mathcal{H}_0, \mathcal{H}_1, \ldots, \mathcal{H}_{D-1}$, one corresponding to each disk. At any point of time, $\mathcal{H}_i[j]$ stores $K_{i,j}$, where $K_{i,j}$ is the smallest key value in the smallest block of run j on disk i.

At any time, in order for SRM to read in the smallest block from disk i, it merely needs to read in the smallest block of run j on disk i, where j is the run with the smallest key in $\mathcal{H}_i$.

4 The SRM Merging Procedure

In this section, we discuss how SRM merges R runs striped across D disks, with the randomized layout described in Section 2. Our main focus is the I/O schedule, which describes, at any point of time, the next parallel I/O read operation to be executed, primarily using the forecasting data structure FDS. Internal memory management of merge data is done at block level granularity so internal fragmentation within internal memory blocks is not an issue. By an internal block, we mean a set of contiguous internal memory locations with fixed boundaries, large enough to hold B records.

4.1 Internal Memory Management

The number M/B of blocks in internal memory, expressed in terms of R and D, is $2R + 4D + RD/B$. Of these, SRM maintains a *dynamically changing partition* $\{M_L, M_R, M_D\}$ of $2R + 2D$ physical blocks of the internal memory used for unprocessed records.

Definition 3 The set M_L of R internal blocks is maintained such that at any point of time if the leading block of any run is in internal memory, its internal block is in M_L. The set M_D contains D internal blocks maintained such that each I/O read operation of SRM can read in D blocks from disks into the internal memory blocks of M_D. The remaining $R + D$ internal memory blocks comprise the set M_R. The sets are maintained such that as soon as there are D unoccupied internal blocks, a parallel read can be initiated into M_D. In addition, SRM uses a set M_W of $2D$ internal memory blocks as an output buffer.

The D internal memory blocks of M_D are used specifically to initiate read operations bringing in blocks at the earliest possible time, before they begin participating. FDS and related auxiliary data structures occupy not more than about RD/B blocks. The output buffer M_W is large enough to ensure that the output run can be written out in full parallelism in the forecasting format described earlier.

Next we indicate briefly how the dynamic partition of the $2R+2D$ blocks into $\{M_R, M_D, M_L\}$ can be maintained as the merge proceeds.

1. As soon as a leading block of a run r ceases to participate, if r's new leading block occupies an internal block of M_R, then M_R and M_L mutually exchange the internal blocks corresponding to the new and old leading blocks.
2. Whenever an I/O read operation brings in a run r's leading block into M_D, then M_L and M_D exchange internal blocks corresponding to the old and new leading blocks, so M_D gets an unoccupied internal block.
3. Whenever M_R has any unoccupied internal block while M_D has an occupied one, the two are exchanged.

Definition 4 Suppose we order a set A of blocks in ascending order by the blocks' smallest key values. We define $Rank_A(b)$, for $b \in A$, to be the rank of block b in this order. For any time t during the merge, we define the following terms:

- $\mathcal{F}_t$ denotes the set of full blocks in internal memory such that no block b in $\mathcal{F}_t$ is the leading block of any run at time t. All blocks of $\mathcal{F}_t$ occupy internal blocks in the set M_R.

- $\mathcal{S}_t$ denotes the set of D blocks, each of which is the smallest block on a disk i, $0 \le i \le D-1$.
- $Fset_t(l) = \{b \mid b \in \mathcal{F}_t, \; Rank_{\mathcal{F}_t}(b) \ge |\mathcal{F}_t| - l + 1\}$ denotes the set of the l highest ranked blocks of $\mathcal{F}_t$. By definition, no block of $Fset_t(l)$ is a leading block.
- $OutRank_t = \min_{b \in \mathcal{S}_t}\{Rank_{\mathcal{F}_t \cup \mathcal{S}_t}(b)\}$ denotes the rank of the smallest-ranked block of $\mathcal{S}_t$ in the set $\mathcal{F}_t \cup \mathcal{S}_t$.

Definition 5 The operation $ParRead_t$ is executed at a time t only when D unoccupied internal blocks are available in M_D. Using FDS, it reads in the set $\mathcal{S}_t$ of D blocks into M_D, exchanging internal blocks of M_D with M_L as in point 2 above, if necessary. FDS is updated in the obvious way using the implanted information in blocks of $\mathcal{S}_t$.

Definition 6 The operation $Flush_t(j)$ virtually flushes out the j blocks of the set $Fset_t(j)$ occupying internal blocks in M_R at time t. In SRM, we have $j \le D$ whenever $Flush_t(j)$ is invoked. The flushing of blocks is virtual in the sense that the blocks *do not actually have to be written out to disk*, so there is no I/O involved. *Leading blocks of runs are never flushed out.* FDS is updated appropriately to reflect the fact that the flushed out blocks will have to be read back from their original disks when needed.

4.2 Merging Algorithm

In this subsection, we specify the parallel I/O read schedule of SRM. As far as writes go, each write is executed with full write parallelism as soon as the next output stripe of D formatted blocks are ready in the output buffer M_W. The basic idea is to read in D blocks of the set $\mathcal{S}_t$ whenever possible; flushing a few of the highest ranked non-leading blocks in $\mathcal{F}_t$, if necessary.

1. At the beginning, SRM reads the first block from each of the R runs into the R internal blocks of M_L using parallel reads.
2. Until the merge is completed, whenever the I/O system is free at a time t when D unoccupied blocks are available in M_D, SRM does the following:
   (a) If there are D unoccupied internal blocks in the set M_R at time t, a $ParRead_t$ operation is initiated.
   (b) Else if the number of occupied internal blocks in M_R at time t is $R + extra$, where $1 \le extra \le D$, and if $OutRank_t > extra$, a $ParRead_t$ operation is initiated.
   (c) Else if the number of occupied internal blocks in M_R at time t is $R + extra$ where $1 \le extra \le D$, and if $OutRank_t \le extra$, a $Flush_t(extra - OutRank_t + 1)$ operation is invoked, followed by initiation of a $ParRead_t$ operation.

In Step 2c, the flushed out blocks are the ones that will not be used in the "near future", as we will see in the next section. Lemma 1 below, whose proof we defer to the full version of this paper, ensures that SRM runs to completion, needing only the specified amount of memory.

Lemma 1 *Consider any $ParRead_t$ operation invoked by SRM at time t. No block not in internal memory when $ParRead_t$ has completed begins participating in the* *merge until at least D unoccupied physical blocks are available in internal memory. This ensures that there will be enough space for the next read operation to begin before any outside block begins participating.*

Lemma 1 shows that the next I/O read can begin before any block brought in by it begins participation. Thus there is genuine prefetching ability, which is useful in overlapping I/O operations with internal processing.

5 Using Phases to Count the Number of *ParRead* Operations

For the purpose of analysis, we break the process of merging R runs comprising a total of N' records into a sequence of *phases*, and we upper bound the expected number of read operations required during a phase. The overall bound on the merge process can then be computed using the linearity of expectations.

Consider the set of all blocks except blocks in the initial set. Each such block can participate in the merge only after its preceding blocks have ceased to participate, so that if the ith block of a run is currently participating, $(i-1)B$ records from that run have already been output.

Definition 7 Consider the set $\mathcal{R}_0$ of all blocks of all runs excluding the initial block of each run. The *participation index* of any block in $\mathcal{R}_0$ is i if it is the ith block, in chronological order, to begin participating in the merge, where $1 \leq i \leq \frac{N'-RB}{B}$. We denote by set $\mathcal{P}_j$, where $1 \leq j \leq \frac{N'-RB}{BR}$, the subset of $\mathcal{R}_0$ consisting of blocks with participating indices i such that $(j-1)R+1 \leq i \leq jR$.

Using parallel reads, SRM tries to fill the R internal blocks of M_R using FDS. Every incoming block provides new implanted information. When a block with small participation index is on disk while memory is full, we bring it in, by flushing blocks that are *surely not* among the next R blocks to begin participating in the merge, as Lemma 2 (whose proof we omit here) shows.

Lemma 2 *Consider a flush $Flush_t$ at time t. The $R + OutRank_t - 1$ smallest-ranked blocks of $\mathcal{F}_t$ (in M_R) do not get flushed out: they remain in M_R after the flush completes.*

The merge proceeds in a sequence of phases, each contributing the R blocks to the output run.

Definition 8 The first phase begins at the time p_0 of the last $ParRead_{p_0}$ read operation invoked in Step 1 of SRM. The jth phase ends at the time p_j of the read $ParRead_{p_j}$ such that any block $b \in \mathcal{R}_0$ with participation index i_b, where $i_b \leq jR$, has been read into memory (at least once) from disk by some $ParRead_t$ such that $t \leq p_j$. The $ParRead_t$ read operations such that $p_{j-1} < t \leq p_j$ are the *reads forming the jth phase*, and the set $\mathcal{P}_j$ defined above is the *set of blocks forming the jth phase*.

The next step is to show that when the jth phase is over, blocks with participation indices smaller than jR have already been "taken care of."

Lemma 3 *After the read* $ParRead_{p_j}$ *at time* p_j *is complete, we have*

1. *No block still on disk after* $ParRead_{p_j}$ *completes has a participating index less than* $jR + 1$.
2. *No block* b *with participating index* $i_b \leq (j+1)R$ *can be flushed out by any* $Flush_t$, *where* $t > p_j$.

Proof: We prove the claims by induction. For the base case $j = 1$, observe that by time p_1, all blocks having participation indices at most R have been read in at least once. By Lemma 2, none of these could have been flushed out at any time $t \leq p_1$ because they would be among the R smallest ranked blocks of M_R, thus proving the claim. Now consider the second claim for $j = 1$. Suppose there is a flush $Flush_t$ at any time $t > p_1$. By the previous claim, if X is the number of blocks in M_R that have participating indices at most R, then $OutRank_t > X$. By Lemma 2, the $R + OutRank_t - 1$ smallest ranked blocks at time t remain in M_R after the flush. Of these, at most $X \leq OutRank_t - 1$ blocks have ranks smaller than R. Thus if there were any block b with rank at most $2R$ at time t, b must also have been among the $R + OutRank_t - 1$ smallest ranked blocks. Hence the base case is proved for both claims.

Now suppose both claims are true for some $j - 1$. Then by inductive hypothesis for Claim 1, when the read $ParRead_{p_{j-1}}$ at p_{j-1} completes, the smallest participating index of any block still on disk is $(j-1)R + 1$. By the inductive hypothesis for Claim 2 none of the blocks with participating indices at most jR can get flushed at any time t after p_{j-1}. By definition, all blocks with indices at most jR are read in by p_j. Since none of them can get flushed after p_{j-1}, the first claim is true for j. The second claim for j follows by an argument identical to the argument for its base case. □

We are led immediately to the following lemma that gives the number of in the output run of the merge at the end of the first j phases.

Lemma 4 *At least* jRB *records from blocks of the set* $\mathcal{R}_0$ *are already in the merged output of SRM* before *any block still on disk after time* p_j *begins participating in the merge. Thus all reads* $ParRead_t$ *such that* $t \leq p_j$ *and the reads* I_0 *required in the step 1 of SRM can be charged to these* jRB *records in the output of SRM.*

We wish to obtain a handle on the number of read operations $ParRead_t$ where $p_{j-1} < t \leq p_j$ (the reads associated with the jth phase). We will show that this number is the largest number of blocks in $\mathcal{P}_j$ that need to be read in from any particular disk.

Definition 9 Consider the blocks of the set $\mathcal{P}_{j+1}$ just after $ParRead_{p_j}$ is completed. Such a block b is said to be on *level* 0 if it is already in internal memory and level h if it is the hth smallest block on its disk at that time. Let L_{j+1} denote the highest level of any block of $\mathcal{P}_{j+1}$. In the case of the first phase, let L_1 be the highest level of any block of $\mathcal{P}_1$ after step 1 of SRM.

We prove the following characterization of the number of reads associated with the $j+1$st phase.

Lemma 5 *The* L_{j+1}*th read of SRM* after *time* p_j *is done at time* p_{j+1}.

Proof: The proof follows from Lemma 3. No block with participation index at most jR is still on disk after the read at p_j completes. Moreover no block with participation index at most $(j+1)R$ can be flushed out after p_j. Clearly then, since any $ParRead_t$ operation reads in the smallest block from every disk at time t, we have the nice property that *all blocks of $\mathcal{P}_{j+1}$ that are at the hth level will get read by the hth read after p_j*, never to get flushed out. This proves the lemma. □

We can now relate progress in the merge to the number of read operations involved in phases by means of the following lemma that follows from Lemmas 5 and 4.

Lemma 6 *The first $I_0 + \sum_{1\le i\le j} L_i$ reads of SRM can be charged to at least jRB records in the output of the merge, where I_0 is the number of reads SRM incurs in Step 1. The total number of I/O read operations required to complete the merge is given by the above sum with $j = \frac{N'/B-R}{R}$.*

In the next section we give a bound on the expectation of L_i with respect to the randomization involved in the choice of disks for the initial blocks of the runs. Thereafter, Lemma 6 can be used to obtain the overall bound on the number of reads involved in SRM.

6 Probabilistic Analysis

Consider any arbitrary R runs input to SRM. The initial block of each run is placed on a disk chosen with a uniform probability of $1/D$ independently of other runs. Since runs are cycled across disks, the position of the initial block of the run determines the position of the other blocks of the run. As a result, for any j, the jth block of a run is on one of the D disks with probability $1/D$, depending only on the disk containing that run's initial block.

Lemma 7 *Consider the ith phase and the related set $\mathcal{P}_i$ of blocks such that $1 \le i \le \frac{N'-RB}{RB}$. Let n_j denote the number of blocks from the jth run in $\mathcal{P}_i$, so that $\sum_{0\le j\le R-1} n_j = R$. Let C_j denote the ordered list $(b_{j,0}, b_{j,1}, \ldots, b_{j,n_j-1})$ of the n_j blocks of run j in $\mathcal{P}_i$, where the order is the chronological order of participation. Let D_j denote the disk from which the block $b_{j,0}$ originates, for each run j, $0 \le j \le R-1$. Then the disks D_j are independently distributed, each with uniform probability over the set $\{0, 1, \ldots, D-1\}$ of D disks.*

Definition 10 We call each C_j a *chain* of length n_j of contiguous blocks of run j. The disk corresponding to any given block of C_j is determined by the disk of the first block $b_{j,0}$ of C_j.

We recall that the number of reads L_i in the ith phase is the maximum level of any block on any disk at time p_{i-1}.

Definition 11 Consider the set $\mathcal{P}_i$. Let L'_i be the maximum level of any block of $\mathcal{P}_i$ on disk, considering all of $\mathcal{P}_i$'s blocks on their respective original disks.

In the next subsection, we will be able to bound $E(L'_i)$. The following lemma shows that L'_i is an overestimate of L_i, so we get a bound on L_i too.

Lemma 8 *With L_i and L'_i defined as above, we have $L_i \le L'_i$ and $E(L_i) \le E(L'_i)$.*

6.1 The Dependent Occupancy Problem

In this section, we define the dependent occupancy problem, relate it to both the well-studied *classical occupancy* problem and the analysis of SRM and thereby prove Theorem 1.

> In the **classical occupancy** problem of parameters $\{N_b, D\}$, N_b balls are thrown into D bins independently each with uniform probability of $1/D$ of falling into any bin. We are interested in the expectation $\mathcal{C}(N_b, D)$ of the maximum number of balls in any bin [VF90, KSC78].
>
> In the **dependent occupancy** problem, we consider C chains c_j ($1 \leq j \leq C$), each consisting of n_j balls, where $\sum_{1 \leq j \leq C} n_j = N_b$. The chains are thrown into one of D bins $\{0, 1, \ldots, D-1\}$, wherein they get placed in the bins in such a manner that if the first ball $b_{j,0}$ of chain c_j falls in the bin s, the ith ball $b_{j,i-1}$ falls in the bin $(s + i - 1) \bmod D$. The first ball of each of the C chains is placed into a bin independently of the other chains with uniform probability $1/D$. We are interested in the expectation $E(n_1, n_2, \ldots, n_{C-1}, n_C)$ of the maximum number of balls that fall into any bin for arbitrary n_j's and C satisfying the constraint above. When values of the n_j's and C are known, we call this the dependent occupancy problem with parameters $\{N_b, D, C, n_1, n_2, \ldots, n_C\}$. We use $\overline{E(N_b, D)}$ to denote the quantity $\max\{E(n_1, n_2, \ldots, n_{C-1}, n_C) : \sum_{i=1}^{C} n_i = N_b\}$.

Our goal here is to relate $E(n_1, n_2, \ldots, n_{C-1}, n_C)$ to the well-known statistic $\mathcal{C}(N_b, D)$. In fact we have the following conjecture regarding these quantities.

Conjecture 1 *With $E(n_1, n_2, \ldots, n_{C-1}, n_C)$ and $\mathcal{C}(N_b, D)$ as defined above, $E(n_1, n_2, \ldots, n_{C-1}, n_C) \leq \mathcal{C}(N_b, D)$.*

Intuition: The simple intuition is that when a chain consists of some $1 < l \leq D$ balls, we are guaranteed that all those l balls fall into separate bins. If instead of the l-long chain, l balls are distributed independently and with uniform probability among the D bins, the expected maximum number of balls in any bin is certainly greater than 1. This observation and the fact that we could prove Theorem 2 below are the basis of our conjecture. □

Recently Knuth [Knu98a] made the following stronger conjecture.

Conjecture 2 *Assuming $n_1 > 1$, we have*

$$E(n_1, n_2, \ldots, n_{C-1}, n_C) \leq E(n_1 - 1, 1, n_2, n_3, \ldots, n_{C-1}, n_C)$$

It may be observed that if Conjecture 2 is true then a dependent occupancy problem can be transformed into a classical occupancy problem via a sequence of occupancy problems, with their expected maximum occupancies remaining non-decreasing over the sequence.

However since the above observations remain conjectures, in order to obtain bounds on our algorithm we proved the following bound on the quantity $\overline{E(N_b, D)}$, which is, by definition, an upper bound even on $\mathcal{C}(N_b, D)$.

Theorem 2 *Let $N_b = kD$, for $k > 0$. Then*

1. *If k is a constant as $D \to \infty$,*

$$\overline{E(N_b, D)} \leq \frac{\ln D}{\ln \ln D}\left(1 + \frac{\ln \ln \ln D}{\ln \ln D} + \frac{1 + \ln k}{\ln \ln D} + O\left(\frac{(\log \log \log D)^2}{(\log \log D)^2}\right)\right).$$

2. *In the case where $k = r \ln D$ and $r = \Omega(1)$, we have*

$$\overline{E(N_b, D)} \leq \left(1 + \sqrt{\frac{2}{r}} + \frac{\ln r}{\sqrt{2r} \ln D} + O\left(\frac{1}{r} + \frac{\log r}{r \log D} + \frac{1}{D\sqrt{r}}\right)\right)\frac{N_b}{D}.$$

Note that the right hand side is of the form $\frac{N_b}{D}(1 + o(1))$ when $r \to \infty$. In the case when $r = \Theta(1)$, we have $\overline{E(N_b, D)} \leq cN_b/D$ for some constant $c > 0$.

Notes: We refer the reader to Theorem 2 of [BGV97] for details of the proof. It may be observed that the leading few terms in the bound we proved here on $\overline{E(N_b, D)}$ are the same as those in the bound proved in [KSC78] for $\mathcal{C}(N_b, D)$, which bolsters the case for Conjecture 1 above. Our proof for the upper bound necessarily required an approach different from the one used in [KSC78] to bound $\mathcal{C}(N_b, D)$. The proof in [KSC78] uses the independence of distribution of the classical occupancy problem to exploit the fact that the occupancy distribution of balls in the D bins is asymptotically equivalent to the D distribution of D appropriate independent Poisson variables. Our proof uses a different technique that overcomes the difficulties imposed by the inherent dependence in the occupancy problem we consider. The proof in [KSC78] also serves as a lower bound on classical expected maximum occupancy but our techniques can be modified to do the same. □

Using linearity of expectations, Lemmas 6, and 8 and recognizing that $E(L_i') \leq \overline{E(R, D)}$ and that $E(I_0) = \mathcal{C}(R, D) \leq \overline{E(R, D)}$, we bound the expected number of reads of one merge step of SRM in the following theorem.

Theorem 3 *The expected number of I/O read operations during one merge step of SRM is bounded by $\frac{N'}{RB}\overline{E(R, D)}$.*

This means that to proceed by R blocks in the merge output, the number of reads SRM needs is less than the maximum occupancy when R balls are randomly thrown into D bins.

Our main result, Theorem 1, follows from Theorem 3 and the bounds for $overline E(R, D)$ from Theorem 2.

7 A Deterministic Variant

As mentioned earlier, when the inputs are randomly uniformly distributed we expect a version of our algorithm that uses a staggered distribution of starting blocks of runs on disks to give an average performance comparable to the bounds we show here. Intuitively, this effect occurs because with long, random input runs, the R leading blocks of the R input runs at any time during the merge tend to be spread no worse than randomly among the D disks.

When the R input runs are staggered on the D disks, the first R/D runs have their first blocks on disk 0 with subsequent blocks cyclically stored on the disks, the next R/D blocks have their first blocks stored on disk 1 and so on. With random inputs, it might be beneficial to exploit the stagger "while it lasts" since the stagger guarantees an even distribution of leading blocks on disks. If the runs are "short enough," we can expect the runs to be more or less staggered on the disks *throughout* the duration of the merge on the average, ensuring very efficient I/O. On the other hand, as observed in the previous paragraph, even when runs are "long," we can expect I/O to be reasonably efficient. Hence, if the technique of having runs staggered deterministically during initial merge stages (when runs are "short") is combined with the technique that exploits the randomness in maximum occupancy situations during later merge stages (when runs are "long"), we might be able to obtain an improvement in overall I/O performance of the mergesort. We might thus be able to prove theoretical optimality (within a constant factor) for an increased range of M, B, D values for our mergesort, compared to the range in Theorem 1.

8 Comparisons between SRM and DSM in Practice

In DSM, the popularly-used merging technique with disk striping, there are $O(M/DB)$ runs merged in each merge step with perfect read parallelism. The price that DSM pays for not using the disks independently is that it can merge only $\Theta(M/DB)$ rather than $\Theta(M/B)$ runs at a time. However, DSM is often more efficient in practice than the previously developed asymptotically optimal sorting algorithms, since the latter algorithms have larger overheads.

When the amount M of internal memory is small or the number D of disks is large, our SRM method is clearly superior to DSM. For example, if $M = \Theta(DB)$, DSM is suboptimal by a factor of $\Theta(\log D)$ whereas SRM is suboptimal by a factor of $\Theta(\log D/\log\log D)$. The improvement of SRM over DSM shows up in practice even when M is substantially larger or the number of disks D is small, which is when DSM has better I/O performance than previously developed external sorting algorithms.

In this section, we compare the performance of the DSM and SRM mergesorts on existing parallel disk systems. In Subsection 8.2 we make a comparison based on *an estimate of the upper bound on the expected worst-case* number of I/O operations of an SRM mergesort with the exact number of I/O operations for the standard DSM mergesort with disk striping, using parameters of presently existing or feasible parallel disk systems. In Subsection 8.3 we make a similar comparison based on actual simulations of SRM to estimate its average-case I/O performance on random inputs. Both the comparisons show that SRM's I/O performance is always better than DSM's. Moreover, simulations indicate that SRM's actual I/O performance is much better than that implied by the estimate of the analytical bound in Subsection 8.2.

8.1 Expressions for the number of I/O operations

In this subsection and the following ones, we assume that SRM is able to merge $R = kD$ runs at a time, for interesting values of k and D. We consider DSM using the same amount of memory and estimate the relative performances of SRM and DSM.

We will first consider SRM, which merges $R = kD$ runs at a time. The amount of internal memory needed to support the merging is $M = (2kD + 4D)B + kD^2$. The total number of writes SRM requires is $\frac{N}{DB}(1 + \frac{\ln(N/M)}{\ln(kD)})$ since it has perfect write parallelism. In terms of reads, SRM requires N/DB reads corresponding to the initial run formation pass. It subsequently requires $\frac{\ln(N/M)}{\ln(kD)}$ passes to perform the merging, each incurring $v\frac{N}{DB}$ reads, where $v = v(k, D)$ is the *overhead factor* that represents a multiplicative overhead over and above the minimum number $\frac{N}{DB}$ of reads for a single pass. The total number of I/O operations SRM takes to sort N records is thus

$$\frac{N}{DB}\left(2 + \frac{\ln(N/M)}{\ln(kD)}(1+v)\right) = \frac{N}{DB}\left(2 + C_{SRM}\ln\frac{N}{M}\right),$$

where

$$C_{SRM} = \frac{1+v}{\ln(kD)}. \tag{1}$$

Now let us compute in a similar way the number of I/O operations needed by DSM to sort N records using the same amount of memory as SRM above. We assume that DSM uses $2D$ blocks per run for I/O read buffers and $2D$ blocks for I/O write buffers. DSM merges $(M/B - 2D)/2D = k + 1 + kD/2B$ runs at a time. It can be verified that the number of I/O operations required by DSM to sort N records is

$$\frac{N}{DB}\left(2 + 2\frac{\ln(N/M)}{\ln(k+1+kD/2B)}\right) = \frac{N}{DB}\left(2 + C_{DSM}\ln\frac{N}{M}\right),$$

where

$$C_{DSM} = \frac{2}{\ln(k+1+kD/2B)}. \tag{2}$$

8.2 Comparison based on expected worst-case performance of SRM

Our goal in this subsection is to compare SRM and DSM based on the *expected worst-case performance* of SRM. Our comparison uses k and D values corresponding to realistic systems. To make the comparison possible, we need to obtain an estimate of the overhead factor $v(k, D)$ defined in the previous subsection corresponding to these values of k and D.

Note that for a pair of given finite values of k and D, the estimate of the expected worst-case value of v using Theorem 1 is lax because of contributions from lower-order terms. As explained in Section 6, the expected number of reads done by SRM is bounded by an expression involving expected maximum occupancies of the classical and dependent cases. We get more useful comparisons between SRM and DSM by replacing the maximum dependent occupancy by the maximum

	$D = 5$	$D = 10$	$D = 50$	$D = 100$	$D = 1000$
$k = 5$	1.6	1.7	2.2	2.3	2.7
$k = 10$	1.4	1.5	1.8	1.9	2.2
$k = 20$	1.3	1.4	1.5	1.6	1.8
$k = 50$	1.2	1.2	1.3	1.4	1.5
$k = 100$	1.11	1.16	1.22	1.26	1.3
$k = 1000$	1.04	1.05	1.08	1.08	1.1

Table 1: The overhead $v(k, D)$, computed by estimating $\mathcal{C}(kD, D)/k$ using computer simulations.

classical occupancy in the bound for the expected number of reads required by SRM. This approach can be justified on two counts. First, the upper bound in Theorem 2 that we proved for the expected maximum dependent occupancy is the same as the expression for the expected maximum of the classical occupancy problem.

For given values of k and D, we simulate throwing kD balls into D bins repeatedly to estimate $\mathcal{C}(kD, D)$. Thus the expected worst-case overhead $v(k, D)$ is estimated by repeated ball-throwing experiments to estimate $\frac{\mathcal{C}(kD,D)}{k}$. Table 1 shows the values of v based on such experiments.

For every k, D pair considered, we use (1) to estimate an expected worst-case value for C_{SRM} using the abovementioned estimates for v. In Table 2 we present the C_{SRM}/C_{DSM} ratio for several values of k and D. The ratio C_{SRM}/C_{DSM} represents the relative I/O advantage of SRM over DSM, neglecting the $2N/DB$ I/Os that both methods require during the initial run formation. We used block size $B = 1000$ records for all k, D pairs. (The choice of B is not significant so long as it is reasonable.) In our representation, $2k$ is roughly the number of internal memory blocks available per disk.

Table 2 shows that the SRM uses signficantly fewer I/Os than does DSM. For example, for $D = 50$, $k = 100$, which translates into $M = 10.5$ million records of internal memory, SRM uses 0.60 times as many I/Os as does the slower DSM, not counting the initial run formation pass that they share in common. When D is small, as k increases relative to D, the C_{SRM}/C_{DSM} ratio gradually increases toward 1, which indicates a lessening advantage of SRM over DSM when there are few disks and a huge amount of internal memory. As we show in the next subsection, since we overestimate the number of reads required by SRM in our analysis, SRM actually exhibits even better performance than indicated in Tables 1 and 2.

8.3 Using simulations to count SRM's I/O operations

In the previous subsection we showed by using classical maximum occupancy to estimate the overhead term v that SRM performs well compared to DSM on arbitrary inputs. In this section we use *simulations of the algorithm itself* to estimate v on *average-case inputs* and make a similar comparison.

Our experiments consist of simulating SRM while merging $R = kD$ sorted runs, and each of length L, for a large range of k and D values. The runs input

	$D = 5$	$D = 10$	$D = 50$	$D = 100$	$D = 1000$
$k = 5$	0.71	0.62	0.51	0.48	0.46
$k = 10$	0.72	0.66	0.54	0.50	0.48
$k = 20$	0.75	0.68	0.56	0.53	0.49
$k = 50$	0.77	0.71	0.59	0.55	0.50
$k = 100$	0.78	0.72	0.61	0.57	0.51
$k = 1000$	0.83	0.77	0.67	0.63	0.56

Table 2: The performance ratio C_{SRM}/C_{DSM} for memory size $M = (2k+4)DB + kD^2$, with block size $B = 1000$. (Both M and B are expressed in units of records.) The overhead factor v in C_{SRM} is based on computer simulations of $\mathcal{C}(kd, D)/k$.

	$D = 5$	$D = 10$	$D = 50$
$k = 5$	1.0	1.0	1.2
$k = 10$	1.00	1.0	1.1
$k = 50$	1.00	1.00	1.00

Table 3: The overhead factor $v(k, D)$ for memory size $M = (2k + 4)DB + kD^2$ obtained from simulations.

to the merge were generated such that *each set of input runs was equally likely.* SRM's actions while merging depend only on the relative order of the input keys. Thus, there is an obvious one-to-one correspondence between the set of all possible input runs to the merge and the set of partitions of the set $I = \{1, 2, \ldots, LkD\}$, each partition splitting I into kD disjoint subsets of size L. We generate average-case inputs to the merge by generating partitions of the set I, with each partition being equally likely.

Our simulations indicate that SRM's I/O overhead is noticeable only when k is small compared with the number of disks D. We ran our simulations for several different sets of parameters. Not only did we vary k and D, but for each k, D pair we also varied B and L (where L is as defined above). We present here an illustrative sample of typical simulation outcomes, with our main focus on the parameters k and D. Table 3 shows the multiplicative overhead v corresponding to simulations for interesting k, D pairs. It can be seen that the estimates of v based on average-case inputs are smaller than the corresponding estimates of the expectation of worst-case values of v in Table 1. (For values of k larger than the ones shown here, the simulations gave values v that were practically equal to 1.) The simulations show in the average-case that SRM has little or no overhead when k is reasonably large and that the number of I/O read operations required to carry out the merge is practically N'/DB, where $N' = LkD$ is the number of records in the merged output. In our simulations, N' was always 1000 times bigger than kDB. Longer simulations tend to be time consuming.

The average-case simulations indicate when k is reasonably large that there is little or no flushing of blocks by SRM. Intuitively, the rate of consumption (due to merging) of the blocks in internal memory is such that there is almost always space to read in the next "smallest" D blocks without the need to flush.

In order to make a similar comparison as the previous subsection, but based

	$D = 5$	$D = 10$	$D = 50$
$k = 5$	0.56	0.47	0.37
$k = 10$	0.61	0.52	0.40
$k = 50$	0.71	0.63	0.51

Table 4: The performance ratio C'_{SRM}/C_{DSM} for memory size $M = (2k+4)DB + kD^2$ where C'_{SRM} is computed using the overhead value $v(k, D)$ obtained from simulations.

on results of simulations of the algorithm itself, we computed the term C'_{SRM} analogously to C_{SRM} using (1), by using the average-case values for v obtained from simulations of the algorithm.

Table 4 below shows the C'_{SRM}/C_{DSM} ratios so obtained, for various k, D pairs. We note that the entries in Table 4 are smaller than the corresponding entries in Table 2, indicating that SRM's performance is indeed better than that implied by the more pessimistic upper bound. Moreover, SRM's low overhead in the average-case indicates that for all practical purposes it is an optimal external sorting algorithm.

9 Realistic Values for Parameters *k*, *D*, and *B*

We have shown in the previous section for a wide range of k, D, and B values that SRM performs very efficiently. Since not every k, D, B triplet corresponds to that of a realistic machine, we try in this section to obtain a crude approximation of the relative magnitudes of k, D, and B in typical computers. We argue that k is generally much larger than D in realistic machines, which implies that SRM is the method of choice on such machines, still noticeably faster than DSM.

In our terminology, where the internal memory size is $M = (2kD+4D)B+kD^2$ and the merge order is $R = kD$, the expression $2kD$ is roughly equal to M/B, the number of blocks that fit in main memory, under the realistic assumption that $D = O(B)$. Let us consider a fast, uniprocessor workstation attached to, say, $D = 5$ independent disks for parallel I/O. We are likely to find internal memories of the order of 100 megabytes on such machines and disk block or track sizes of the order of 10–50 kilobytes. This would mean that k may be on the order of 200–1000 when $D = 5$ on such workstations. If the same amount of main memory is used with instead 10 disks and 50-kilobyte disk blocks, k would be on the order of 100. On the other hand when the number of disks is relatively high as in large-scale multiprocessor computing systems, we would still expect k to be large, even after factoring in the increased block sizes that such machines have. This is primarily because large-scale computing systems tend to have huge internal memories. For instance, in a system with 100 parallel disks and 100-kilobyte disk blocks, one would expect on the order of at least 5–10 gigabytes or more of aggregate internal memory. This would correspond to k being on the order of 500–1000 or more. The relative magnitudes of k, D, and B cited here are meant to represent likely scenarios; it is conceivable that there are systems with different relationships among the values of k, D, and B.

For most values of of k, D, and B, the previous section shows that SRM is extremely efficient. Looking back at the occupancy analysis, the reason for SRM's efficiency is that throwing a large number (say, kD) of balls uniformly and independently into a small number of bins (say, D) does result in a more or less balanced distribution. Even DSM will perform well when D is small and k is large, since it will merge k runs at a time, which is not much worse than merging the optimal kD runs at a time. However, SRM's extremely low overhead still gives it an advantage over DSM.

10 Conclusions

In this paper, we presented a simple, efficient mergesort algorithm for parallel disks that makes a limited use of randomization. The analysis of the I/O performance involved a reduction to certain maximum occupancy problems. We demonstrated the practical merit of the algorithm by showing that it incurs fewer I/O operations than the commonly used disk-striped mergesort, even on realistic parallel disk systems with a small number of disks. We did so analytically by estimating the maximum bucket occupancy values for several k, D pairs and empirically by using simulations to count the number of I/O operations needed by SRM in the average case. We argued that SRM may be considered an optimal external sorting algorithm in practice. The technique of staggering runs might yield further gains in practice if combined with our general approach.

References

[AP94] Alok Aggarwal and C. Greg Plaxton. Optimal parallel sorting in multi-level storage. *Proc. Fifth Annual ACM-SIAM Symp. on Discrete Algorithms*, pages 659–668, 1994.

[AV88] Alok Aggarwal and Jeffrey S. Vitter. The input/output complexity of sorting and related problems. *Communications of the ACM*, 31(9):1116–1127, 1988.

[BGV97] Rakesh D. Barve, Edward F. Grove, and Jeffrey Scott Vitter. Simple randomized mergesort on parallel disks. 23(4), 1997. Special issue on parallel I/O. An earlier version appears in Proc. of the 8th Annual ACM Symposium on Parallel Algorithms and Architectures (SPAA '96), Padua, Italy, June 1996, 109–118.

[Flo72] R. W. Floyd. Permuting information in idealized two-level storage. In R. Miller and J. Thatcher, editors, *Complexity of Computer Computations*, pages 105–109. Plenum, 1972.

[GS84] D. Gifford and A. Spector. The TWA reservation system. *Comm. ACM*, 27(7):650–665, July 1984.

[HGK+94] L. Hellerstein, G. Gibson, R. M. Karp, R. H. Katz, and D. A. Patterson. Coding techniques for handling failures in large disk arrays. *Algorithmica*, 12(2–3), 1994.

[HK81] J. W. Hong and H. T. Kung. I/O complexity: The red-blue pebble game. *Proc. 13th Annual ACM Symp. on Theory of Computation*, pages 326–333, may 1981.

[Knu98a] D. E. Knuth. Personal communication, 1998.

[Knu98b] D. E. Knuth. *Sorting and Searching*, volume 3 of *The Art of Computer Programming*. Addison-Wesley, Reading MA, second edition, 1998.

[KSC78] V. F. Kolchin, B. A. Sevastyanov, and V. P. Chistyakov. *Random Allocations.* Winston & Sons, Washington, 1978.

[LV85] E. E. Lindstrom and J. S. Vitter. The design and analysis of bucketsort for bubble memory secondary storage. *IEEE Transactions on Computers*, C-34:218–233, March 1985.

[Mag87] N. B. Maginnis. Store more, spend less: Mid-range options abound. *Computerworld*, pages 71–82, November 1987.

[NV90] M. H. Nodine and J. S. Vitter. Large-scale sorting in parallel memories. In *Proc. 3rd ACM Symp. on Parallel Algorithms and Architectures*, pages 29–39, 1990.

[NV93] M. H. Nodine and J. S. Vitter. Deterministic distribution sort in shared and distributed memory multiprocessors. *Proc. 5th Annual ACM Symp. on Parallel Algorithms and Architectures*, pages 120–129, 1993.

[PGK88] D. A. Patterson, G. Gibson, and R. H. Katz. A case for redundant arrays of inexpensive disks (raid). *Proc. 1988 ACM-SIGMOD Conf. on Management of Data*, pages 109–116, 1988.

[PSV94] V. S. Pai, A. A. Schaffer, and P. J. Varman. Markov analysis of multiple-disk prefetching strategies for external merging. *Theoretical Computer Science*, 128(2):211–239, June 1994.

[Uni89] University of California at Berkeley. *Massive Information Storage, Management, and Use (NSF Institutional Infrastructure Proposal)*, January 1989. Technical Report No. UCB/CSD 89/493.

[VF90] J. S. Vitter and Ph. Flajolet. Average-case analysis of algorithms and data structures. In Jan van Leeuwen, editor, *Handbook of Theoretical Computer Science, Volume A: Algorithms and Complexity*, chapter 9, pages 431–524. North-Holland, 1990.

[VS94] J. S. Vitter and E. A. M. Shriver. Algorithms for parallel memory I: Two-level memories. *Algorithmica*, 12(2–3):110–147, 1994.

DEPT. OF COMPUTER SCIENCE, DUKE UNIVERSITY, DURHAM, NC 27708-0129
E-mail address: `rbarve@cs.duke.edu`

E-mail address: `efg@cs.duke.edu`

E-mail address: `jsv@cs.duke.edu`

DIMACS Series in Discrete Mathematics
and Theoretical Computer Science
Volume **43**, 1999

Randomized Greedy Algorithms for the Hypergraph Partitioning Problem

R. Battiti, A. Bertossi, and R. Rizzi

ABSTRACT. We propose a series of randomized greedy construction schemes for the hypergraph partitioning problem. While the final results are inferior to those obtained by recent multi-level methods, the advantages of our greedy schemes are their simplicity and low computational complexity. The best greedy algorithms considered obtain low cut values and large standard deviations of the results. Therefore, when independent repetitions are considered, the quality of the best solution greatly improves and, in some cases, it is superior to the variable-depth Fiduccia-Mattheyses (FM) algorithm, for smaller CPU times. Furthermore, the algorithms can be used as building blocks in more complex schemes. For example, we successfully employ our greedy schemes to produce initial partitions for *improvement-based* heuristics. In particular, if FM is run starting from partitions generated by our greedy schemes, instead of random initial solutions, a significant improvement of the average solution quality is achieved for comparable computation times.

1. Introduction

The hypergraph partitioning problem is defined as follows (in this paper we consider only the case of partitioning into two sets, also called bisection). Let $H(V,E)$ be an hypergraph, that is a set V of *nodes* plus a set E of subsets of V called *hyperedges* (or simply *edges*), see also Fig. 1. One is required to partition the nodes in V into two sets, say Set_0 and Set_1, aiming at minimizing the number of *broken* edges, i.e, edges in E containing nodes on both sides of the partition, while obeying a constraint on the smallest possible sizes (min_size) of Set_0 and Set_1. Equivalently, one is given a percentage value, for example 45%, and is required to return a partition in which the cardinalities of Set_0 and Set_1 both exceed $0.45 * |V|$. Because of its physical realization (nodes are modules and a hyperedge is an electrical network connecting a set of modules) one often uses the term "circuit partitioning."

Hypergraph partitioning has recently emerged as a central issue in VLSI design [**1, 9**]. First of all, in VLSI placement, a divide and conquer approach is taken

1991 *Mathematics Subject Classification.* Primary 68M10 , 68R10 ; Secondary 68P05 , 90B12, 90C35, 90C27.

Key words and phrases. greedy algorithms, hypergraph-partitioning, data structures, computer implementation .

where the circuit is hierarchically divided into smaller components using hypergraph partitioning [**16**]. The formulation also finds applications in *rapid prototyping* where the goal typically is to partition to a minimal number of FPGAs (Field Programmable Gate Arrays) under different constraints such as available number of pins and routing resources [**4, 12, 13, 14, 22**]. Another application area is the *design for testability* of VLSI circuits where the circuit is partitioned into smaller parts to facilitate testing [**21**]. However the importance of the hypergraph partitioning problem goes beyond VLSI design. For example, electrical circuits with multiple-pin nets are readily modeled as hypergraphs. Other applications include data mining [**18**], efficient storage of large databases on disks [**19**], clustering and partitioning of the roadmap database for routing applications [**20**], de-clustering data in parallel databases [**19**].

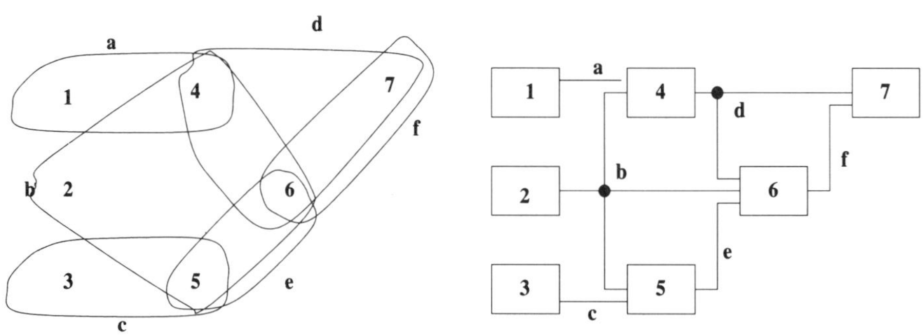

FIGURE 1. A hypergraph (left) and a concrete realization as electrical circuit (right).

In the last years one has observed a blossoming of graph and hypergraph partitioning algorithms and software packages (METIS, MELO, Paraboli, SCOTCH). The solution quality and reliability improvements which have come along is remarkable and would have been unpredictable only a few years ago [**1, 9, 5, 6**]. We refer to [**1**] for a recent survey of the problem and for detailed references.

Because graph bisection is a special case of hypergraph bisection and is NP-hard, hypergraph bisection is NP-hard as well. However, because of the importance of the problem, many heuristic algorithms have been developed [**1**]. A widely used class of *iterative improvement* partitioning algorithms start by randomly generating an initial partition and then attempt to progressively reduce the cut value of the partition by repeatedly moving vertices between the two parts of the partition. The most used among these algorithms is the Fiduccia-Mattheyses [**7, 8**] (FM) algorithm, derived from the seminal Kernighan-Lin algorithm for graph partitioning [**10**]. In the FM algorithm a vertex (that is not *locked*) is moved that results in the greatest reduction in the edge-cuts, which is also called the *gain* for moving the vertex. Because more nodes can have the same gain, Krishnamurty [**11**] proposed to look ahead up to r levels of gains before making moves as a tie breaking rule to be used when more nodes achieve the same reduction of the edge-cuts, a frequent case of real-world partitioning problems.

In this paper we analyze the performance of greedy strategies in the context of hypergraph bisection. The basic motivation of our research is to investigate whether one could extend to hypergraphs the successful performance of greedy schemes previously obtained on graphs [**2**]. The results obtained are not competitive with state-of-the-art multi-level algorithms [**9**]. Nonetheless, our analysis shows greedy schemes to be a practical and viable choice for applications characterized by strong trade-offs between solution quality and computation time, like for example declustering data in parallel databases and other applications in parallel computing. Moreover, our schemes can be used as building blocks of more complex schemes. For example, we compared the classical *improvement-based* Fiduccia-Mattheyses heuristic, starting from a random initial solution, against the same heuristic, but starting from solutions produced by a greedy scheme. While the computation time of the two algorithms is comparable, the improvement in terms of average solution quality is significant.

The following part of this paper is organized as follows. The benchmark graphs considered for the VLSI applications are described in Sec. 2, some simple greedy approaches are presented in Sec. 3, while the more effective Likelihood Greedy algorithm is presented in Sec. 4. The results obtained when Fiduccia-Mattheyses starts from our greedy solutions are discussed in Sec. 5. Finally, the implementation details, computational complexity, and measured CPU times are presented in Sec. 6.

2. Benchmarks

The computational tests in this paper are executed on the collection of hypergraphs (nets) proposed as benchmarks for the hypergraph partitioning problem by Chuck Alpert[1]. All nets are in *netlist* format.

The nets are collected into three major groups:

- **Cheng's nets**, made available by C.K. Cheng at UCSD.
- **VPNR nets**, obtained by translating in netlist format some instances first represented in VPNR format.
- **Large nets**, made available by Lars Hagen (lars@cadence.com).

The characteristics of these hypergraphs are shown in Table 1. The column labeled "Best" lists the heuristically best value from Table 7 of [**9**]. Most of these results are obtained with a finely tuned multi-level algorithms (hMETIS) that is considered the state of the art for hypergraph problems of interest in the VLSI community. The two additional columns list quantities involved in the computational complexity of some greedy schemes ($|h|$ is the cardinality of hyperarc h), see Sec 6.

Chuck Alpert has made available an efficient implementation of the Fiduccia-Mattheyses Algorithm [2]. The implementation uses a LIFO (last-in-first-out) bucket structure which many researchers discovered to be significantly more effective than either FIFO (first-in-first-out) or random bucket structures [**8**]. This FM implementation provides a standard and basic benchmark against which other partitioning codes can be measured in terms of runtimes and solution quality.

We have run our Greedy algorithms 100 times on each single net. In the tables one reports the minimum value found among all runs (Min), the average values obtained (Ave) and the standard deviation of the values (Sdev). The standard

[1]These nets are publically available on the "Circuit Partitioning Page" at URL `http://vlsicad.cs.ucla.edu/~cheese/benchmarks.html` .

[2]Available on WWW at URL `http://vlsicad.cs.ucla.edu/~cheese/codes.html`.

graph	vertices	hyperedges	pins	"Best"	$\Sigma_{h\in E(H)}\|h\|^2$	$\Sigma_{h\in E(H)}\|h\|^2 \log \|V\|$
balu	801	735	2697	27	33637	97669
p1	833	902	2908	47	15422	45042
bm1	882	903	2910	47	15426	45436
t4	1515	1658	5975	48	258673	822686
t3	1607	1618	5807	57	136691	438233
t2	1663	1720	6134	87	217016	698985
t6	1752	1541	6638	60	251541	815881
structP	1952	1920	5471	33	22555	74216
t5	2595	2750	10076	71	558422	1906529
19ks	2844	3282	10547	104	382443	1320931
p2	3014	3029	11219	139	85723	298242
s9234P	5866	5844	14065	40	46076	173630
biomedP	6514	5742	21040	83	2582834	9850535
s13207P	8772	8651	20606	53	77706	306402
s15850P	10470	10383	24712	42	90199	362595
industry2	12637	13419	48158	168	1786670	7328284
industry3	15406	21923	65791	241	425483	1781790
s35932	18148	17828	48145	41	2476306	10546163
s38584	20995	20717	55203	47	342383	1479819
avq.small	21918	22124	76231	127	63318301	274852139
s38417	23849	23843	57613	50	230903	1011190
avq.large	25178	25384	82751	127	63331341	278722576
golem3	103048	144949	338419	1424	940085	4712683

TABLE 1. The characteristics of the hypergraphs used to evaluate the algorithms. The "Best" column lists the heuristically best value from Table 7 of [**9**].

deviation of the estimated averages can be obtained by dividing the above Sdev value by the square root of the number of tests, i.e. by 10.

The computational tests discussed in this paper have been executed on a machine with the following characteristics:

- CPU: Pentium Pro
- Clock: 200 MHz
- Configuration: 160 MByte
- Operating System: Red Hat Linux release 4.1 (Vanderbilt)
- Compiler: Gnu **g++** version 2.7.2.1 with maximum optimization

In this preliminary investigation, the focus of attention is on the average results and standard deviations obtained by the different approaches and on their computational complexity more than on the the obtained computing times. Nonetheless, CPU times for the given machines are reported in Sec. 6.3. The standard deviation is of interest when independent repetitions of the algorithms are considered and the minimum is returned.

3. Simple greedy approaches

Motivated by the previous success of some greedy heuristics for the graph partitioning problem [**3, 2**], we investigate in this paper the behavior of a greedy approach for the hypergraphs partitioning problem.

Our general scheme is the following: At the beginning both Set_0 and Set_1 are empty and all nodes are marked as *free*. Then free nodes are inserted one by one into either Set_0 or Set_1. Insertions are definitive: once a node has been added to

Set_i ($i = 0, 1$) it will stay there until the end of the algorithm. A node inserted into Set_i is no longer free but *fixed* to i.

To ease the comparisons among different strategies, we have adopted the following rule. First, nodes are added alternately to Set_0 and Set_1 until the sizes of both sets reach the *min_size* value. Finally, the remaining free nodes are inserted one by one into either set, aiming at obtaining the smallest increase of newly broken edges.

The algorithms considered in this paper correspond to the above scheme, differing only in the way they choose the next node to be inserted. Each greedy scheme can be described with the help of a *gain* function which expresses the *pricing policy* of the heuristic. For each node v, $gain(v, 0)$ is the heuristic's estimate of the advantage of assigning v to Set_0. Similarly $gain(v, 1)$ is the gain for assigning v to Set_1. At each move the scheme greedily strives to maximize the gain.

All nodes of a certain gain are stored into a list and the following criterion [**8**] is used for inserting and deleting nodes. When the gain of a node is increased the node is inserted at the beginning of the corresponding list. When the gain of a node is decreased the node is inserted at the end. At each move the greedy scheme chooses the first node in the list of nodes of maximum gain.

A common characteristic of the schemes considered here is that the gain of node v is the sum of many terms, one for every hyperedge containing v:

$$gain(v; i) = \sum_{h \ni v} g_h(v; i)$$

A scheme is defined by specifying $g_h(v; i)$ for a generic hyperedge h. When $g(v; 0) = -g(v; 1)$ for each node v, the scheme is said *consistent.* Consistent schemes turned out to be most effective. When a scheme is consistent we can define the gain of a node as an absolute value (not depending on a specific side of the bipartition), precisely as $g(v) = g(v; 1)$. When choosing a node to be inserted into Set_1, the heuristic strives to maximize the gain. When searching for a node to be inserted into Set_0, a node of minimum gain is chosen. More precisely, all consistent schemes obey the following rules:

(i) when we seek for a node to be inserted into Set_1, then we choose randomly and with uniform probability a node of maximum gain.

(ii) when we seek for a node to be inserted into Set_0 then we choose randomly and with uniform probability a node of minimum gain.

(iii) Assume we have to choose a node to be inserted in whichever set of the partition. If max gain $> -$min gain then we decide to insert into Set_1 and go to *(ii)*. If max gain $= -$min gain then we choose at random and with equal probability whether to insert into Set_1 and go to *(ii)* or to insert into Set_0 and go to *(i)*. Otherwise we decide to insert into Set_0 and go to *(i)*.

We are interested in heuristics which are robust and can be executed repeatedly to improve the solution quality. Therefore *randomization* is desirable in our greedy schemes. At the beginning, when all nodes are still free and all gains are zero, all nodes are inserted into the zero-gain list in a random order. The experiments show that this source of randomness suffices to derive a robust heuristic. The standard deviation of the cut value returned for a given input hypergraph is significant, so that much better results can be achieved by repeating the greedy construction with a different random seed.

Let us now describe how the gains are calculated in the different schemes.

3.1. Δf Greedy. The method is the most straightforward greedy approach, based only on the consideration of the function to be minimized. In each step of the algorithm, let f be the number of broken edges. At termination f is the value of the partition obtained. An edge is said to be *broken* when it contains nodes both fixed to 0 and fixed to 1.

The idea behind the Δf Greedy algorithm is to choose randomly between those nodes which produce the smallest possible increase of the number of broken edges (the value of f) when they are added to the current partition.

Let v be a free node and h an hyperedge. If $v \notin h$ then $g_h(v;i) = 0$ for $i = 0, 1$. Otherwise $g_h(v;i)$ is given as follows.

$$g_h(v;i) = \begin{cases} -1 & \text{if } |h|_{1-i} > 0 \text{ and } |h|_i = 0 \\ 0 & \text{otherwise} \end{cases}$$

where $|h|_i = |h \cap Set_i|$ (for $i = 0, 1$). When we look for a node to be added to Set_i we search for a free node v which maximizes $gain(v;i)$. When i is not given, that is, when $|Set_0| \geq min\_size$ and $|Set_1| \geq min\_size$, then one maximizes the gain both over v and over $i \in \{0, 1\}$. If there are ties, a random choice among the winning nodes is executed. The implementation uses a standard "buckets" structure.

As expected, and as noted independently by many previous researchers, the results obtained by this immediate greedy algorithm are very unsatisfactory, see Table 2. The problem is caused by the large number of ties that are present during the greedy construction: the algorithm is forced to make a "blind" choice among a very large set of winning nodes.

3.2. Critical Edges Greedy. In our previous work about graph partitioning [**2**] a greedy approach with a gain function given by the *difference* between connections of node v to Set_0 and connections of node v to Set_1 was the most successful option. Therefore we were motivated to modify the method for the case of hypergraphs.

The modifications must take into account the fact that different hyperedges h containing a node v should not be given the same weight when deciding about an addition. In particular, if h contains already nodes in both sets (see situation (a) in Fig. 2) it should not be considered: there is no way to undo its cut. If at least two nodes of h are free (situations (b) and (c) in Fig. 2) and all other nodes, if any, are fixed to the same set, then one cannot decide the fate of h during a single addition. For example, in situation (c), even if node u is added to Set_0, node v may be added to Set_1 in the future steps, therefore breaking the hyperedge. The cases in which the fate can be decided in one step are those corresponding to situation (d), where only one node is free and all other nodes have been added to the same set. If the free node is added to the same set, h is not broken, otherwise it is. An edge is said 0-*critical* when all of its nodes are fixed to 0, except for a single free node. An edge is 1-*critical* when all of its nodes are fixed to 1, except for a single free node.

Critical Edges Greedy is our first example of a consistent scheme resulting from the adaptation of the Differential-Greedy algorithm [**2**] to hypergraphs. In each step of the algorithm, the *gain* of a free node v is defined as the number of 1-*critical* edges containing v minus the number of 0-*critical* edges containing v. The implementation uses the "buckets" data structure.

When the results are compared with those of Δf Greedy, see Table 2, one observes a much better performance for most graphs. In particular, macroscopic

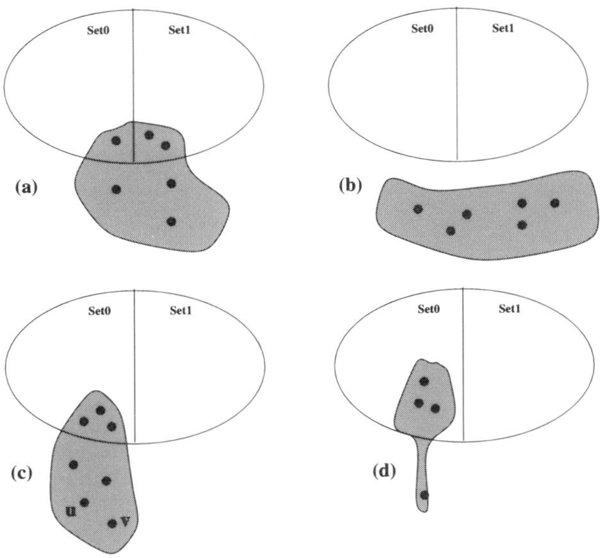

FIGURE 2. Critical and non-critical hyperedges (see text for explanation).

differences are observed for the larges graphs. For example, when the minimum value is considered, one passes from 5,748 to 1,168 broken edges for `industry3`, from 32,034 to 2,383 for `golem3`. Nonetheless, the average and minimum results are very far from those of the FM algorithm.

3.3. Weighted Greedy. The Weighted Greedy algorithm is a second consistent scheme with the aim of considering in a more gradual way the effect of non-critical edges. The previous approach does not give any preference information when the weights obtained by considering only critical edges is the same for different nodes.

Let v be a free node and h an hyperedge. If $v \notin h$ or if h is broken then $g_h(v) = g_h(v;1) = 0$. Otherwise $g_h(v) = \frac{|h|_1}{|h|} - \frac{|h|_0}{|h|}$, where $|h|$ is the number of nodes in h and $|h|_i = |h \cap Set_i|$ for $i = 0, 1$. (In particular either $|h|_1 = 0$ or $|h|_0 = 0$ because h is not broken). The implementation uses binary heaps to store the gain values.

The last column in Table 2 presents the results obtained by the Weighted Greedy algorithm. A notable improvement is obtained with respect to the Critical Edges Greedy. In particular, both the average and the minimum results become roughly comparable to those obtained by FM. In some cases FM results are beaten (these cases are written in boldface in the table). In the other cases the difference is often small.

4. The Likelihood Greedy algorithm

Weighted Greedy clearly outperforms all competitors examined in Table 2. Critical Edges Greedy suffers from the following weakness: only critical edges play a role in the choices of the algorithm. Our first attempt to consider non-critical edges was represented by the Weighted Greedy algorithm. However the policy of Weighted Greedy can be ameliorated. In particular the gain calculation can be

Graph	Fiduccia		Δf Greedy		Critical Edges Greedy		Weighted Greedy	
	Min	Ave (Sdev)	Min	Ave (Sdev)	Min	Ave (Sdev)	Min	Ave (Sdev)
baluP	30	66.3 (18.6)	152	198.1 (17.0)	160	196.8 (11.8)	**28**	**63.1 (23.1)**
p1	47	74.9 (13.7)	202	249.1 (18.6)	147	192.4 (15.2)	57	90.2 (16.9)
bm1	49	75.8 (14.9)	192	242.5 (18.8)	164	197.4 (15.7)	59	87.9 (14.6)
t4	80	135.6 (25.4)	352	426.3 (28.8)	259	307.7 (20.3)	**55**	**114.6 (28.0)**
t3	62	108.7 (22.1)	305	422.0 (33.8)	237	308.4 (22.4)	67	128.1 (34.8)
t2	124	175.5 (24.2)	395	472.9 (26.5)	258	324.1 (17.3)	**100**	**149.3 (26.7)**
t6	60	91.0 (14.8)	417	475 (26.7)	183	212.6 (17.4)	62	117.6 (30.0)
structP	41	55.4 (8.6)	406	444.4 (18.4)	346	418.6 (26.9)	**41**	106.6 (35.6)
t5	104	179.4 (35.6)	521	688.9 (43.7)	371	446.8 (33.4)	**86**	**151.6 (36.8)**
19ks	130	178.9 (26.8)	583	696.3 (40.8)	416	479.2 (32.0)	**124**	221.6 (55.1)
p2	182	278.5 (40.3)	832	928.5 (47.0)	752	808.9 (21.5)	232	319.9 (54.6)
s9234P	51	89.9 (25.2)	1122	1182.9 (25.4)	326	372.2 (16.9)	60	123.9 (44.9)
biomedP	83	126.5 (40.1)	1259	1569.4 (172.2)	890	944.7 (23.2)	128	223.1 (34.6)
s13207P	78	126.2 (21.3)	1553	1670.2 (34.3)	456	499.4 (16.4)	118	181.0 (37.7)
s15850P	104	184.0 (29.4)	1934	2030.4 (34.3)	558	627.9 (20.9)	**104**	**167.9 (33.0)**
industry2	264	602.1 (161.8)	2815	3233.0 (157.1)	1702	1885.1 (73.3)	306	715.3 (295.7)
industry3	263	484.6 (161.0)	5748	6167.1 (112.6)	1168	2648.3 (815.6)	555	961.5 (335.0)
s35932	85	203.5 (53.7)	3170	3302.0 (52.1)	1137	1245.6 (42.2)	**45**	**187.8 (92.9)**
s38584	63	258.8 (120.3)	4300	4476.6 (64.7)	1788	1889.9 (42.4)	97	**253.5 (106.7)**
avq.small	343	620.3 (116.3)	4904	5573.6 (233.7)	3454	3637.0 (70.2)	**328**	**590.7 (120.3)**
s38417	147	368.0 (92.3)	4747	4870.3 (58.4)	1546	1647.3 (36.8)	**95**	**198.2 (48.6)**
avq.large	373	759.3 (123.9)	5335	6166.7 (295.0)	3449	3625.1 (79.6)	**325**	**694.9 (157.5)**
golem3	2175	3217.9 (304.0)	32034	32379.5 (160.1)	2383	3449.6 (714.2)	5634	6109.6 (214.5)

TABLE 2. Simple greedy schemes compared with Fiduccia-Mattheyses (FM)

unfair if two hyperedges h_1 and h_2 have widely different numbers of *nodes already added.* For example, suppose that $|h_1| = 3$, $|h_2| = 6$, while $|h_1|_1 = 2$ and $|h_2|_1 = 4$ (and no node in Set_0). The contribution to the gain from h_1 and from h_2 will be equal, while h_1 is a critical edge and the addition of the single remaining free node to Set_1 should be preferred to the addition of a free node in h_2, with the danger of a subsequent cut of h_2 when its second node is added. In other words, the number of nodes already added to hyperedge should *not* influence the addition of the remaining nodes. Our first greedy approaches have been modified to take these comments into account and the experimental results obtained confirm the intuitive expectations.

Likelihood Greedy is a consistent scheme. Let v be a free node and h an hyperedge. If $v \notin h$ or if h is broken then $g_h(v) = g_h(v;1) = 0$. Otherwise $g_h(v)$ is given as follows.

$$g_h(v) = \begin{cases} \frac{1}{2^{|h|_f - 1}} & \text{if } |h|_1 > 0 \text{ and } |h|_0 = 0 \\ -\frac{1}{2^{|h|_f - 1}} & \text{if } |h|_0 > 0 \text{ and } |h|_1 = 0 \\ 0 & \text{otherwise.} \end{cases}$$

where $|h|_i = |h \cap Set_i|$ (for $i = 0, 1$) and $|h|_f$ denotes the number of free nodes of h.

The rationale for such a choice of gains is the following. Let u be any free node of h. The *a priori* probability of node u being assigned to Set_0 is $\frac{1}{2}$ and it is the same as the probability of node u being assigned to Set_1. Let us assume that $|h|_1 > 0$ and $|h|_0 = 0$. If v is assigned to Set_0 then hyperedge h gets broken. To favor additions to Set_1 the gain $g_h(v)$ is positive. However, assigning node v to Set_1 is not sufficient to guarantee that h will not be in the final cut set

Graph	Fiduccia		(3)-Likelihood Greedy		(10)-Likelihood Greedy		Likelihood Greedy	
	Min	Ave (Sdev)	Min	Ave (Sdev)	Min	Ave (Sdev)	Min	Ave (Sdev)
baluP	30	66.3 (18.6)	33	62.2 (15.1)	29	51.6 (13.8)	**29**	**51.4 (14.1)**
p1	47	74.9 (13.7)	65	89.2 (11.3)	67	83.7 (10.0)	63	82.9 (11.0)
bm1	49	75.8 (14.9)	61	88.8 (11.3)	59	87.0 (9.9)	62	86.2 (9.9)
t4	80	135.6 (25.4)	74	128.0 (19.9)	69	112.2 (17.3)	**67**	**104.4 (19.8)**
t3	62	108.7 (22.1)	70	123.4 (30.2)	60	100.3 (27.6)	**60**	**93.9 (25.7)**
t2	124	175.5 (24.2)	108	142.2 (16.7)	104	138.6 (19.3)	**103**	**141.4 (18.1)**
t6	60	91.0 (14.8)	93	122.3 (12.9)	64	89.8 (11.7)	67	**84.9 (9.6)**
structP	41	55.4 (8.6)	42	82.9 (19.5)	46	84.2 (19.3)	46	82.3 (21.7)
t5	104	179.4 (35.6)	109	145.0 (22.0)	85	123.6 (22.9)	**76**	**127.0 (22.4)**
19ks	130	178.9 (26.8)	163	248.2 (36.7)	166	225.4 (27.6)	165	225.1 (27.4)
p2	182	278.5 (40.3)	252	339.1 (31.0)	232	313.0 (36.6)	222	307.5 (40.1)
s9234P	51	89.9 (25.2)	67	130.3 (35.1)	61	114.6 (33.1)	63	129.5 (28.1)
biomedP	83	126.5 (40.1)	156	214.2 (36.1)	123	194.5 (33.7)	91	159.6 (44.1)
s13207P	78	126.2 (21.3)	115	168.0 (12.4)	98	158.7 (16.9)	97	151.0 (17.3)
s15850P	104	184.0 (29.4)	90	150.7 (29.4)	89	133.6 (25.4)	**92**	**131.7 (26.2)**
industry2	264	602.1 (161.8)	601	881.0 (114.0)	499	762.7 (112.6)	310	768.2 (166.4)
industry3	263	484.6 (161.0)	618	1047.7 (308.4)	454	873.7 (274.1)	334	819.0 (341.1)
s35932	85	203.5 (53.7)	85	155.2 (39.8)	74	105.0 (25.8)	**73**	**103.6 (33.9)**
s38584	63	258.8 (120.3)	165	275.4 (64.5)	93	202.1 (73.1)	73	**202.6 (74.6)**
avq.small	343	620.3 (116.3)	312	535.6 (82.6)	228	516.3 (74.5)	**241**	**513.3 (73.1)**
s38417	147	368.0 (92.3)	91	180.8 (38.9)	82	144.6 (28.1)	**74**	**143.6 (24.8)**
avq.large	373	759.3 (123.9)	376	584.8 (87.7)	348	567.0 (89.3)	382	**567.1 (87.4)**
golem3	2175	3217.9 (304.0)	5084	5736.0 (229.1)	1935	3077.8 (312.3)	2405	**3009.5 (313.4)**

TABLE 3. Different versions of the Likelihood Greedy scheme

returned by the algorithm. In fact, this will require that each free node of h is assigned to Set_1. Therefore, if node v is assigned to Set_1, then the probability of h not being broken is $\frac{1}{2^{|h|_f-1}}$, in the crude assumption that each node has equal probability of being added to either set and that the individual additions are done independently. In fact, let x_h be a random variable associated to our random heuristic and whose value is 1 if edge h is in the final cutset returned by the algorithm and 0 otherwise. Because assigning v to Set_0 breaks h then the expected value $E[x_h|v \to Set_0] = 1$. On the other side $E[x_h|v \to Set_1] = 1 - \frac{1}{2^{|h|_f-1}}$. Thus $g_h(v) = g_h(v;1) = E[x_h|v \to Set_0] - E[x_h|v \to Set_1] = \frac{1}{2^{|h|_f-1}}$.

This greedy scheme is not to be confused with the probability-based approach proposed by Dutt and Deng [**5**] in the framework of iterative-improvement methods. It is of interest to observe that rough estimates of probabilities to end up in the different sets are useful in both algorithms to greatly improve the performance.

In order to consider in more detail the effect of this probabilistic strategy we consider also a version called (k)-Likelihood Greedy, in which the gain is set to zero if the number of free nodes in an hyperedge is greater than k.

In Table 3 the three columns in addition to that for FM (that is always reported to ease the comparison) list the results obtained by the (3)-Likelihood Greedy, (10)-Likelihood Greedy, and the Likelihood Greedy without limitations on the size of the hyperarcs. When k increases in (k)-Likelihood Greedy the performance improves rapidly at the beginning ($k = 2, 3$) and then reaches a "performance plateau", explained by the fact that hyperedges with large number of nodes are rare in the benchmark. When one considers the average values, the performance of the Likelihood Greedy scheme without limitations is in general better than that

Graph	Fiduccia		Greedy (Likelihood)		Greedy + Fiduccia		Perf. Ratio
	Min	Ave (Sdev)	Min	Ave (Sdev)	Min	Ave (Sdev)	Min / "Best"
baluP	30	66.3 (18.6)	29	51.4 (14.1)	**27**	**32.9 (7.5)**	1
p1	47	74.9 (13.7)	63	82.9 (11.0)	**47**	**69.5 (8.4)**	1
bm1	49	75.8 (14.9)	62	86.2 (9.9)	**48**	**72.9 (9.0)**	1.02
t4	80	135.6 (25.4)	67	104.4 (19.8)	**60**	**89.0 (14.5)**	1.25
t3	62	108.7 (22.1)	60	93.9 (25.7)	**60**	**82.2 (17.9)**	1.05
t2	124	175.5 (24.2)	103	141.4 (18.1)	**94**	**127.0 (16.8)**	1.08
t6	60	91.0 (14.8)	67	84.9 (9.6)	**60**	**76.5 (8.7)**	1
structP	41	55.4 (8.6)	46	82.3 (21.7)	**41**	57.1 (9.0)	1.24
t5	104	179.4 (35.6)	76	127.0 (22.4)	**74**	**102.2 (16.6)**	1.04
19ks	130	178.9 (26.8)	165	225.1 (27.4)	**127**	181.3 (21.9)	1.22
p2	182	278.5 (40.3)	222	307.5 (40.1)	**171**	**246.7 (34.4)**	1.23
s9234P	51	89.9 (25.2)	63	129.5 (28.1)	**49**	106.5 (28.5)	1.22
biomedP	83	126.5 (40.1)	91	159.6 (44.1)	84	**120.5 (33.9)**	1.01
s13207P	78	126.2 (21.3)	97	151.0 (17.3)	91	128.2 (16.2)	1.71
s15850P	104	184.0 (29.4)	92	131.7 (26.2)	**76**	**117.3 (24.7)**	1.80
industry2	264	602.1 (161.8)	310	768.2 (166.4)	**216**	**517.2 (128.0)**	1.28
industry3	263	484.6 (161.0)	334	819.0 (341.1)	**253**	555.9 (223.2)	1.04
s35932	85	203.5 (53.7)	73	103.6 (33.9)	**73**	**101.6 (29.5)**	1.78
s38584	63	258.8 (120.3)	73	202.6 (74.6)	**56**	**138.8 (59.3)**	1.19
avq.small	343	620.3 (116.3)	241	513.3 (73.1)	**220**	**474.6 (79.7)**	1.73
s38417	147	368.0 (92.3)	74	143.6 (24.8)	**68**	**128.8 (22.5)**	1.36
avq.large	373	759.3 (123.9)	382	567.1 (87.4)	**270**	**514.1 (70.4)**	2.12
golem3	2175	3217.9 (304.0)	2405	3009.5 (313.4)	**1633**	**2367.2 (265.7)**	1.14

TABLE 4. FM starting from Likelihood Greedy solution

of Weighted Greedy. On the contrary, when one considers the *minimum* values, some of the advantage is lost, and Weighted Greedy obtains superior results in about 50% of the cases. This effect is caused by the smaller standard deviation of results obtained by Likelihood Greedy: a more accurate selection of the node to be added improves the average results but decreases the amount of "diversification." In the last columns of Table 3 a bold value indicates that FM results are beaten or duplicated. Let us note that the performance is particularly significant if one considers that the CPU times requires by the Likelihood Greedy scheme are less than those required by FM, in some cases by a large factor, see Table 5.

5. Fiduccia-Mattheyses starting from our greedy solutions

Because greedy schemes are quite effective and fast, it is worth considering the solution quality obtained if one runs the Fiduccia-Mattheyses algorithm starting from a solution generated by a greedy scheme instead that from a random partition.

Table 4 displays the results obtained by executing the Fiduccia-Mattheyses procedure starting from a solution generated by the *Likelihood Greedy* heuristic. In addition, the last column shows the "performance ratio" of the combination, defined as the "Min" value (obtained in the 100 runs) divided by the "Best" heuristic value of Table 1.

6. Implementation and computational complexity

A main contribution of Fiduccia and Mattheyses [7] was the use of the now standard *buckets* data structure: nodes with a same gain are collected into the same bucket, so that the nodes with maximum (or minimum) gain values are immediately available. Motivated by the results in [**8**], our buckets are implemented as doubly linked lists and we confirmed experimentally that the following policy is most convenient: when $g(v;i)$ increases then v is inserted at the beginning of the bucket corresponding to the new value; when $g(v;i)$ decreases then v is inserted at the end of the new bucket. When we are extracting a node of maximum gain (for insertion into Set_0 or Set_1) then we take the first node in the corresponding bucket. For a consistent scheme the above policy translates as follows: when $g(v)$ is increasing then v is inserted at the beginning of the bucket corresponding to the new value; when $g(v)$ is decreasing then v is inserted at the end of the new bucket. When we are extracting a node of maximum gain (for insertion into Set_1) then we take the first node in the bucket of maximum gain; when we extract a node of minimum gain (for insertion into Set_0) then we take the last node in the bucket of minimum gain nodes. As suggested in [**8**], the randomization is introduced by permuting the nodes randomly at the beginning of the algorithm, just before insertion into the buckets. The remaining steps of the algorithms are deterministic with only one exception: when both sides of the partition contain at least *min_size* nodes and the maximum gains achievable by inserting into Set_0 and into Set_1 are the same, then we choose at random and with equal probability whether to perform the next insertion into Set_0 or into Set_1.

For the additional implementational details it is convenient to distinguish between the various schemes considered, as it is done in the next two subsections.

6.1. Δf Greedy and Critical Edges Greedy. Given an hypergraph $H(V,E)$, the cardinality $|e|$ of an edge $e \in E$ is the number of nodes contained in e. We denote by $card(H)$ the maximum cardinality of an edge in E. Let p denote the number of pins in the hypergraph, that is $p = \sum_{e \in E} |e|$. The degree of node v is the number of hyperedges containing v. We denote by $deg(H)$ the maximum degree of a node in H.

Storing an hypergraph requires an amount of memory which is linear in p, the number of pins. The time needed to input the hypergraph is hence $\Theta(p)$. We will show that Δf and Critical Edges Greedy require $\Theta(p)$ time to return a partition together with its cut value. One cannot ask for more, because computing the cut value of a given partition requires $\Theta(p)$ time on its own.

Let us denote by γ the range of different possible gain values during execution. Let us note that the value of γ depends on the single greedy scheme under consideration. In particular, both in Δf Greedy and Critical Edges Greedy, the gain of node v is an integer with absolute value bounded by the degree of node v, hence $\gamma \leq 2\ deg(H) + 1$. This allows for the standard buckets data structure, the one introduced by Fiduccia and Mattheyses [**7**]. In particular, buckets for all possible gain values are allocated, together with two additional arrays. The first stores for each node v the gain of v (and hence which bucket contains v). The second records the position of v inside the bucket associated to $g(v)$. This ensemble can be thought of as a two-coordinate system which results in an efficient (constant time) implementation of the needed operations. We refer to [**7, 8, 2**] and [**1**] for a detailed introduction to the standard use of buckets. Let us only mention that the data structure permits to update the gains of the nodes in constant time while keeping the buckets of minimum and maximum gains always at disposal. In particular, choosing the next node to be fixed (the first or last node into the minimum or maximum gain bucket) is a constant time operation whose cost can be considered as $O(|V|)$ for the overall execution. Therefore it remains to bound the complexity of updating the gains. As we said above the gain of node v can be seen as the sum of many terms, one for every hyperedge containing v:

$$gain(v) = \sum_{h \ni v} g_h(v)$$

Moreover, let us note that $g_h(v_1)$ and $g_h(v_2)$ have the same value for any two free nodes $v_1, v_2 \in h$. When this common value $g_h(.)$ changes, we say that edge h *changes status.* Each time a node is fixed we must look at all edges h containing v and check if they change their status. If this happens then the gain values of all free nodes in h must be updated. In detail, the following operations are executed:

```
1.  for each node v
2.      fix v either to 0 or 1
3.      for each edge h containing v
4.          if h changes status by fixing v then
5.          update the gain of each free node in h.
```

In the case of Δf Greedy each edge h undergoes one or two status changes: when a first node of h is fixed (either to 0 or 1) and when h gets eventually broken. Thus the total number of gain updates is $\Theta(p)$.

In the case of Critical Edges Greedy each edge h undergoes at most one status change, if it eventually becomes 0-critical or 1-critical. However when this happens h contains precisely one free node and hence the total number of gain updates is $O(|E|)$.

Let us now consider the computational complexity cause by the control structure in lines 1. — 4. For each edge h we store $|h|_0$, $|h|_1$ and $|h|_f$. Each time we test *"if h changes status"* we first update the values of $|h|_0$, $|h|_1$ and $|h|_f$. From the knowledge of $|h|_0$, $|h|_1$ and $|h|_f$ we can readily decide if h gets broken or fixed to either 0 or 1 (Δf Greedy). We can also check if h becomes critical (Critical Edges Greedy). We conclude that testing *"if h changes status"* requires constant time in all three schemes. The cost caused by the control structure of lines 1. — 4. is therefore $\Theta(p)$. To summarize, the complexity of Δf and Critical Edges Greedy is $\Theta(p)$.

The above asymptotic complexity analysis is complemented by the measured average PU times of the different greedy schemes in Table 5, where the effects of non-asymptotic terms is also visible.

6.2. Weighted Greedy and Likelihood Greedy. The situation is significantly different when considering Weighted Greedy and Likelihood Greedy. In fact, standard buckets permit to update the gains of the nodes in constant time if the gain values are integers of bounded value. If this is not the case, then binary heaps can be used to store the gain values. Our implementation of Weighted and Likelihood Greedy employs a binary heap structure. In particular, the binary heap template "sortseq" from LEDA [**17**] has been used. Here, the amortized time cost for a single bucket operation (node update, extract max or min) is $O(\log b)$, where b is the maximum number of buckets ever contained in the list. Obviously $b \leq |V|$.

Graph	Fiduccia Ave	Δf Ave	Critical Ave	Weighted Ave	Likely(3) Ave	Likely(10) Ave	Likely Ave	Likely + Fid Ave
baluP	0.049	0.012	0.008	0.014	0.007	0.042	0.054	0.083
p1	0.071	0.013	0.005	0.016	0.008	0.040	0.053	0.082
bm1	0.072	0.011	0.008	0.017	0.008	0.043	0.054	0.091
t4	0.182	0.027	0.010	0.045	0.016	0.209	0.151	0.226
t3	0.259	0.027	0.013	0.085	0.015	0.190	0.237	0.302
t2	0.200	0.027	0.012	0.042	0.018	0.207	0.129	0.211
t6	0.270	0.028	0.013	0.051	0.017	0.049	0.212	0.288
structP	0.208	0.027	0.014	0.028	0.019	0.043	0.059	0.172
t5	0.433	0.048	0.017	0.079	0.024	0.238	0.277	0.406
19ks	0.648	0.049	0.023	0.116	0.027	0.084	0.428	0.701
p2	0.619	0.050	0.021	0.081	0.029	0.088	0.327	0.583
s9234P	1.302	0.100	0.049	0.073	0.054	0.093	0.175	0.605
biomedP	1.750	0.127	0.049	0.140	0.057	0.123	0.460	1.081
s13207P	1.960	0.159	0.081	0.121	0.088	0.138	0.294	0.966
s15850P	2.207	0.194	0.101	0.144	0.113	0.179	0.377	0.895
industry2	4.323	0.310	0.128	0.945	0.164	0.343	3.134	4.263
industry3	4.202	0.429	0.171	0.407	0.229	0.444	1.493	3.748
s35932	6.052	0.387	0.169	0.385	0.193	0.315	0.578	1.097
s38584	6.284	0.489	0.226	0.455	0.247	0.421	1.386	2.995
avq.small	7.943	0.547	0.216	1.530	0.250	0.422	5.772	7.079
s38417	5.977	0.511	0.261	0.409	0.262	0.415	1.014	2.430
avq.large	8.632	0.612	0.259	1.686	0.292	0.490	8.181	8.283
golem3	43.129	3.072	1.676	2.254	2.806	4.422	5.250	39.729

TABLE 5. Average CPU times obtained for the different algorithms

Not only the cost of updating a single gain value is now $O(\log|V|)$ as discussed before, but the total number of gain updates increases.

The crucial point is that in these greedy schemes edges change status each time one of their nodes is fixed. For any edge h, this happens precisely $|h|$ times, as many as the nodes in h. When edge h changes status, we update the gain of at most $|h|$ nodes. The total number of gain updates is therefore $\Theta(\sum_{h\in E(H)}|h|^2)$. The complexity of these greedy schemes is therefore $O(\sum_{h\in E(H)}|h|^2\log|V|)$.

6.3. CPU times. The above described computational complexities agree in a qualitative way with the measured CPU times, although the non-asymptotic terms in the complexity functions are still observable for the small size problems. In particular, the terms proportional to $|V|$, the number of nodes, are not negligible with respect to the higher order terms.

When one analyzes the computation times, see Table 6.3, one observes a roughly comparable (and in some cases smaller) average computation times for the combined option (Likelihood Greedy followed by Fiduccia-Mattheyses) than for MF starting from random solutions, especially for the larger graphs. This phenomenon is due to the fact that, while our greedy schemes are not much more expensive than what is required for a random assignment and function evaluation, the time subsequently spent in the Fiduccia-Mattheyses refinement phase tends to be smaller if one starts from a reasonably good initial partition. Let us remember that, when the average and minimum values of the combined option are considered, the results are better than those of FM in most cases, as it is shown in Table 4.

7. Conclusions

Motivated by a previous success with some greedy heuristics for the graph partitioning problem [**3, 2**], we investigated in this paper the behavior of a greedy approach to the hypergraph partitioning problem.

The conclusions to be derived from our experimental tests are that these techniques are *not* competitive with the state of the art, that is given by multi-level algorithms especially tuned to the hypergraph partitioning instances of interest for VLSI applications. The best values obtained are in most cases within 20-30% but in some cases up to two times larger than the best heuristic values.

Nonetheless, two advantages of our greedy schemes against more sophisticated methods are simplicity and low computational complexity. More surprisingly, if we look at the quality of the best solution over independent repetitions and within a same amount of allotted CPU time, some of our greedy schemes compare well against heuristics which have been the state of the art until some years ago, in particular the Fiduccia-Mattheyses algorithm. The use of independent repetitions is also at the basis of the popular "greedy randomized adaptive search procedure" (GRASP) approach [**15**]. The algorithms considered in this paper are simpler, because no criteria for the construction of candidate lists is considered, beyond the criterion given by having the same gain values (ties). Even the average quality of solutions returned after a single run could be of interest for some applications, considering that the time requirements are in some cases comparable with those required for a random assignment plus objective function (cut value) evaluation. Moreover, our schemes can be employed to produce initial partitions for *improvement-based* heuristics. For example, we compared the classical Fiduccia-Mattheyses heuristic, starting from a random initial solution, against the same heuristic, but starting from solutions produced by a greedy scheme. While the computation time of the two algorithms is roughly the same, the improvement in terms of solution quality is significant.

Of course, the present work is a preliminary investigation. The next step is to integrate our proposed greedy techniques into state-of-the-art multi-level techniques.

References

[1] C.J. Alpert and A.B. Kahng, *Recent directions in netlist partitioning: A survey*, UCLA Computer Science Department, Los Angeles, 1996.

[2] R. Battiti and A. A. Bertossi, *Differential greedy for the 0–1 equicut problem*, Network Design: Connectivity and Facilities Location (D.Z. Du and P.M. Pardalos, eds.), DIMACS: Series in Discrete Mathematics and Theoretical Computer Science, vol. 40, American Mathematical Society, 1997, pp. 3–22.

[3] ———, *Greedy and prohibition–based heuristics for graph-partitioning*, Tech. Report UTM-97-512, Dip. di Matematica, Univ. di Trento, 1997, revised version to appear in IEEE Transactions on Computers.

[4] N.C. Chou, L.T. Liu, and C.K. Cheng, *Circuit partitioning for huge logic emulation systems*, Proc. 31th ACM/IEEE Design Automation Conference, 1994, pp. 244–249.

[5] S. Dutt and W. Deng, *A probability-based approach to VLSI circuit partitioning*, Proc. 31th ACM/IEEE Design Automation Conference, June 1996.

[6] ———, *VLSI circuit partitioning by cluster-removal using iterative improvement techniques*, Proc. IEEE/ACM International Conference on CAD, November 1996.

[7] C.M. Fiduccia and R.M. Mattheyses, *A linear time heuristic for improving network partitions*, Proc. ACM/IEEE Design Automation Conference, 1982, pp. 175–181.

[8] L.W. Hagen, D.J.-H. Huang, and A.B. Kahng, *On implementation choices for iterative improvement partitioning algorithms*, IEEE Trans. Computer-Aided Design (1995), submitted.
[9] G. Karypis, R. Aggarwal, V. Kumar, and S. Shekhar, *Multilevel hypergraph partitioning: Applications in VLSI domain*, Tech. report, University of Minnesota, Department of Computer Science, April 1997.
[10] B. Kernighan and S. Lin, *An efficient heuristic procedure for partitioning graphs*, Bell Systems Technical J. **49** (1970), 291–307.
[11] B. Krishnamurthy, *An improved min-cut algorithm for partitioning VLSI networks*, IEEE Trans. Computers **33** (1984), no. 5, 438–446.
[12] R. Kuznar, F. Brglez, and K. Kozminski, *Cost minimization of partitions into multiple devices*, Proc. 30th ACM/IEEE Design Automation Conference, 1993, pp. 315–320.
[13] ———, *Multi-way netlist partitioning into heterogeneus: fpgas and minimization of total device cost and interconnect*, Proc. 31th ACM/IEEE Design Automation Conference, 1994, pp. 238–243.
[14] ———, *A unified cost model for min-cut partitioning with replication applied to optimization of large heterogeneous fpga partitions*, Proc. 31th ACM/IEEE Design Automation Conference, 1994, pp. 271–276.
[15] M. Laguna, T. A. Feo, and H. C. Elrod, *A greedy randomized adaptive search procedure for the two-partition problem*, Oper. Res. **42** (1994), 677.
[16] T. Lengauer, *Combinatorial algorithms for integrated circuit layout*, John Wiley and Sons, 1990.
[17] K. Mehlhorn and S. Naeher, *LEDA, a platform for combinatorial and geometric computing*, Communications of the ACM **38** (1995), no. 1, 96–102.
[18] B. Mobasher, N. Jain, E.H. Han, and J. Srivastava, *Web mining: Pattern discovery from world wide web transactions*, Tech. Report TR-96-050, University of Minnesota, Department of Computer Science, 1996.
[19] S. Shekhar and R. Aggarwal, *Partitioning similarity graphs: A framework for declustering problems*, Tech. Report TR-94-18, University of Minnesota, Department of Computer Science, 1994.
[20] ———, *Clustering roadmaps for routing: A hypergraph approach*, Tech. report, University of Minnesota, Department of Computer Science, 1997.
[21] S. Tragoudas, R. Farrell, and F. Makedon, *Circuit partitioning into small sets: A tool to support testing with further applications*, Proc. 28th ACM/IEEE Design Automation Conference, 1991, pp. 518–521.
[22] N.S. Woo and J. Kim, *An efficient method of partitioning circuits for multiple-fpga implementation*, Proc. 30th ACM/IEEE Design Automation Conference, 1993, pp. 202–207.

(R. Battiti) Dipartimento di Matematica, Università di Trento, Via Sommarive 14, 38050 Povo (Trento) Italy
E-mail address, R. Battiti: `battiti@science.unitn.it`

(A. Bertossi) Dipartimento di Matematica, Università di Trento, Via Sommarive 14, 38050 Povo (Trento) Italy
E-mail address, A. Bertossi: `bertossi@science.unitn.it`

(R. Rizzi) Dipartimento di Matematica, Università di Trento, Via Sommarive 14, 38050 Povo (Trento) Italy
E-mail address, R. Rizzi: `rrizzi@rtm.science.unitn.it`

DIMACS Series in Discrete Mathematics
and Theoretical Computer Science
Volume **43**, 1999

Elementary Algebra Revisited: Randomized Algorithms

Gene Cooperman and George Havas

ABSTRACT. We look at some simple algorithms for elementary problems in algebra that yield dramatic efficiency improvements over standard methods through randomization. The randomized algorithms are, in a sense, "obvious". Their formal statement was delayed partly by the need for rigorous analysis, but more so by the need to re-think traditional approaches to elementary algorithms. We illustrate this philosophy with some basic problems in computational number theory (GCD of many integers), linear algebra (low-rank Gaussian elimination) and group theory (random subproducts for subgroup construction), along with a brief survey of other areas.

1. Introduction

Because the textbook algorithms of elementary algebra seem to be so efficient, researchers often neglect to examine still more efficient randomized algorithms. Nevertheless, there is scope for some interesting and potentially very important applications of randomization.

A paper of this length cannot hope to provide a survey of the subject. So, it is our intention, rather, to illustrate the utility of randomization in several recent innovations that are related to our own work. The topics chosen reflect elementary algorithms that are either known or can easily be taught to high school students. Often, it is not difficult to see a candidate for a randomized algorithm, but a formal analysis may be more difficult. It is hoped that these examples will inspire more people to look at randomization as a source of greater efficiency. We would also like to point out the excellent survey paper [**13**] in this volume, concerning generation of randomized non-singular matrices, k-subspaces and other structures of interest.

In keeping with our philosophy, the ideas for randomization are simple. The "interesting bits" are in the analysis. In this paper, we discuss the GCD problem for many integers, one of the oldest algorithms of elementary algebra, in detail, while surveying several other algorithms. In addition to GCD's, we give examples of some applications of randomizations to low-rank Gaussian elimination, and to computing canonical forms of matrices.

1991 *Mathematics Subject Classification.* Primary 68Q25; Secondary 11-04, 15-04, 20-04.
Supported in part by NSF Grant CCR-9509783.
Supported in part by the Australian Research Council.

For GCD's, one wishes to find the GCD of k integers, $n_1, \dots, n_k$, with few GCD computations. The key idea of the algorithm is to form two random linear sums of the form $\sum_{i=1}^k a_i n_i$, and take the GCD of those two random linear sums. It is clear that this computation yields a multiple of the true GCD. Hence, we describe an "error" as occurring when the two random linear sums each include an additional prime factor that is not present in the k integers. The issue is to determine a maximum magnitude for the randomly chosen coefficients, a_i, sufficient to guarantee low probability of error. A refined version of such an algorithm is given in [**3**] together with a rigorous analysis and some experimental results.

For low-rank Gaussian elimination, one uses a similar trick. If one can rapidly determine the rank, r, of a matrix and form r independent linear combinations of matrix rows, then the Gaussian elimination algorithm admits an obvious optimization. The algorithm consists of considering the rows, $v_1, \dots, v_k$, of a matrix, and forming r random linear sums, $\sum_{i=1}^k a_i v_i$, where the random coefficients, a_i, are now chosen independently from the set $\{0, 1\}$.

The ideas above are specified in the context of integers and vectors, respectively. The ideas can be generalized in the direction of mathematical groups by considering random subproducts for groups. This yields some surprisingly strong results.

Finally, the use of randomization in computing canonical forms of matrices shows performance gains over deterministic algorithms. The randomized algorithms achieve close to theoretically optimal performance in some practical circumstances.

2. GCD's of many integers

Consider a model of computation where calculating GCDs is relatively expensive in comparison to other arithmetic. This is a reasonably realistic model. Thus, we wish to do few GCD computations in comparison with the naive algorithm (iterative computation of pairwise GCDs), which in the worst case requires $k-1$ GCD calculations.

THEOREM 1. *For positive integers $n_1, \dots, n_k$, one can probabilistically find $\gcd(n_1, \dots, n_k)$ at the cost of $4k$ arithmetic operations, one GCD calculation, and the computation of $2k$ random integers chosen in the range from 1 to M, where $M = \max(1024\lceil \frac{\log_2 k}{150} + 0.1\rceil, (n_1)^2, \dots, (n_k)^2)$. All operations involve quantities of size at most $M\sum_{i=1}^k n_i \leq kM\max(n_1, \dots, n_k)$. The probability of error is bounded above by 0.622.*

PROOF. Without loss of generality, we assume that $n_1 \leq n_2 \leq \cdots \leq n_k$.

The reduction algorithm consists of computing $z = \gcd(\sum_{i=1}^k a_i n_i, \sum_{i=1}^k b_i n_i)$, where $a_i, b_i \in [1, M]$ are randomly chosen integers in the specified range. The value of z will be a multiple of $\gcd(n_1, \dots, n_k)$.

We now consider $x = z/\gcd(n_1, \dots, n_k)$. We bound the probability that $x \neq 1$. The analysis consists of finding an upper bound on the probability that x has a prime factor, p, for each of three cases: (A) $M < p \leq kM^{3/2}$, (B) $M/10 < p \leq M$, and (C) $p \leq M/10$. (Note $x \leq kM^{3/2}$, and so no other cases are possible.) The sum of these three probability bounds will then be shown to be less that the probability stated in the theorem.

CASE A: $(M < p \leq kM^{3/2})$

In this case, p is co-prime to n_k since $p > n_k$ and p is prime. Since $M < p$ and $\sum_{i=1}^k b_i n_i \leq kM^{3/2}$, there are at most $\lceil \log(kM^{3/2})/\log M \rceil$ distinct primes, p, that

divide $\sum_{i=1}^{k} b_i n_i$. Consider $\sum_{i=1}^{k} a_i n_i$ as a function of a_k. Since $a_k \leq M < p$, for each fixed p there is at most one value of a_k for which p divides $sum_{i=1}^{k} a_i n_i$. Since a_k is chosen uniformly from the range $[1, M]$, the probability that a prime $p > M$ divides x is bounded above by $\lceil \log(kM^{3/2})/\log M \rceil / M$. So, the probability that p divides x is bounded above by
$\lceil \log(kM^{3/2})/\log M \rceil / M \leq (\frac{\log_2 k}{10} + \frac{3}{2})/M \leq 15\lceil \frac{\log_2 k}{150} + 0.1 \rceil / M$.

CASE B: $(M/10 < p \leq M)$
Choose s to be the smallest positive integer such that p^s does not divide $\gcd(n_1, \ldots, n_k)$. (Hence, $s \geq 1$ always.) Choose j such that p^s does not divide n_j. Fix the a_i's arbitrarily except for a_j, and consider $\sum_{i=1}^{k} a_i n_i$ as a function of a_j. If p^s divides $\sum_{i=1}^{k} a_i n_i$ for a given value of a_j, then the next value of a_j for which p^s divides $\sum_{i=1}^{k} a_i n_i$ will be $a_j + p$. Since a_j is chosen uniformly from the range $[1, M]$, the probability that p^s divides $\sum_{i=1}^{k} a_i n_i$ is at most $\lceil M/p \rceil / M \leq \frac{1}{p} + \frac{1}{M} \leq 2/p$. Similarly, the probability that p^s divides $\sum_{i=1}^{k} b_i n_i$ is at most $2/p$. Since the a_i and b_i were chosen independently, the probability that p divides x is at most $4/p^2$. So, the overall probability that p divides x for some prime in the given range is bounded above by $\sum_{p=(M/10)+1}^{M} (4/p^2) \leq \int_{p=M/10}^{M} (4/p^2)\, dp = 36/M$.

CASE C: $(p \leq M/10)$
Choose s and j as in the previous case. One follows the argument of the previous case. Since a_j is chosen uniformly from the range $[1, M]$ while now $p \leq M/10$, one can conclude for this case that the probability that p^s divides $\sum_{i=1}^{k} a_i n_i$ is at most $\lceil M/p \rceil / M \leq \frac{1}{p} + \frac{1}{M} \leq \frac{1}{p} + \frac{1}{10p} = \frac{11}{10}/p$. Similarly, the probability that p^s divides $\sum_{i=1}^{k} b_i n_i$ is at most $\frac{11}{10}/p$. Since the a_i and b_i were chosen independently, the probability that p divides x is at most $(\frac{11}{10}/p)^2$. So, the overall probability that p divides x for some prime in the given range is bounded above by $(\frac{11}{10}/2)^2 + (\frac{11}{10}/3)^2 + (\frac{11}{10}/5)^2 + (\frac{11}{10}/7)^2 + \sum_{p=11}^{M/10} (\frac{11}{10}/p)^2 < 0.51004 + \int_{p=10}^{M/10} (\frac{11}{10}/p)^2\, dp = 0.51004 + (\frac{11}{10})^2(\frac{1}{10} - 10/M) < 0.571$.

Hence, the sum of the probabilities of error for the three cases is

$$0.571 + 36/M + 15\lceil \frac{\log_2 k}{150} + 0.1 \rceil / M \leq 0.622$$

based on the choice of M in the hypothesis. This completes the proof. □

REMARK 2. *A different approach to solving this problem is implicit in* [**8**, Proof of Lemma 1.2].

As is always the case with Monte Carlo algorithms, the probability can be improved by repeating the algorithm. Specifically, the probability that r invocations all return an erroneous answer is at most 0.622^r.

2.1. Las Vegas conversion. Any Monte Carlo algorithm which has an easy check as to whether the solution is correct may be simply converted to a Las Vegas algorithm: run the algorithm, check the solution, and repeat till the solution is confirmed.

In this case checking the solution is very easy: check that the hypothetical GCD exactly divides each of the k input numbers. Thus, adding k more arithmetic operations to the algorithm, we get a Las Vegas algorithm for which the expected number of invocations needed to achieve a correct answer is at most $1 + 0.622 + 0.622^2 + 0.622^3 + \cdots < 2.65$. Multiplying 2.65 by the appropriate complexity estimate yields the following corollary.

COROLLARY 3. *One can find* $\gcd(n_1, \ldots, n_k)$ *for arbitrary positive integers* $n_1, \ldots, n_k$ *with an expected cost of at most* $13.25k$ *arithmetic operations,* $5.3k$ *random integers in the range* $[1, M]$*, and* 2.65 *GCDs. All operations involve quantities of size at most* $kM^{3/2}$*, with* M *defined as in Theorem 1.*

Notice that $\lceil \frac{\log_2 k}{150} + 0.1 \rceil = 1$ for $k \leq 10^{40}$. It is worth remarking that the only uses of division in this algorithm are in the GCD algorithm and in the Las Vegas verification of the answer. All other operations consist of multiplication and addition, which are usually faster.

2.2. Improved Reliability. The above analysis can be strengthened to yield a probability of error of at most 0.2. Define a *spurious prime factor* as a prime factor of x. (Recall that x was defined as $\gcd(\sum_{i=1}^k a_i n_i, \sum_{i=1}^k b_i n_i)/\gcd(n_1, \ldots, n_k)$.)

REMARK 4. *The largest contribution to the above probability comes from small primes. For example, the possibility of additional prime factors* 2 *or* 3 *occurring spuriously in the answer contributed a probability of error of more than* 0.4369. *If one separately checks for factors of* 2 *and* 3 *in the answer and replaces* 0.622 *by the finer estimate of* 0.6219, *the probability of error is reduced to at most* $0.6219 - 0.4369 = 0.185$.

In implementations, this cost saving may be worthwhile. Since the answer returned, y, will be a multiple of the true answer, it requires one to directly find the largest power of 2 that divides $\gcd(y, n_1, \ldots, n_k)$ and similarly for 3. Since $y \leq kM$, there are at most $\log_2 kM$ additional spurious prime factors in y. So, the additional computation costs $O(k \log(kM))$ using an obvious algorithm.

However, the worst case computational cost occurs only if y has many prime factors corresponding to 2 and 3. In typical cases, there will be at worst a small constant number, r, of such factors, and so the additional computation costs will typically involve only $k + 5r$ additional divisions. This is seen by first finding how many times 2 and 3 occur as factors of y and then carrying out the k divisions into $n_1, \ldots, n_k$ and up to r adjustments to y as spurious prime factors corresponding to 2 or 3 are found. This operation should be carried out before the Las Vegas test at the end to see if y divides n_i for all i. If one can be assured that r is small (for example, if the inputs, $n_1, \ldots, n_k$ are drawn from a "random" distribution for appropriate definition of random), then this yields a still faster GCD algorithm. Taking $r < k/5$ yields a Monte Carlo algorithm with probability of error at most 0.2. The expected number of invocations is then reduced from 2.65 to 1.25 at the cost of raising the $5k$ arithmetic operations per invocation to at most $7k$.

Still further heuristics are possible. As just one example, consider including the result of a previous invocation as an additional input, n_{k+1} in the next invocation.

3. Random Subproducts for Group Theory

Random subproducts provide a useful unifying framework for both the randomized GCD algorithm, discussed above, and for the low-rank Gaussian elimination algorithm, discussed in the next section. Random subproducts were first used in a more limited setting (for orbits in permutation groups [**2**]). A survey of many of the implications of random subproducts is found in [**4**]. The more general description here has surprising consequences.

Definition. A *random subproduct* of a sequence of group elements $(g_1, g_2, \dots, g_k)$ is an instance of a product $g_1^{e_1} g_2^{e_2} \cdots g_k^{e_k}$ where the e_i are independent random variables uniformly distributed over $\{0, 1\}$.

We also refer to a *random subproduct on a set of group elements*, instead of a random subproduct of a sequence, when an arbitrary ordering of the set suffices.

THEOREM 5. [**1**, Lemma 2.4] *Let H be a non-trivial, proper subgroup of the finite group G, and let G have a generating set S of size s. If $g \in G$ is a random subproduct on S, then one can compute g at the cost of at most s group multiplications, and* $\Pr(g \notin H) \geq 1/2$.

PROOF. For some i, $g_i \notin H$, (since $H \neq G$). Let $w = ug_i^{\epsilon_i}v$, with $g_i \notin H$, $g_{i+1}, \dots, g_r \in H$. So, $v \in H$.
Case I: $u \in H$. With probability $1/2$, $\epsilon_i = 1$, and so $w = ug_iv \notin H$.
Case II: $u \notin H$. With probability $1/2$, $\epsilon_i = 0$, and so $w = uv \notin H$. □

Note that this algorithm makes no mention of a particular group representation. This "representation-free" character of the result also provides a strong result for the case of matrix groups, which are notoriously difficult to analyze. The ideas of this section yield an almost trivial algorithm to find generators for the normal closure of a group, first described in [**4**]. The normal closure algorithm also illustrates the general ideas of this philosophy.

Definition. Let $H \subseteq G$, with $H = \langle U \rangle$ and $G = \langle S \rangle$. A *random normal subproduct with respect to U and S* is a conjugate of the form h^g where h is a random subproduct on U and g is a random subproduct on S.

THEOREM 6. [**1**, Lemma 3.6] *Let $H = \langle U \rangle$ and $G = \langle S \rangle$ be subgroups of a (possibly infinite) group J, with U and S finite. Suppose that H is not normalized by G. Let h^g be a random normal subproduct with respect to U and S. Then*

$$\Pr(h^g \in \langle H^G \rangle \setminus H) \geq 1/4.$$

The proof of this theorem is clear by analogy with Theorem 5. Its usage is more interesting. In looking for a normal closure, $K = \langle H^G \rangle$, we initially set $K = \langle U \rangle$. Our goal is to add generators to K, until $K = \langle H^G \rangle$. Theorem 6 asserts that a random normal subproduct will provide a "new" generator in $\langle K^G \rangle \setminus K$ with probability at least $1/4$. The random normal subproduct is added to the generators of K, and the previous random normal subproduct construction is repeated.

One can show that the algorithm operates on $O(l^2)$ generators, where l is an upper bound on the length of a subgroup chain from its representation and assuming $\max(|S|, |U|) = O(l)$. One can estimate l based either on the representation as a permutation or matrix group, or one can take $l = \log_2 |G|$ for $H < G$. So $4l$ random normal subproducts suffice in some averaged sense for full generation of $\langle H^G \rangle$. Formally, one uses Chernoff's theorem to show cl random normal subproducts suffice with high probability, for c a constant that is larger than 4 by some small factor.

Other similar ideas are described in [**1**], where random subproducts are used for fast algorithms that test solvability and nilpotency of groups in $O(l^2 \log^4 l)$ group operations, that find a normal closure in $O(l \log^4 l)$ (by a different algorithm), and that find a generating set of size $O(\log |G|)$ from an initial generating set, S, in $O(|S| \log l)$ group operations.

The last result is important for the next section, and so we highlight it.

THEOREM 7. [**1**, Theorem 2.3] *Let $G = \langle S \rangle$ be a finite group. Let l be a known upper bound on the length of all subgroup chains in G. Then for any fixed parameter p such that $0 < p < 1$, with probability at least p one can find a generating set S' with $|S'| = O(l)$, using $O(|S| \log l)$ group operations.*

4. Low-rank Gaussian elimination

It is well known that Gaussian elimination operates in $O(n^3)$ time or operations in a given field, for n the matrix dimension. For low-rank matrices of rank $r < n$, this time can be reduced to $O(nr^2 + n^2 \log^2 r)$ field operations. This should be compared with the standard upper bound (Gaussian elimination) of $O(n^2 r)$ and the lower bound of $O(n^2)$ required simply to look at all matrix entries.

THEOREM 8. *A system of n linear equations of rank r (where r is not known a priori) can be solved (or found inconsistent) (with a small probability of error) using $O(nr^2 + n^2 \log^2 r)$ field operations.*

A useful way to understand this construction is by using the ideas of random subproducts, discussed in the previous section. The idea is to consider an additive group on vectors in dimension n. Consider the group generated by the row vectors of the matrix. A maximum subgroup chain in this group is of length $l = r$. If one could efficiently construct $O(r)$ generators for this group, it is clear that $O(nr^2)$ field operations would then suffice for the Gaussian elimination. Theorem 7 allows us to assume $l = r$, and find $O(r)$ generators in $O(n^2 \log r)$ field operations.

In this specialized setting, $l = r$ and a randomized subproduct of the row vectors v_i becomes $\sum_{i=1}^{n} e_i v_i$ for random $e_i \in \{0, 1\}$. We initially set $K = \mathrm{span}(0)$ for 0 the zero vector. We add cr random subproducts to the generators of K for suitable $c > 4$, where each random subproduct has probability at least $1/4$ of being "new", by Theorem 5. This requires $crn = O(nr)$ operations. So if the rank, r, is known a priori, we form a rectangular $cr \times n$ matrix and carry out Gaussian elimination on that matrix in $O(nr^2)$ field operations.

We now describe the full algorithm. The only additional idea is to guess the rank and invoke "repeated doubling". One initially guesses a rank of $\tilde{r} = 1$ and carries out Gaussian elimination on the resulting $c\tilde{r} \times n$ matrix, and determines an estimated rank, $\tilde{r}'$, of the matrix with $c\tilde{r}$ row vectors. One then doubles the guessed rank, $\tilde{r}$, repeats the Gaussian elimination, and compares the newly estimated rank, $\tilde{r}'$, with the old value. If the values are the same, then we have discovered the rank with high probability, and we have successfully carried out a Gaussian elimination. If the ranks are different, we double the guessed rank, $\tilde{r}$, again, and repeat the process. It is clear that $\lceil \log_2 r \rceil$ iterations will be required (where r is the true rank), with each iteration requiring $O(n\tilde{r}^2)$ field operations. The time is dominated by the last iteration with $\tilde{r} \leq 2r$. So, the overall time is $O(nr^2)$, if one is given the $c\tilde{r}$ row vectors. As we have seen, one can form the $c\tilde{r}$ row vectors using $O(n^2 \log \tilde{r})$ field operations for each of the $O(\log r)$ iterations, yielding the overall estimate of $O(nr^2 + n^2 \log^2 r)$.

If partial information is known, a priori, about r, then one can often achieve a better complexity. If the rank, r, is known exactly, then the time can be reduced to $O(nr^2 + n^2 \log r)$ in an obvious manner.

If $r < \epsilon m$ for some small $\epsilon > 0$ but r itself is not bounded, we can achieve considerable savings by first replacing the m rows by a set of $O(r)$ rows using our Monte Carlo algorithms. With large probability, the rows obtained will generate the same row space. The new rows are obtained as sums of subsets of the original rows and will require $O(m \log r)$ row operations to compute. Subsequently, we can complete the work by Gaussian elimination, using $O(r^2)$ row operations.

5. Computing canonical forms of matrices

There has been much recent work on finding theoretically efficient and practical algorithms for computing canonical forms of matrices over various computational domains. In a well-defined sense finding optimal ways to obtain the canonical forms is NP-hard, and worst-case complexity for standard algorithms is exponential in terms of both space and time [**6, 12**]. Randomization has proved to be a powerful tool in addressing such difficult problems. Recent progress includes [**7, 8, 9, 10, 11, 15, 17, 18**], which cover a comprehensive range of algorithms including: heuristically driven with excellent practical performance; deterministic with guaranteed performance; and randomized. In the randomized context both Monte Carlo and Las Vegas algorithms have been developed.

Interesting applications of randomization to Smith normal form computation problems for integer matrices are given by Giesbrecht [**8, 9**] as Monte Carlo algorithms using black-box methods. The GCD of many integers problem of Section 2 arises in this context. The upshot is algorithms with improved space and time complexity for both sparse and dense input. However no comparably efficient Las Vegas versions are known.

In a polynomial matrix context, Storjohann and Labahn [**17**] present a Las Vegas probabilistic algorithm which finds the Smith normal form in $\mathbf{Q}[x]^{n\times n}$ of a nonsingular input matrix in $\mathbf{Z}[x]^{n\times n}$. Their complexity results improve significantly on previous algorithms in both theory and practice.

References

[1] L. Babai, G. Cooperman, L. Finkelstein, E.M. Luks, and Á. Seress, "Fast Monte Carlo Algorithms for Permutation Groups", *J. Computer and System Sciences* **50** (1995) 296–308.

[2] L. Babai, E. Luks, and Á. Seress, "Fast Management of Permutation Groups", *Proc. 29th IEEE FOCS* (1988) 272–282.

[3] G. Cooperman, S. Feisel, J. von zur Gathen, G. Havas, "GCD of many integers", *Northeastern University Tech. Rept. NU-CCS-97-12* (1997).

[4] G. Cooperman and L. Finkelstein, "Combinatorial Tools for Computational Group Theory", *Groups and Computation*, DIMACS Ser. Discrete Math. Theoret. Comput. Sci. **11** (1993) 53–86.

[5] G. Cooperman, L. Finkelstein and B. York, "Parallel Implementations of Group Membership and the Method of Random Subproducts", *Proc. of 1992 Dartmouth Institute for Advanced Graduate Studies in Parallel Computation Symposium (DAGS'92)*, D. Johnson, F. Makedon, and P. Metaxas (eds.), Department of Mathematics and Computer Science, Dartmouth College, Hanover, N.H., pp. 94–100.

[6] X.G. Fang and G. Havas, "On the worst-case complexity of integer gaussian elimination", *ISSAC'97* (Proc. 1997 Internat. Sympos. Symbolic Algebraic Comput.) ACM Press (1997) 28–31.

[7] M. Giesbrecht, "Nearly optimal algorithms for canonical matrix forms", *SIAM Journal on Computing* **24** (1995) 948–969.
[8] M. Giesbrecht, "Fast computation of the Smith normal form of an integer matrix", *ISSAC'95* (Proceedings of the 1995 International Symposium on Symbolic and Algebraic Computation), ACM Press (1995) 110–118.
[9] M. Giesbrecht, "Probabilistic computation of the Smith normal form of a sparse integer matrix", *Algorithmic Number Theory*, Springer Lecture Notes in Computer Science **1122** (1996) 173–186.
[10] G. Havas and B.S. Majewski, "Integer matrix diagonalization", *J. Symbolic Computation* **24** (1997) 399–408.
[11] G. Havas, B.S. Majewski and K.R. Matthews, "Extended gcd and Hermite normal form algorithms via lattice basis reduction", *Experimental Mathematics* **7** (1998) 125–135.
[12] G. Havas and C. Wagner, "Matrix reduction algorithms for Euclidean rings", *Proc. 3rd Asian Symposium Computer Mathematics* (to appear).
[13] I. Pak, "When and How n Choose k", DIMACS Ser. Discrete Math. Theoret. Comput. Sci. (this volume).
[14] A. Storjohann, "Near optimal algorithms for computing Smith normal forms of integer matrices", *ISSAC'96* (Proceedings of the 1996 International Symposium on Symbolic and Algebraic Computation), ACM Press (1996) 267–274.
[15] A. Storjohann, "Computing Hermite and Smith Normal Forms of Triangular Integer Matrices", *Linear Algebra and its Applications* (to appear).
[16] A. Storjohann and G. Labahn, "Asymptotically fast computation of Hermite normal forms of integer matrices", *ISSAC'96* (Proceedings of the 1996 International Symposium on Symbolic and Algebraic Computation), ACM Press (1996) 259–266.
[17] A. Storjohann and G. Labahn, "A Fast Las Vegas Algorithm for Computing Smith Normal Form of a Polynomial Matrix", *Linear Algebra and its Applications* **253** (1997) 155–173.
[18] C. Wagner, "Normalformberechnung von Matrizen über euklidischen Ringen", Dissertation, Fachbereich Mathematik und Informatik der Universität/GH Essen (1997).

College of Computer Science, Northeastern University, Boston, MA 02115, USA
E-mail address: gene@ccs.neu.edu
URL: http://www.ccs.neu.edu/home/gene

Centre for Discrete Mathematics and Computing, Department of Computer Science and Electrical Engineering, The University of Queensland, Queensland 4072, Australia
E-mail address: havas@csee.uq.edu.au
URL: http://www.it.uq.edu.au/~havas

DIMACS Series in Discrete Mathematics
and Theoretical Computer Science
Volume **43**, 1999

Combinatorial Property Testing (a survey)

Oded Goldreich

ABSTRACT. We consider the question of determining whether a given object has a predetermined property or is "far" from any object having the property. Specifically, objects are modeled by functions, and distance between functions is measured as the fraction of the domain on which the functions differ. We consider (randomized) algorithms which may query the function at arguments of their choice, and seek algorithms which query the function at relatively few places.

We focus on combinatorial properties, and specifically on graph properties. The two standard representations of graphs – by adjacency matrices and by incidence lists – yield two different models for testing graph properties. In the first model, most appropriate for dense graphs, distance between N-vertex graphs is measured as the fraction of edges on which the graphs disagree over N^2. In the second model, most appropriate for bounded-degree graphs, distance between N-vertex d-degree graphs is measured as the fraction of edges on which the graphs disagree over dN.

To illustrate the two models, we survey results regarding the complexity of testing whether a graph is Bipartite. For a constant distance parameter, a constant number of queries suffice in the first model, whereas $\widetilde{\Theta}(\sqrt{N})$ queries are necessary and sufficient in the second model.

1. Introduction and Summary of Known Results

The following general formulation of Property Testing was suggested in [**17**]:

> Let P be a fixed property of functions, and f be an unknown function. The goal is to determine (possibly probabilistically) if f has property P or if it is far from any function which has property P, where distance between functions is measured with respect to some distribution D on the domain of f. Towards this end, one is given examples of the form $(x, f(x))$, where x is distributed according to D. One may also be allowed to query f on instances of one's choice.

The above formulation is inspired by the PAC learning model [**37**]. In fact, property testing is related to variants of PAC learning as has been shown in [**17**]. The general formulation above allows the consideration of arbitrary distributions rather than uniform ones, and of testers which utilize only randomly chosen instances (rather than being able to query instances of their own choice). However, we do

1991 *Mathematics Subject Classification.* Primary 68Q22; Secondary 68Q25.

Currently visiting LCS, MIT. Supported by DARPA grant DABT63-96-C-0018. This survey is based on joint work with Shafi Goldwasser and Dana Ron [**17, 18**], and joint works with Dana Ron [**19, 20**].

not consider these latter generalizations here, but rather focus on the special case (formulated previously in [**34, 33**]) where the distribution is uniform on the domain of the function, and testers are allowed to query the function on instances of their choice. Thus, the above formulation simplifies to the following definition, in which we associate a property with the class of functions satisfying it.

DEFINITION 1. (property tester): *Let S be a finite set, and P a subset of functions mapping S to $\{0,1\}^*$. A* (property) **tester** *for P is a probabilistic oracle machine*[1]*, M, which given a* distance parameter $\epsilon > 0$ *and oracle access to an arbitrary function $f : S \mapsto \{0,1\}^*$ satisfies the following two conditions:*

1. (the tester accepts f if it is in P)
   If $f \in \mathrm{P}$ then $\mathbf{Pr}(M^f(\epsilon)=1) \geq \frac{2}{3}$.
2. (the tester rejects f if it is far from P)
   If $|\{x \in S : f(x) \neq g(x)\}| > \epsilon \cdot |S|$, for every $g \in \mathrm{P}$, then $\mathbf{Pr}(M^f(\epsilon)=1) \leq \frac{1}{3}$.

Property testing (as just defined) emerges naturally in the context of program checking [**10, 29, 16, 34**] and probabilistically checkable proofs (PCP) [**6, 7, 14, 4, 3, 8, 11, 5, 9, 22, 24**]. Specifically, in the context of program checking, one may choose to test that the program satisfies certain properties before checking that it computes a specified function. This paradigm has been followed both in the theory of program checking [**10, 34**], and in practice where often programmers first test their programs by verifying that the programs satisfy properties that are known to be satisfied by the function they compute. In the context of probabilistically checkable proofs, the property tested is being a codeword with respect to a specific code. This paradigm, explicitly introduced in [**7**], has shifted from testing codes defined by low-degree polynomials [**6, 7, 14, 4, 3**] to testing Hadamard codes [**3, 8, 11, 5, 28, 36**], and recently to testing the "long code" [**9, 22, 24, 36**].

Much of the work cited above deals with the development and analysis of testers for algebraic properties; specifically, linearity, multi-linearity, and low-degree polynomials [**10, 29, 6, 7, 14, 16, 34, 4, 3, 8, 11, 5**]. In contrast, following [**17**] we focus on testing combinatorial properties, and specifically on testing graph properties such as Bipartiteness.

THE RELEVANT PARAMETERS. The main parameter relevant to property testing is the permitted *distance parameter*, denoted ϵ. In addition, one may consider a confidence (or error bound) parameter, denoted δ, rather than fixing an error bound of $\frac{1}{3}$ (as done in Definition 1 above). Standard amplification techniques do apply here, and thus we choose to fix the error bound. The complexity measures we focus on are the *query complexity* and the *running time* of the tester. All testers discussed in this survey have running-time which is substantially smaller than the full description of the function.

1.1. Motivation. The definition of property testing is a relaxation of the standard definition of a decision task: The tester is allowed arbitrary behavior when the object does not have the property, and yet is "close" to an object having the property. Thus, a property tester may be far more efficient than a standard decision procedure (for the same property).

[1] Alternatively, one may consider a RAM model of computation, in which trivial manipulation of domain and range elements (e.g., reading/writing an element and comparing elements) is performed at unit cost.

In case the object is huge, as in case one thinks of a function and algorithms which operate in time polynomial in the length of the arguments to the function, there is actually no other alternative to property testing. That is, it is typically infeasible (i.e., requires exponential time in the length of the arguments) to decide whether such a function has the desired property. A property testing algorithm which operates in time polynomial in the length of the arguments thus offers a feasible approximation to a problem which is intractable in the exact formulation.

Property testers are valuable also in case one deals with objects of feasible size (i.e., size for which scanning the entire object is feasible): If a property tester is much faster than the exact decision procedure then it makes sense to run it before running the decision procedure. In case the object is far from having the property, we may obtain an indication towards this fact, and save the time we might have used running the decision procedure. In case the tester supplies proofs of violation of the property (as in some of the testers discussed below), we have obtain an absolutely correct answer without running the decision procedure at all. Thus, we may only need to run the decision procedure on objects which are close to having the property. In some natural setting where *typical* objects are either good (i.e., have the property) or very bad (i.e., are very far from objects having the property), we may gain a lot. Furthermore, *if* it is *guaranteed* that objects are either good or very bad then we may not even need to run the decision procedure at all. The gain in such a setting is enormous.

Being close to an object which has the property is a notion of approximation which, in certain applications, may be of great value. In some cases, being close to an object having the property translates to a standard notion of approximation. In other cases, it translates to a notion of "dual approximation". This point is clarified and exemplified below (by referring to specific properties). In both cases, a fast property tester which is more efficient than the decision procedure is of value, both if the decision procedure is feasible and more so if it is not.

Alternatively, we may be forced to take action, without having time to run a decision procedure, while given the option of modifying the object in the future, at a cost proportional to the number of added/omitted edges. For example, suppose you are given a graph which represents some design problem, where Bipartite graphs corresponds to a good design and changes in the design correspond to edge additions/omissions. Using a Bipartiteness tester you always accept a good design, and reject with high probability designs which will cost a lot to modify. You may still accept bad designs, but then you know that it will not cost you much to modify them later.

1.2. Testing Graph Properties – An Overview. As stated above, this survey focuses on testing graph properties. Two natural representations of a graph are offered by its adjacency matrix and by its incidence list. Correspondingly, we consider two representations of graphs by functions.

1. An N-vertex graph, $\mathrm{G} = (\mathrm{V}, \mathrm{E})$, can be represented by the *adjacency predicate*, $f : \mathrm{V} \times \mathrm{V} \mapsto \{0, 1\}$, so that $(u, v) \in \mathrm{E}$ if and only if $f(u, v) = 1$.
2. An N-vertex graph of degree bound d, $\mathrm{G} = (\mathrm{V}, \mathrm{E})$, can be represented by the *incidence function*, $g : \mathrm{V} \times [d] \mapsto \mathrm{V} \cup \{0\}$, so that $g(u, i) = v$ if v is the ith vertex incident at u, and $g(u, i) = 0 \notin \mathrm{V}$ if u has less than i neighbors.

As usual, the choice of representation has a fundamental impact on the potential algorithm. Here the impact is even more dramatic since we seek algorithms which only inspect a relatively small fraction of the object (graph represented by a function). Furthermore, there is another fundamental impact of the choice of representation on the task of property testing. This has to do with our definition of distance, which is relative to the size of the domain of the function. In particular, distance ϵ in the first representation means a symmetric difference of $2\epsilon \cdot N^2$ edges, whereas in the second representation this means a symmetric difference of $2\epsilon \cdot dN$ edges. (In both cases, the extra factor 2 is due to the redundant representation which is adopted for sake of simplicity.)

As usual, the first representation (i.e., adjacency predicate) is most appropriate for dense graphs (i.e., $|E| = \Omega(|V|^2)$), whereas the second representation (i.e., incidence function) is applicable and most appropriate for graphs of degree bound d. We demonstrate the difference between the two representations by considering the task of testing whether a graph is Bipartite.

1.2.1. *Some known results in the first (i.e., adjacency predicate) representation.* Testers of complexity which depends only on the distance parameter, ϵ, are known for several natural graph properties [**17**]. In particular, the following properties can be tested in query-complexity poly$(1/\epsilon)$ and time complexity exp(poly$(1/\epsilon)$):

- k-Colorability, for any fixed $k \geq 2$. The query-complexity is poly(k/ϵ), and for $k = 2$ the running-time is $\widetilde{O}(1/\epsilon^3)$. The Bipartite Tester is presented in Section 2.
- ρ-Clique, for any $\rho > 0$. That is, does the N-vertex graph have a clique of size ρN.
- ρ-CUT, for any $\rho > 0$. That is, does the N-vertex graph have a cut of size at least ρN^2. A generalization to k-way cuts works within query-complexity poly$((\log k)/\epsilon)$.
- ρ-Bisection, for any $\rho > 0$. That is, can the vertices of the N-vertex graph be partitioned into two equal parts with at most ρN^2 edges going between them.

REMARKS:

1. For all the above properties, in case the graph has the desired property, the testing algorithm outputs some auxiliary information which allows to construct, in poly$(1/\epsilon) \cdot N$-time, a partition which approximately obeys the property. For example, for ρ-CUT, we can construct a partition with at least $(\rho - \epsilon)N^2$ crossing edges.
2. Except for Bipartite testing, running-time of poly$(1/\epsilon)$ is unlikely, as it will imply $\mathcal{NP} \subseteq \mathcal{BPP}$.
3. The k-Colorability tester has one-sided error: It always accepts k-colorable graphs. Furthermore, when rejecting a graph, this tester always supplies a poly$(1/\epsilon)$-size subgraph which is not k-colorable. All other algorithms have two-sided error, and this is unavoidable within $o(N)$ query-complexity.

All the above property testing problems are special cases of the *General Graph Partition Testing Problem*, parameterized by a set of lower and upper bounds. In this problem one needs to determine whether there exists a k-partition of the vertices so that the number of vertices in each part as well as the number of edges between each pair of parts falls between the corresponding lower and upper bounds (in the set of parameters). A tester for the general problem has been presented in [**17**] too. The algorithm uses $\widetilde{O}(k^2/\epsilon)^{2k+O(1)}$ queries, runs in time exponential in its query-complexity, and has two-sided error.

Going beyond the General Graph Partition Problem, we remark that there are graph properties which are very easy to test in this model (e.g., Connectivity, Hamiltonicity, and Planarity) [**17**]. The reason being that for these properties either every N-vertex graph is at distance at most $O(1/N)$ from a graph having the desired property (and so for $\epsilon = \Omega(1/N)$ the trivial algorithm which always accepts will do), or the property holds only for sparse graphs (and so for $\epsilon = \Omega(1/N)$ one may reject any non-sparse graph). On the other hand, there are ("unnatural") graph properties in $\mathcal{NP}$ which are extremely hard to test; namely, any testing algorithm must inspect at least $\Omega(N^2)$ of the vertex pairs [**17**]. In view of the above, we believe that providing a characterization of graph properties, according to the complexity of testing them, may be very challenging.

RELATION TO RECOGNIZING GRAPH PROPERTIES: Our notion of testing a graph property P is a *relaxation* of the notion of *deciding the graph property* P which has received much attention in the last two decades [**30**]. In the classical problem there are no margins of error, and one is required to accept all graphs having property P and reject all graphs which lack it. In 1975 Rivest and Vuillemin [**35**] resolved the Aanderaa–Rosenberg Conjecture [**32**], showing that any deterministic procedure for deciding any non-trivial monotone N-vertex graph property must examine $\Omega(N^2)$ entries in the adjacency matrix representing the graph. The query complexity of randomized decision procedures was conjectured by Yao to be $\Omega(N^2)$. Progress towards this goal was made by Yao [**38**], King [**27**] and Hajnal [**21**] culminating in an $\Omega(N^{4/3})$ lower bound. This stands in striking contrast to the testing results of [**17**] mentioned above, by which some non-trivial monotone graph properties can be *tested* by examining a constant number of locations in the matrix.

APPLICATION TO THE STANDARD NOTION OF APPROXIMATION: The relation of testing graph properties to the standard notions of approximation is best illustrated in the case of Max-CUT. Any tester for the class ρ-cut, working in time $T(\epsilon, N)$, yields an algorithm for approximating the maximum cut in an N-vertex graph, up to additive error ϵN^2, in time $\frac{1}{\epsilon} \cdot T(\epsilon, N)$. Thus, for any constant $\epsilon > 0$, using the above tester of [**17**], we can approximate the size of the max-cut to within ϵN^2 in constant time. This yields a constant time approximation scheme (i.e., to within any constant relative error) for dense graphs, improving over previous work of Arora *et. al.* [**2**] and de la Vega [**13**] who solved this problem in polynomial-time (i.e., in $O(N^{1/\epsilon^2})$–time and $\exp(\widetilde{O}(1/\epsilon^2)) \cdot N^2$–time, respectively). In the latter works the problem is solved by actually constructing approximate max-cuts. Finding an approximate max-cut does not seem to follow from the mere existence of a tester for ρ-cut; yet, as mentioned above, the tester in [**17**] can be used to find such a cut in time linear in N.

RELATION TO "DUAL APPROXIMATION" (cf., [**25, 26**]): To illustrate this relation, we consider the ρ-Clique Tester mentioned above. The traditional notion of approximating Max–Clique corresponds to distinguishing the case in which the max-clique has size at least ρN from, say, the case in which the max-clique has size at most $\rho N/2$. On the other hand, when we talk of testing "ρ-Cliqueness", the task is to distinguish the case in which an N-vertex graph has a clique of size ρN from the case in which it is ϵ-far from the class of N-vertex graphs having a clique of size ρN. This is equivalent to the "dual approximation" task of distinguishing the case in which an N-vertex graph has a clique of size ρN from the case in which any ρN subset of the vertices misses at least ϵN^2 edges. To demonstrate that these two tasks are vastly different we mention that whereas the former task is NP-Hard, for $\rho < 1/4$ (see [**9, 22, 23**]), the latter task can be solved in $\exp(O(1/\epsilon^2))$-time, for any $\rho, \epsilon > 0$. We believe that there is no absolute sense in which one of these approximation tasks is more important than the other: Each of these tasks may be relevant in some applications and irrelevant in others.

1.2.2. *Some known results in the second (i.e., incidence function) representation.* Testers of complexity which depends only on the distance parameter, ϵ, are known for several natural graph properties [**19**]. In particular, the following properties can be tested in time (and thus query-complexity) $\text{poly}(d/\epsilon)$:

- *Connectivity.* The tester runs in time $\widetilde{O}(1/\epsilon)$. In case the graph is connected the algorithm always accepts, whereas in case the graph is ϵ-far from being connected the algorithm rejects with probability at least $\frac{2}{3}$ and furthermore supplies a small counter-example to connectivity (in the form of an induced subgraph which is disconnected from the rest of the graph).
- *k-edge-connectivity.* The algorithms run in time $\widetilde{O}(k^3 \cdot \epsilon^{-3})$. For $k = 2, 3$ improved algorithms have running-times $\widetilde{O}(\epsilon^{-1})$ and $\widetilde{O}(\epsilon^{-2})$, respectively. Again, k-edge-connected graphs are always accepted, and rejection is accompanied by a counter-example.
- *k-vertex-connectivity, for $k = 2, 3$.* The algorithms run in time $\widetilde{O}(\epsilon^{-k})$.
- *Planarity.* The algorithm runs in time $\widetilde{O}(d^4 \cdot \epsilon^{-1})$.
- *Cycle-Freeness.* The algorithms run in time $\widetilde{O}(\epsilon^{-3})$. Unlike all other algorithms, this algorithm has two-sided error probability, which is unavoidable for testing this property (within $o(\sqrt{N})$ queries).

The complexity of Bipartiteness testing is considered in Section 3. We survey an $\Omega(\sqrt{N})$ lower bound on the query complexity of any tester [**19**] and a recent result of [**20**] by which a natural algorithm of running time (and query complexity) $\widetilde{O}(\text{poly}(1/\epsilon) \cdot \sqrt{N})$ is a good tester. The lower bound stands in sharp contrast to the situation in the first model (i.e., representation by adjacency predicates), where Bipartite testing is possible in $\text{poly}(1/\epsilon)$-time. We note that the $\widetilde{O}(\text{poly}(1/\epsilon) \cdot \sqrt{N})$-time tester (for the incidence function representation) is significantly faster than the linear time decision procedure (provided that ϵ is not too small).

1.3. Combinatorics beyond Graph Theory – Testing Monotonicity.

In this subsection we mention a partial result referring to testing a combinatorial property which is seemingly unrelated to graph theory. A function $f : \{0,1\}^n \mapsto \{0,1\}$ is called monotone if $f(x) \leq f(y)$, for every $x < y$, where the partial order between strings is defined analogously to the set inclusion relation. That is, $x_1 x_2 \cdots x_n < y_1 y_2 \cdots y_n$ if $x_i \leq y_i$ for all i's, and $x \neq y$. Let $w(x)$ denote the *weight* of x (i.e., the number of 1's in x). The following natural test of Monotonicity was was suggested and partially analyzed in [**18**].

ALGORITHM 1. (Candidate Monotonicity Tester):
Given n, ϵ and oracle access to a function $f : \{0,1\}^n \mapsto \{0,1\}$, repeat $2n^2/\epsilon$ times:

1. *Uniformly select $x \in \{0,1\}^n$, and obtain the value $f(x)$.*
2. *In case $f(x) = 1$, obtain the values of $f(y)$ for all y's satisfying $y > x$ and $w(y) = w(x) + 1$. If one of these $f(y)$'s is 0 then* reject.
3. *Analogously, in case $f(x) = 0$, obtain the values of $f(y)$ for all y's satisfying $y < x$ and $w(y) = w(x) - 1$. If one of these $f(y)$'s is 1 then* reject.

If all iterations were completed without rejecting then accept.

Clearly, Algorithm 1 accepts every monotone function with probability 1. The conjecture that it rejects, with probability at least 2/3, any function which is ϵ-far from being monotone was reduced in [**18**] to the following combinatorial conjecture referring to the Boolean Lattice (cf., background in [**12**]). For each i, $0 \leq i \leq n$, let $\mathrm{L}_i \subset \{0,1\}^n$ denote the set of n-bit long strings of weight i. Let G_n be the leveled *directed* (acyclic) graph over the vertex set $\{0,1\}^n$, where there is a directed edge from y to x if and only if $x < y$ and $w(x) = w(y) - 1$ (i.e., x and y are in adjacent L_i's).

CONJECTURE 1. *Let r and s be integers satisfying, $0 \leq r < s \leq n$, and let $\mathrm{R}, \mathrm{S} \subset \{0,1\}^n$, be sets such that $\mathrm{R} \subseteq \mathrm{L}_r$, and $\mathrm{S} \subseteq \mathrm{L}_s$, and $|\mathrm{R}| = |\mathrm{S}| = m$. Suppose that there exists a 1-to-1 mapping ψ from S to R such that for every $y \in \mathrm{S}$, there is a directed path in G_n from y to $\psi(y)$. Then there exist m vertex-disjoint directed paths from S to R in G_n.*

We stress that these vertex-disjoint paths do not have to respect ψ. In fact, if one requires these paths to respect ψ then the conjecture becomes false (cf., [**18**]).

THEOREM 1. [**18**]: *If Conjecture 1 holds then Algorithm 1 is a property tester for monotonicity. In particular, if f is monotone then Algorithm 1 always accepts, whereas if f is ϵ-far from being monotone then Algorithm 1 rejects with probability at least* 2/3.

1.4. Rest of this survey. In Sections 2 and 3 we consider the complexity of testing Bipartiteness in the two graph representations discussed above: In Section 2 we consider representation by an adjacency predicate, and in Section 3 by an incidence function. Concluding remarks appear in Section 4.

2. Testing Bipartiteness in the First Representation

In this section we consider the representation of N-vertex graphs by adjacency predicates mapping pairs $\{1,2,...,N\} \times \{1,2,...,N\}$ to $\{0,1\}$. The bipartite tester is extremely simple: It selects a tiny, random set of vertices and checks whether the induced subgraph is bipartite.

ALGORITHM 2. (Bipartite Tester in the first model [**17**]):
On input N, d, ϵ and oracle access to an adjacency predicate of an N-vertex graph, $\mathrm{G} = (\mathrm{V}, \mathrm{E})$:

1. *Uniformly select a subset of $\widetilde{\mathrm{O}}(1/\epsilon^2)$ vertices of* V.
2. *Accept if and only if the subgraph induced by this subset is Bipartite.*

Step (2) amounts to querying the predicate on all pairs of vertices in the subset selected at Step (1). As will become clear from the analysis, it actually suffice to query only $\widetilde{O}(1/\epsilon^3)$ of these pairs.

THEOREM 2. [**17**]: *Algorithm 2 is a Bipartite Tester* (in the adjacency predicate representation)*. Furthermore, the algorithm always accepts a Bipartite graph, and in case of rejection it provides a witness of length* poly$(1/\epsilon)$ (that the graph is not bipartite).

Proof: Let R be the subset selected in Step (1), and G_R the subgraph induced by it. Clearly, if G is bipartite then so is G_R, for any R. The point is to prove that if G is ϵ-far from bipartite then the probability that G_R is bipartite is at most 1/3. Thus, from this point on we assume that at least ϵN^2 edges have to be omitted from G to make it bipartite.

We view R as a union of two disjoint sets U and S, where $t \stackrel{\text{def}}{=} |U| = O(\epsilon^{-1} \cdot \log(1/\epsilon))$ and $m \stackrel{\text{def}}{=} |S| = O(t/\epsilon)$. We will consider all possible partitions of U, and associate a partial partition of V with each such partition of U. The idea is that in order to be consistent with a given partition, (U_1, U_2), of U, all neighbors of U_1 (resp., U_2) must be placed opposite to U_1 (resp., U_2). We will show that, with high probability, most high-degree vertices in V do neighbor U and so are forced by its partition. Since there are relatively few edges incident to vertices which do not neighbor U, it follows that with very high probability each such partition of U is detected as illegal by G_R. Details follow, but before we proceed let us stress the key observation: It suffices to rule out relatively few (partial) partitions of V (i.e., these induced by partitions of U), rather than all possible partitions of V.

We use the notations $\Gamma(v) \stackrel{\text{def}}{=} \{u : (u,v) \in E\}$ and $\Gamma(X) \stackrel{\text{def}}{=} \cup_{v \in X}\Gamma(v)$. Given a partition (U_1, U_2) of U, we define a (possibly partial) partition, (V_1, V_2), of V so that $V_1 \stackrel{\text{def}}{=} \Gamma(U_2)$ and $V_2 \stackrel{\text{def}}{=} \Gamma(U_1)$ (assume, for simplicity that $V_1 \cap V_1$ is indeed empty). As suggested above, if one claims that G can be "bipartited" with U_1 and U_2 on different sides then $V_1 = \Gamma(U_1)$ must be on the opposite side to U_2 (and $\Gamma(U_1)$ opposite to U_1). Note that the partition of U places no restriction on vertices which have no neighbor in U. Thus, we first ensure that *most* "influential" (i.e., "high-degree") vertices in V have a neighbor in U.

DEFINITION 2.1. (high-degree vertices and good sets): *We say that a vertex $v \in V$ is of* high-degree *if it has degree at least $\frac{\epsilon}{3}N$. We call* U good *for* V *if all but at most $\frac{\epsilon}{3}N$ of the high-degree vertices in* V *have a neighbor in* U.

Note that NOT insisting that U neighbors *all* high-degree vertices allows us to show that a random U of size unrelated to the size of the graph has this feature. (If we were to insist that U neighbors *all* high-degree vertices then we would have had to use $|U| = \Omega(\log N)$.)

CLAIM 2.2. *With probability at least* 5/6, *a uniformly chosen set* U *of size* t *is good.*

PROOF. For any high-degree vertex v, the probability that v does not have any neighbor in a uniformly chosen U is at most $(1 - \epsilon/3)^t < \frac{\epsilon}{18}$ (since $t = \Omega(\epsilon^{-1}\log(1/\epsilon))$). Hence the expected number of high-degree vertices which do not have a neighbor in a random set U is less than $\frac{\epsilon}{18} \cdot N$, and the claim follows by Markov's Inequality. ∎

DEFINITION 2.3. (disturbing a partition of U): *We say that an edge* disturbs *a partition* (U_1, U_2) *of* U *if both is end-points are in the same* $\Gamma(U_i)$, *for some* $i \in \{1, 2\}$.

CLAIM 2.4. *For any good set* U *and any partition of* U, *at least* $\frac{\epsilon}{3}N^2$ *edges disturb the partition.*

PROOF. Each partition of V has at least ϵN^2 violating edges (i.e., edges with both end-points on the same side). We upper bound the number of these edges which are not disturbing. Actually, we upper bound the number of edges which have an end-point not in $\Gamma(U)$.

- The number of edges incident to high-degree vertices which do not neighbor U is bounded by $\frac{\epsilon}{3}N \cdot N$ (at most $\frac{\epsilon}{3}N$ such vertices each having at most N incident edges).
- The number of edges incident to vertices which are not of high-degree vertices is bounded by $N \cdot \frac{\epsilon}{3}N$ (at most N such vertices each having at most $\frac{\epsilon}{3}N$ incident edges).

This leaves us with at least $\frac{\epsilon}{3}N^2$ violating edges connecting vertices in $\Gamma(\mathrm{U})$ (i.e., edges disturbing the partition of U). ■

The theorem follows by observing that $\mathrm{G_R}$ is bipartite only if either (1) the set U is not good; or (2) the set U is good and there exists a partition of U so that none of the disturbing edges occurs in $\mathrm{G_R}$. Using Claim 2.2 the probability of event (1) is bounded by 1/6; and using Claim 2.4 the probability of event (2) is bounded by the probability that there exists a partition of U so that none of the corresponding $\geq \frac{\epsilon}{3}N^2$ disturbing edges has both edge-point in S. Actually, we pair the m vertices of S, and consider the probability that none of these pairs is a disturbing edge for a partition of U. Thus the probability of event (2) is bounded by

$$2^{|\mathrm{U}|} \cdot \left(1 - \frac{\epsilon}{3}\right)^{m/2} < \frac{1}{6}$$

where the inequality is due to $m = \Omega(t/\epsilon)$. The theorem follows. ■

COMMENT: The procedure employed in the proof yields a poly$(1/\epsilon) \cdot N$-time algorithm for 2-partitioning a bipartite graph so that at most ϵN^2 edges lie within the same side. This is done by running the tester, determining a partition of U (defined as in the proof) which is consistent with the bipartite partition of R, and partitioning V as done in the proof (with vertices which do not neighbor U, or neighbor both $\mathrm{U_1, U_2}$, placed arbitrarily). Thus, the placement of each vertex is determined by inspecting at most $\widetilde{O}(1/\epsilon)$ entries of the adjacency matrix.

3. Testing Bipartiteness in the Second Representation

In this section we consider the representation of N-vertex graphs of degree bound d by incidence functions mapping pairs

$$\{1, 2, ..., N\} \times \{1, 2, ..., d\} \text{ to } \{0, 1, 2, ..., N\}.$$

3.1. A lower bound. In contrast to Theorem 2, under the incidence function representation there exists no Bipartite tester of complexity independent of the graph size.

THEOREM 3. [**19**]: *Testing Bipartiteness* (with constant ϵ and d) *requires* $\Omega(\sqrt{N})$ *queries* (in the incidence function representation).

Proof Idea: For any (even) N, we consider the following two families of graphs:

1. The first family, denoted $\mathcal{G}_1^N$, consists of all degree-3 graphs which are composed by the union of a Hamiltonian cycle and a perfect matching. That is, there are N edges connecting the vertices in a cycle, and the other $N/2$ edges are a perfect matching.
2. The second family, denoted $\mathcal{G}_2^N$, is the same as the first *except* that the perfect matchings allowed are restricted as follows: the distance on the cycle between every two vertices which are connected by an perfect matching edge must be odd.

Clearly, all graphs in $\mathcal{G}_2^N$ are bipartite. One first proves that almost all graphs in $\mathcal{G}_1^N$ are far from being bipartite. Afterwards, one proves that a testing algorithm that performs less than $\alpha\sqrt{N}$ queries (for some constant $\alpha < 1$) is not able to distinguish between a graph chosen randomly from $\mathcal{G}_2^N$ (which is always bipartite) and a graph chosen randomly from $\mathcal{G}_1^N$ (which with high probability will be far from bipartite). Loosely speaking, this is done by showing that in both cases the algorithm is unlikely to encounter a cycle (among the vertices it has inspected). ∎

3.2. An algorithm. The lower bound of Theorem 3 is essentially tight. Furthermore, the following natural algorithm constitutes a Bipartite tester of running time $\text{poly}((\log N)/\epsilon) \cdot \sqrt{N}$.

ALGORITHM 3. (Bipartite Tester in the second model [**20**]):
On input N, d, ϵ and oracle access to an incidence function for an N-vertex graph, $\mathrm{G} = (\mathrm{V}, \mathrm{E})$, of degree bound d, repeat $T \stackrel{\text{def}}{=} \Theta(\frac{1}{\epsilon})$ times:

1. *Uniformly select s in* V.
2. (Try to find an odd cycle through vertex s):
   (a) *Perform $K \stackrel{\text{def}}{=}$* $\text{poly}((\log N)/\epsilon) \cdot \sqrt{N}$ *random walks starting from s, each of length $L \stackrel{\text{def}}{=}$* $\text{poly}((\log N)/\epsilon)$.
   (b) *Let R_0* (resp., R_1) *denote the vertices set reached from s in an even* (resp., odd) *number of steps in any of these walks.*
   (c) *If $R_0 \cap R_1$ is not empty then* reject.

If the algorithm did not reject in any one of the above T iterations, then it accepts.

THEOREM 4. [**20**]: *Algorithm 3 is a Bipartite Tester* (in the incidence function representation). *Furthermore, the algorithm always accepts a Bipartite graph, and in case of rejection it provides a witness of length* $\text{poly}((\log N)/\epsilon)$ (that the graph is not bipartite).

MOTIVATION – THE SPECIAL CASE OF RAPID MIXING GRAPHS. The proof of Theorem 4 is quite involved. As a motivation, we consider the special case where the graph has a "rapid mixing" feature. It is convenient to modify the random walks so that at each step each neighbor is selected with probability $1/2d$, and otherwise (with probability at least $1/2$) the walk remains in the present vertex. Furthermore, we will consider a single execution of Step (2) starting from an arbitrary vertex, s, fixed in the rest of the discussion. The rapid mixing feature we assume is that, for every vertex v, a (modified) random walk of length L starting at s reaches v with probability approximately $1/N$ (say, up-to a factor of 2). Note that if the graph is an expander then this is certainly the case (since $L \geq O(\log N)$).

The key quantities is the analysis are the following probabilities, referring to the parity of the length of a path obtained from the random walk by omitting the self-loops (transitions which remain at current vertex). Let $p^0(v)$ (resp., $p^1(v)$) denote the probability that a (modified) random walk of length L starting at s reaches v while making an even (resp., odd) number of real (i.e., non-self-loop) steps. By the rapid mixing assumption we have (for every $v \in \mathrm{V}$)

$$\frac{1}{2N} < p^0(v) + p^1(v) < \frac{2}{N} \tag{1}$$

We consider two cases regarding the sum $\sum_{v \in \mathrm{V}} p^0(v)p^1(v)$ – In case the sum is (relatively) "small", we show that V can be 2-partitioned so that there are relatively few edges between vertices placed in the same side, which implies that G is close to be bipartite. Otherwise (i.e., when the sum is not "small"), we show that with significant probability, when Step (2) is started at vertex s it is completed by rejecting G. The two cases are presented in greater detail in the following (corresponding) two claims.

CLAIM 3.1. *Suppose* $\sum_{v\in \mathrm{V}} p^0(v)p^1(v) \leq \epsilon/50N$. *Let* $\mathrm{V}_1 \stackrel{\mathrm{def}}{=} \{v \in \mathrm{V} : p^0(v) < p^1(v)\}$ *and* $\mathrm{V}_2 = \mathrm{V} \setminus \mathrm{V}_1$. *Then, the number of edges with both end-points in the same* V_σ *is bounded above by* ϵdN.

Proof Sketch: Consider an edge (u, v) where, without loss of generality, both u and v are in V_1. Then, both $p^1(v)$ and $p^1(u)$ are greater than $\frac{1}{2}\cdot\frac{1}{2N}$. However, one can show that $p^0(v) > \frac{1}{3d}\cdot p^1(u)$: Observe that a walk of length $L-1$ with path-parity 1 ending at u is almost as likely as such a walk having length L, and that once such a walk reaches u it continues to v in the next step with probability exactly $1/2d$. Thus, such an edge contributes at least $\frac{(1/4N)^2}{3d}$ to the sum $\sum_{v\in\mathrm{V}} p^0(v)p^1(v)$. The claim follows. ■

CLAIM 3.2. *Suppose* $\sum_{v\in\mathrm{V}} p^0(v)p^1(v) \geq \epsilon/50N$, *and that Step (2) is started with vertex* s. *Then, with probability at least* 2/3, *the set* $R_0 \cap R_1$ *is not empty* (and rejection follows).

Proof Sketch: Consider the probability space defined by an execution of Step (2) with start vertex s. We define random variables $\zeta_{i,j}$ representing the event that the vertex encountered in the Lth step of the ith walk equals the vertex encountered in the Lth step of the jth walk, and that the ith walk corresponds to an even-path whereas the jth to an odd-path. Then

$$\begin{aligned}
\mathbf{E}(|R_0 \cap R_1|) &> \sum_{i\neq j} \mathbf{E}(\zeta_{i,j}) \\
&= K(K-1)\cdot \sum_{v\in\mathrm{V}} p^0(v)p^1(v) \\
&> \frac{500N}{\epsilon}\cdot \sum_{v\in\mathrm{V}} p^0(v)p^1(v) \\
&\geq 10
\end{aligned}$$

where the second inequality is due to the setting of K, and the third to the claim's hypothesis. Intuitively, we expect that with high probability $|R_0 \cap R_1| > 0$. This is indeed the case, but proving it is less straightforward than it seems, the problem being that the $\zeta_{i,j}$'s are not pairwise independent. Yet, since the sum of the co-variances of the dependent $\zeta_{i,j}$'s is quite small, Chebyshev's Inequality is still very useful (cf., [**1**, Sec. 4.3]). Specifically, letting $\mu \stackrel{\mathrm{def}}{=} \sum_{v\in\mathrm{V}} p^0(v)p^1(v)$ $(= \mathbf{E}(\zeta_{i,j}))$, and $\overline{\zeta}_{i,j} \stackrel{\text{def}}{=} \zeta_{i,j} - \mu$, we get:

$$\begin{aligned}
\mathbf{Pr}\left(\sum_{i\neq j} \zeta_{i,j} = 0\right) &< \frac{\mathbf{V}(\sum_{i\neq j}\zeta_{i,j})}{(K^2\mu)^2} \\
&= \frac{1}{K^4\mu^2}\cdot\left(\sum_{i,j}\mathbf{E}(\overline{\zeta}_{i,j}^2) + 2\sum_{i,j,k}\mathbf{E}(\overline{\zeta}_{i,j}\overline{\zeta}_{i,k})\right) \\
&< \frac{1}{K^2\mu} + \frac{2}{K\mu^2}\cdot\mathbf{E}(\zeta_{1,2}\zeta_{1,3})
\end{aligned}$$

For the second term, we observe that $\mathbf{Pr}(\zeta_{1,2} = \zeta_{2,3} = 1)$ is upper bounded by the probability that $\zeta_{1,2} = 1$ times the probability that the Lth vertex of the first walk appears as the Lth vertex of the third path. Using the rapid mixing hypothesis, we upper bound the latter probability by $2/N$, and obtain

$$\begin{aligned}
\mathbf{Pr}(|R_0 \cap R_1| = 0) &< \frac{1}{K^2\mu} + \frac{2}{K\mu^2}\cdot\mu\cdot\frac{2}{N} \\
&< \frac{1}{3}
\end{aligned}$$

where the last inequality uses $K < N/4$, $\mu \geq \epsilon/50N$ and $K^2 \geq 6\cdot 50N/\epsilon$. The claim follows. ■

BEYOND RAPID MIXING GRAPHS. The proof in [**20**] refers to a more general sum of products; that is, $\sum_{v\in\mathrm{V}} p_{\mathrm{odd}}(v)p_{\mathrm{even}}(v)$, where $\mathrm{U}\subseteq\mathrm{V}$ is an appropriate set of vertices, and $p_{\mathrm{odd}}(v)$ (resp., $p_{\mathrm{even}}(v)$) is the probability that a random walk (starting at s) passes through v after more than $L/2$ steps and the corresponding path to v has odd (resp., even) parity. Much of the analysis in [**20**] goes into selecting the appropriate U (and an appropriate starting vertex s), and pasting together many such U's to cover all of V. Loosely speaking, U and s are selected so that there are few edges from U and the rest of the graph, and $p_{\mathrm{odd}}(u) + p_{\mathrm{even}}(u) \approx 1/\sqrt{|\mathrm{V}|\cdot|\mathrm{U}|}$, for every $u \in \mathrm{U}$. The selection is based on the "combinatorial treatment of expansion" of Mihail [**31**]. Specifically, we use the counterpositive of the standard analysis, which asserts that rapid mixing occurs when all cuts are relatively large, to assert the existence of small cuts which partition the graph so that vertices reached with relatively high probability (in a short random walk) are on one side and the rest of the graph on the other. The first set corresponds to U above and the cut is relatively small with respect to U. A start vertex s for which the corresponding sum is big is shown to cause Step (2) to reject (when started with this s), whereas a small corresponding sum enables to 2-partition U while having few violating edges among the vertices in each part of U.

The actual argument of [**20**] proceeds in iterations. In each iteration a vertex s for which Step (2) accepts with high probability is fixed, and an appropriate set of remaining vertices, U, is found. The set U is then 2-partitioned so that there are few violating edges inside U. Since we want to paste all these partitions together, U may not contain vertices treated in previous iterations. This complicates the analysis, since it must refer to the part of G, denoted H, not treated in previous iterations. We consider walks over an (imaginary) Markov Chain representing the H-part of the walks performed by the algorithm on G. Statements about rapid mixing are made with respect to the Markov Chain, and linked to what happens in random walks performed on G. In particular, a subset U of H is determined so that the vertices in U are reached with probability $\approx 1/\sqrt{|V| \cdot |U|}$ (in the chain) and the cut between U and the rest of H is small. Linking the sum of products defined for the chain with the actual walks performed by the algorithm, we infer that U may be partitioned with few violating edges inside it. Edges to previously treated parts of the graphs are charged to these parts, and edges to the rest of $H \setminus U$ are accounted for by using the fact that this cut is small (relative to the size of U).

4. Concluding Remarks

Randomness plays a pivotal role in the theory of property testing: A deterministic tester for any "non-degenerate" property (and in particular for any of the properties discussed above) needs to query the function on a constant fraction of its domain, and so is of little interest.

The results regarding property testing, known to date, are rather sporadic. For more than a dozen natural graph properties, testers are known in the adjacency predicate representation, and for some testers are known in the incidence function representation. More than half a dozen of these testers are interesting, and though they share some techniques, no general structure seems to arise. Some negative results in [**17**] seem to indicate that general results may be hard to obtain: For example, it was shown that there exist properties in $\mathcal{NP}$ which require high query complexity for testing. Also some properties are easy to test with one-sided error, whereas other require two-sided error to be tested efficiently. Thus, obtaining "structural" results regarding easily testable properties may be very challenging as well as of great interest.

Acknowledgments

This survey is based on joint work with Shafi Goldwasser and Dana Ron [**17, 18**], and joint works with Dana Ron [**19, 20**].

References

[1] N. Alon and J.H. Spencer, *The Probabilistic Method*, John Wiley & Sons, Inc., 1992.

[2] S. Arora, D. Karger, and M Karpinski. Polynomial time approximation schemes for dense instances of NP-hard problems. In *Proceedings of the Twenty-Seventh Annual ACM Symposium on the Theory of Computing*, pages 284–293, 1995.

[3] S. Arora, C. Lund, R. Motwani, M. Sudan, and M. Szegedy. Proof verification and intractability of approximation problems. In *Proceedings of the Thirty-Third Annual Symposium on Foundations of Computer Science*, pages 14–23, 1992.

[4] S. Arora and S. Safra. Probabilistic checkable proofs: A new characterization of NP. In *Proceedings of the Thirty-Third Annual Symposium on Foundations of Computer Science*, pages 1–13, 1992.

[5] M. Bellare, D. Coppersmith, J. Håstad, M. Kiwi, and M. Sudan. Linearity testing in characteristic two. In *Proceedings of the Thirty-Sixth Annual Symposium on Foundations of Computer Science*, pages 432–441, 1995.

[6] L. Babai, L. Fortnow, and C. Lund. Non-deterministic exponential time has two-prover interactive protocols. *Computational Complexity*, 1(1):3–40, 1991.

[7] L. Babai, L. Fortnow, L. Levin, and M. Szegedy. Checking computations in polylogarithmic time. In *Proceedings of the Twenty-Third Annual ACM Symposium on Theory of Computing*, pages 21–31, 1991.

[8] M. Bellare, S. Goldwasser, C. Lund, and A. Russell. Efficient probabilistically checkable proofs and applications to approximation. In *Proceedings of the Twenty-Fifth Annual ACM Symposium on the Theory of Computing*, pages 294–304, 1993.

[9] M. Bellare, O. Goldreich, and M. Sudan. Free bits, pcps and non-approximability – towards tight results. Extended abstract in *Proceedings of the Thirty-Sixth Annual Symposium on Foundations of Computer Science*, pages 422–431, 1995. To appear in *SICOMP*.

[10] M. Blum, M. Luby, and R. Rubinfeld. Self-testing/correcting with applications to numerical problems. *Journal of Computer and System Sciences*, 47:549–595, 1993.

[11] M. Bellare and M. Sudan. Improved non-approximability results. In *Proceedings of the 26th Annual ACM Symposium on the Theory of Computing*, pages 184–193, 1994.

[12] B. Bollobás. *Combinatorics*. Cambridge University Press, 1986.

[13] W. F. de la Vega. MAX-CUT has a randomized approximation scheme in dense graphs. To appear in *Random Structures and Algorithms*, 1994.

[14] U. Feige, S. Goldwasser, L. Lovász, S. Safra, and M. Szegedy. Approximating clique is almost NP-complete. In *Proceedings of the Thirty-Second Annual Symposium on Foundations of Computer Science*, pages 2–12, 1991.

[15] A. Frieze and R. Kanan. The regularity lemma and approximation schemes for dense problems. In *Proceedings of the Thirty-Seventh Annual Symposium on Foundations of Computer Science*, pages 12–20, 1996.

[16] P. Gemmell, R. Lipton, R. Rubinfeld, M. Sudan, and A. Wigderson. Self-testing/correcting for polynomials and for approximate functions. In *Proceedings of the Twenty-Third Annual ACM Symposium on Theory of Computing*, pages 32–42, 1991.

[17] O. Goldreich, S. Goldwasser, and D. Ron. Property testing and its connection to learning and approximation. To appear in *JACM*. Extended abstract in *Proc. of the 37th IEEE Symp. on Foundation of Computer Science*, pages 339–348, 1996. Available as TR96-057 of *ECCC*, `http://www.eccc.uni-trier.de/eccc/`, 1996.

[18] O. Goldreich, S. Goldwasser, and D. Ron. A note on testing monotinicity. Unpublished working draft, 1996. Manuscript, 1997. Available from `http://theory.lcs.mit.edu/~oded/test.html`.

[19] O. Goldreich and D. Ron. Property Testing in Bounded Degree Graphs. In *Proc. of the 29th ACM Symp. on Theory of Computing*, pages 406–415, 1997.

[20] O. Goldreich and D. Ron. A sublinear Bipartite Tester for Bounded Degree Graphs. To appear in *Proc. of the 30th ACM Symp. on Theory of Computing*, 1998. Available from `http://theory.lcs.mit.edu/~oded/test.html`.

[21] P. Hajnal. An $\Omega(n^{4/3})$ lower bound on the randomized complexity of graph properties. *Combinatorica*, 11(2):131–144, 1991.

[22] J. Håstad. Testing of the long code and hardness for clique. In *Proc. of the 28th ACM Symp. on Theory of Computing*, pages 11–19, 1996.

[23] J. Håstad. Clique is hard to approximate within $n^{1-\epsilon}$. In *Proc. of the 37th IEEE Symp. on Foundation of Computer Science*, pages 627–636, 1996.

[24] J. Håstad. Getting optimal in-approximability results. In *Proc. of the 29th ACM Symp. on Theory of Computing*, pages 1–10, 1997.

[25] D. S. Hochbaum and D. B. Shmoys. Using dual approximation algorithms for scheduling problems: Theoretical and practical results. *Journal of the Association for Computing Machinery*, 34(1):144–162, January 1987.

[26] D. S. Hochbaum and D. B. Shmoys. A polynomial approximation scheme for machine scheduling on uniform processors: Using the dual approximation approach. *SIAM Journal on Computing*, 17(3):539–551, 1988.

[27] V. King. An $\Omega(n^{5/4})$ lower bound on the randomized complexity of graph properties. *Combinatorica*, 11(1):23–32, 1991.

[28] M. Kiwi. *Probabilistically Checkable Proofs and the Testing of Hadamard-like Codes*. PhD thesis, Massachusetts Institute of Technology, 1996.

[29] R. J. Lipton. New directions in testing. Unpublished manuscript, 1989.

[30] L. Lovász and N. Young. Lecture notes on evasiveness of graph properties. Technical Report TR–317–91, Princeton University, Computer Science Department, 1991.

[31] M. Mihail. Conductance and convergence of Markov chains – A combinatorial treatment of expanders. In *Proceedings 30th Annual Symp. on Foundations of Computer Science*, pages 526–531, 1989.

[32] A. L. Rosenberg. On the time required to recognize properties of graphs: A problem. *SIGACT News*, 5:15–16, 1973.

[33] R. Rubinfeld. Robust functional equations and their applications to program testing. In *Proceedings of the 35th Annual Symposium on Foundations of Computer Science*, 1994. To appear in *SIAM Journal on Computing.*

[34] R. Rubinfeld and M. Sudan. Robust characterization of polynomials with applications to program testing. *SIAM Journal on Computing*, 25(2):252–271, 1996.

[35] R. L. Rivest and J. Vuillemin. On recognizing graph properties from adjacency matrices. *Theoretical Computer Science*, 3:371–384, 1976.

[36] L. Trevisan. Recycling queries in PCPs and in linearity tests. To appear in *Proc. of the 30th ACM Symp. on Theory of Computing*, 1998.

[37] L. G. Valiant. A theory of the learnable. *Communications of the ACM*, 27(11):1134–1142, November 1984.

[38] A. C. C. Yao. Lower bounds to randomized algorithms for graph properties. In *Proceedings of the Twenty-Eighth Annual Symposium on Foundations of Computer Science*, pages 393–400, 1987.

DEPARTMENT OF COMPUTER SCIENCE AND APPLIED MATHEMATICS, WEIZMANN INSTITUTE OF SCIENCE, REHOVOT, ISRAEL.

E-mail address: oded@wisdom.weizmann.ac.il

DIMACS Series in Discrete Mathematics
and Theoretical Computer Science
Volume **43**, 1999

Randomized and Deterministic Local Search for SAT and Scheduling Problems

Jun Gu

ABSTRACT. Random techniques are simple and effective for a wide range of algorithm design. In this research, we study local search algorithms for solving large size satisfiability (SAT) problem and scheduling problem. Several randomized and deterministic local search algorithms have been developed. We compared their performance with a variety of SAT and scheduling problem instances. Our experimental studies indicate that randomized local search algorithms are particularly efficient for large size NP-complete problems that have many solutions. We give several efficient local search algorithms and show their performance for solving large size satisfiability problems. For an NP-hard optimization problem with one or fewer optimum solutions, randomized local search algorithms perform satisfactorily well and, in some cases, a deterministic local search algorithm can find higher quality solutions much more efficiently. Following this, we give two efficient local search algorithms and compare their performance for solving large size scheduling problem.

1. Introduction

Random techniques are simple and effective for a wide range of algorithm design. Many efficient randomized algorithms have been developed [**72, 71, 91, 89, 87**]. An interesting question lies to know when and how to use random techniques for what type of algorithm design. In this research, we study randomized and deterministic local search algorithms for solving large size satisfiability (SAT) problem and scheduling problem. We select these two significant problems for case study since they provide deep insight to search algorithm design.

The goal of the satisfiability problem is to determine whether there exists an assignment of truth values to variables that makes a conjunctive normal form (*CNF*) formula satisfiable [**15, 27**]. The SAT problem is a *core* of a large family of computationally intractable NP-complete problems. Such NP-complete problems have

---

1991 *Mathematics Subject Classification.* Primary 03B05, 90A05, 90B35, 68Q25, 68M20, 90C27, 90B40; Secondary 90A80, 90B80.

Key words and phrases. Combinatorial optimization, constraint satisfaction problem, local search, randomized algorithms, satisfiability problem, scheduling, search.

This research was supported in part by NSERC Research Grant OGP0046423, NSERC Strategic Grants MEF0045793 and STR0167029, 1996 DIMACS Workshop on the satisfiability (SAT) problems, and a DIMACS visitor program.

been identified as central to a number of areas in computing theory and engineering. Since SAT is NP-complete, it is unlikely that any SAT algorithm has a fast worst-case time behavior. However, clever algorithms can rapidly solve many SAT formulas of practical interest. There has been great interest in designing efficient algorithms to solve most SAT formulas [**33**].

A scheduling problem involves determining an optimal schedule that places a set of tasks on a time line, subject to temporal constraints, capacity constraints, and resource constraints. We consider static scheduling algorithms that schedule an edge-weighted *directed acyclic graph* (DAG), or called *task graph*, to a set of homogeneous processors to minimize the completion time. This problem is known to be NP-hard in its general forms [**27**], and optimal solutions are known in restricted cases [**14, 24, 32**]. There has been considerable research effort in this area resulting in various heuristic algorithms [**13, 23, 73, 102, 111, 126, 128**].

Traditional methods treat SAT and scheduling as discrete, constrained decision problems. A variety of enumeration and backtracking methods were developed. As a result, they can solve small size SAT problems having few hundreds of variables or small size scheduling problems with few hundreds of tasks. In this work, we have used *min-conflict* heuristic and developed randomized and deterministic local search algorithms which can solve large size SAT and scheduling problem instances. Our experimental studies indicate that randomized local search algorithms are particularly efficient for large size NP-complete problems that have many solutions. We will give several efficient local search algorithms and show their experimental performance for solving large size satisfiability problems. For an NP-hard optimization problem with one or fewer optimum solutions, randomized local search algorithms perform satisfactorily well and, in some cases, a deterministic local search algorithm can find higher quality solutions much more efficiently. For this case study, we give two efficient local search algorithms and compare their performance for solving large size scheduling problem.

This paper is organized as follows. In the next section, we will briefly discuss the constraint satisfaction problem which is a general model for SAT and scheduling problems. In Section 3, we give a summary of early development of local search algorithms for SAT and scheduling problems. Section 4 describes basic framework and major components in local search. Efficient local search algorithms for solving large size SAT and scheduling problems are discussed in Section 5 and Section 6, respectively. Finally, Section 7 concludes this paper.

2. Constraint Satisfaction Problems

A constraint satisfaction problem (CSP) is to determine whether a set of constraints over discrete variables can be satisfied. Each constraint must have a form that is easy to evaluate, so any difficulty in solving such a problem comes from the interaction between the constraints and the need to find a setting for the variables that simultaneously satisfies all the constraints.

A discrete CSP model consists of the following three components [**34, 48**]:

- *n variables*: $\{x_1, x_2, \ldots, x_n\}$. An assignment is a tuple of n values assigned to the n variables.
- *n domains*: $\{D_1, D_2, \ldots, D_n\}$. Domain D_i contains d possible *values* (also called *labels*) that x_i may be instantiated, *i.e.*, $D_i = \{l_{i,1}, l_{i,2}, \ldots, l_{i,d}\}$.

$$\textbf{CSP}\begin{cases}\begin{array}{c}\text{Discrete}\\ \text{CSP}\end{array}\begin{cases}N\text{-queen problem}\\ \text{Graph coloring problem}\\ \text{Scheduling problem}\end{cases}\\ \\ \begin{array}{c}\text{Binary}\\ \text{CSP}\end{array}\begin{cases}\text{SAT problem}\\ \text{Max-SAT problem}\end{cases}\end{cases}$$

FIGURE 1. Some examples of the constraint satisfaction problem (CSP). SAT problem is a special case of CSP, i.e., a CSP with binary values.

- A subset of $D_1 \times D_2 \times \ldots \times D_n$ is a set of *constraints*. A set of *order-l constraints* ($l \leq n$) imposed on a subset of variables $\{x_{i_1}, x_{i_2}, \ldots, x_{i_l}\} \subseteq \{x_1, x_2, \ldots, x_n\}$ is denoted as

$$C_{i_1, i_2, \ldots, i_l} \subseteq D_{i_1} \times D_{i_2} \times \ldots \times D_{i_l}.$$

An order-l constraint indicates the compatibility (*i.e.*, consistency/inconsistency or conflicting measure) among l variables for a given variable assignment. The variables *conflict* if their values do not satisfy the constraint. In practice, two frequently used constraints are *unary constraints* imposed on a single variable ($C_i \subseteq D_i$) and *binary constraints* imposed on a pair of variables ($C_{i,j} \subseteq D_i \times D_j$).

Solving a CSP entails minimizing local inconsistency and finding a consistent value assignment (*i.e.*, a consistent labeling) to the variables subject to the given constraints.

Constraint satisfaction problems are extremely common. Most NP-complete problems are initially stated as constraint satisfaction problems [**96**]. Indeed, the proof that a problem is NP-complete implies an efficient way to transform the problem into a constraint satisfaction problem. For a few special forms of the constraint satisfaction problem there exist algorithms that solve such formulas in polynomial worst-case time. When no polynomial-time algorithm is known for a particular form of constraint satisfaction problem, it is common practice to solve such formulas with a search algorithm.

Problems that are commonly formulated as constraint satisfaction or satisfiability problems for the purposes of benchmarking include graph coloring and the n-queen problems (see Figure 1). In the case of the n-queen problem, although analytical solutions for the problem exist [**1, 3, 5**], they provide a restricted subset of solutions and can not find a general solution to the n-queen problem. In practical applications, one must use a search algorithm to find a general solution to the n-queen problem. The n-queen problem provided an early CSP model for the SAT and scheduling problem (see Section 3).

In a SAT formula, each constraint is expressed as a clause, making SAT a special case of the constraint satisfaction problem. Due to this close relationship, any CSP algorithm can be transformed into a SAT algorithm, and this can usually be done in a way that maintains the efficiency of the algorithm.

A scheduling problem is itself a CSP with timing, capacity and resource constraints. Each constraint can be expressed as a clause in the SAT problem. Therefore any CSP or SAT algorithm can be directly applied to solve a scheduling problem. Due to a variety of application situations, scheduling problem instances have often been used as benchmarks for CSP and SAT algorithm testing.

3. Early Work on Local Search Algorithms for SAT and Scheduling Problems

Early work on constraint satisfaction, simulated annealing, and complexity theory contributed significantly to the original development of local search algorithms for SAT and scheduling problems. Five notable early developments are: (1) Boolean local relaxation, (2) *min-conflict* formulation for SAT and CSP, (3) the n-queen models and algorithms for scheduling applications, (4) a simulated annealing algorithm, and (5) a 2-SAT algorithm.

3.1. Boolean Local Relaxation. Boolean local relaxation may be viewed as a kind of deterministic local search. The basic idea was to formulate conflicts among truth values in a set of Boolean objective functions and then use a Boolean local search procedure to minimize the conflicts. This Boolean *min-conflict* heuristic was developed for solving the CSP and SAT problems.

For a variable having m values, m Boolean labels are used to indicate the variables' instantiation to a particular value. The conflicts produced by a value assignment are accumulated in a set of Boolean objective functions (one for each label). The objective function for the ith variable and kth label, $f_{i,k}$, is defined as [**34**]:[1]

$$f_{i,q} = \prod_{j=1}^{n} \sum_{p=1}^{m} l_{i,q} \wedge C_{i,j}(q,p) \wedge l_{j,p}, \tag{3.1}$$

where $C_{i,j}(q,p)$ is a constraint between labels $l_{i,q}$ and $l_{j,p}$. Note that the right hand side of Eq. (3.1) is a *CNF* formula with *extended literals*.

In an early implementation, the local conflict minimization procedure was performed using a Boolean relaxation procedure (Figure 2). During each iteration, the algorithm checks each variable for every label and iteratively minimizes the objective functions by flipping truth values assigned to the labels: If the objective function does not change, keep it; If the objective function can be reduced, reduce it, and report the *inconsistency* status. The iteration will terminate once the *inconsistency* signal turns off.

Boolean local relaxation was suitable to VLSI circuit implementation. During 1985 to 1988, several parallel DRA algorithms, i.e., DRA2, DRA3 and DRA5, were implemented in VLSI chips to speed up CSP/SAT computations [**34, 53, 122, 52**]. Furthermore they were used in backtracking search for CSP and SAT applications [**34**].

Because of its iterative local conflict minimization and its direct applications to SAT and CSP, Boolean local search made itself a *predecessor* of several early local search algorithms for CSP, SAT and scheduling problems.

---

[1] In [**34**] the objective function was defined for label $l_{i,k}$ directly.

```
procedure DRA()
boolean inconsistency;
begin
   inconsistency := TRUE;
   k := 0;
   while inconsistency = TRUE do
   begin
      inconsistency := FALSE;
      for variable i := 1 to n
         for label q := 1 to m
         begin
            f_{i,q}^{k+1} := evaluate_objective_function(l, C);
            /* local conflict minimization */
            if f_{i,q}^{k+1} = f_{i,q}^k then continue;
            if f_{i,q}^{k+1} < f_{i,q}^k then
            begin
               f_{i,q}^k := f_{i,q}^{k+1};
               inconsistency := TRUE;
            end;
         end;
      k := k + 1;
   end;
end;
```

FIGURE 2. **DRA:** A local relaxation algorithm.

3.2. *Min-Conflict* **Formulation for SAT and CSP.** Objective functions in DRA algorithms were defined for Boolean labels. During the late eighties, Gu [**34**] observed that if the *conflicts* of all the Boolean objective functions were formulated in *one* objective function, then the global minimum of the objective function would correspond to a conflict-free solution of the given CSP problem. Two first such formulations were developed to solve SAT and n-queen problems.

In the SAT problem, the problem "board" is an n by m rectangle matrix with n variables placed in the horizontal direction and m clauses placed in the vertical direction (Figure 3 (a)). If we evaluate variables in the horizontal direction and count the number of unsatisfiable clauses (i.e., "conflicts") in the vertical direction, we can formulate an objective function as the number of unsatisfied clauses over *all* the variables [**34, 35, 39, 41**]. Thus, the global minimum of the objective function corresponds to the solution of the SAT problem. Reducing the objective function to zero would produce a solution to the SAT problem. For the n-queen problem [**34**], we place n queens in the columns of their row's permutation. This guarantees that no two queens would attack each other on the same row or the same column. The "conflicts" among queens possibly occurring on the diagonal lines are collected to formulate the objective function for solving the n-queen problem (see Figure 3 (b)). Reducing the conflicts to zero would produce a solution to the n-queen problem.

Following conflict formulation of objective function, apparently, the iterative local minimization procedure used in the Boolean relaxation becomes a local search

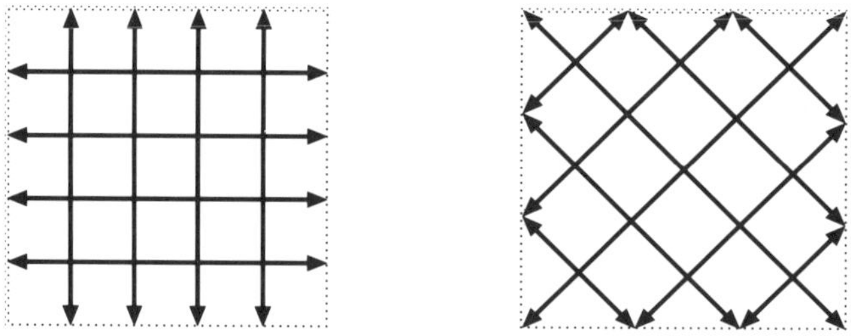

(a) (b)

FIGURE 3. The conflicting direction in (a) the satisfiability problem and (b) the n-queen problem.

procedure to minimize the objective function. This gave the *min-conflict* heuristic in the discrete space. This idea led directly to the early design of *SAT1* algorithms, the first local search algorithm for SAT. Following this, Gu gave a number of randomized local search algorithms for the SAT problem [**35, 39, 41**]. Furthermore efficient heuristics (Section 4.3) were developed to improve the performance of the local search algorithms. Due to the important industrial applications at that time, the effectiveness of the *SAT1* algorithms was tested through two CSP benchmarks, i.e., the SAT problem and the n-queen problems.

An important industrial application for SAT was VLSI engineering. Many VLSI circuit design problems, such as VLSI circuit testing, synthesis, and routing, can be formulated as instances of the SAT or MAX-SAT problems. During the late eighties, the *SAT1* algorithm family was first developed for VLSI applications and was found to be efficient for many VLSI circuit design problems. At that time, however, there was little progress in the theoretical analysis of the *SAT1* algorithms.

Another important application area for the SAT problem was industrial scheduling. During the late eighties, IBM and NASA were working on a number of important scheduling projects. These applications involved solving large size scheduling and task assignment problems under critical timing, spatial, and resource constraints. The scheduling and task assignment problems are well-known as the satisfiability problem. SAT formulas characterize these problems precisely. Significant local search solutions to the scheduling and task assignment problems were derived from the *SAT1* algorithm family. Due to its abstract *CNF* formulation, however, the SAT problem was not able to provide a descriptive geometric model that was able to demonstrate the scheduling and task assignment operations expressively. And it happened to be that the n-queen problem provided an ideal model for these large size scheduling problems.

3.3. N-Queen Scheduling Models and the *QS* Algorithms. The n-queen and scheduling are two problems in constraint satisfaction. During the late eighties, Gu worked on various n-queen problem models for combinatorial optimization [**34, 48**] and found that, by a remarkable coincidence, the n-queen model represents a significant model for scheduling and task assignment problems.

The underlying structure of the n-queen problem, represented by a complete constraint graph, gives a relational model with fully specified constraints among the multiple objects [**34**]. Variations on the dimension, the objects' relative positions, and the weights on the constraints led to a **hyper-queen** problem model which had several simple and basic models:

- **n–queen** problem: the *base* model. N queens are indistinguishable and the constraints among queens are specified by the binary values (i.e., 1 or 0).
- **w–queen** problem: the *weighted* n-queen model. N queens are distinguishable. Each is associated with a cost. The constraints among queens are specified by some weights.
- **3d–queen** problem: queens are to be placed in a 3-dimensional ($l \times m \times n$) rectangular cuboid. A special case, **nm–queen**, is to place queens on an n by m rectangle.
- **q+–queen** problem: more than one queens are allowed to be placed on the same row or the same column.

The *hyper*-queen problem can model the objects/tasks, the performance criteria, the timing, spatial, capacity and resource constraints for a wide range of scheduling and task assignment problems. This made the n-queen problem a general model for many industrial scheduling and task assignment problems. By a remarkable coincidence, the models of several difficult scheduling projects at that time were either the n-queen or the *hyper*-queen problems [**68, 119**]. All of them required efficient solutions to the n-queen or *hyper*-queen problems.

Let's consider a static scheduling problem where we are to schedule an edge-weighted *directed acyclic graph* (DAG), or called *task graph*, to a set of homogeneous processors to minimize the completion time. This problem is known to be NP-hard in its general forms [**27**]. Using the *hyper*-queen model, a set of weighted queens is to be placed on a T by P rectangle (see Figure 4). Here T denote the execution time, P the number of processors, J_i the ith queen (i.e., the ith job and its execution time), and c_{ij} the communication time from the ith job to the jth job, the goal is to place the *job queens* onto the T by P board and minimize the longest execution path, following the given topological constraints. Instead of minimizing conflicts among queens on diagonal lines as in the n-queen problem, here we reduce conflicts among jobs and minimize the longest execution path following the given topological constraints.

We search different geometric patterns in the n-queen and the scheduling problems. These patterns produce different constraints in CSP and do not affect the formulation and the use of *min-conflict* heuristic. In fact, once the constraints are extracted, an algorithm developed for the n-queen problem can be used to solve the scheduling problem with or without minor modifications. The *hyper*-queen models has freed the n-queen problem from its puzzle game background and made many practical scheduling and task assignment applications feasible. These include [**43, 116**] task scheduling, real-time system, task assignment, computer resource management, VLSI circuit design, air traffic control, communication system design, and so on.

Polynomial time, analytical solutions for the n-queen problem exist but they cannot solve the general search problems and have no use in practice [**1, 3, 5, 25, 57, 98**]. Following local conflict minimization [**34, 53**], a $QS1$ algorithm was developed during late 1987 and was implemented during early 1988. It was the

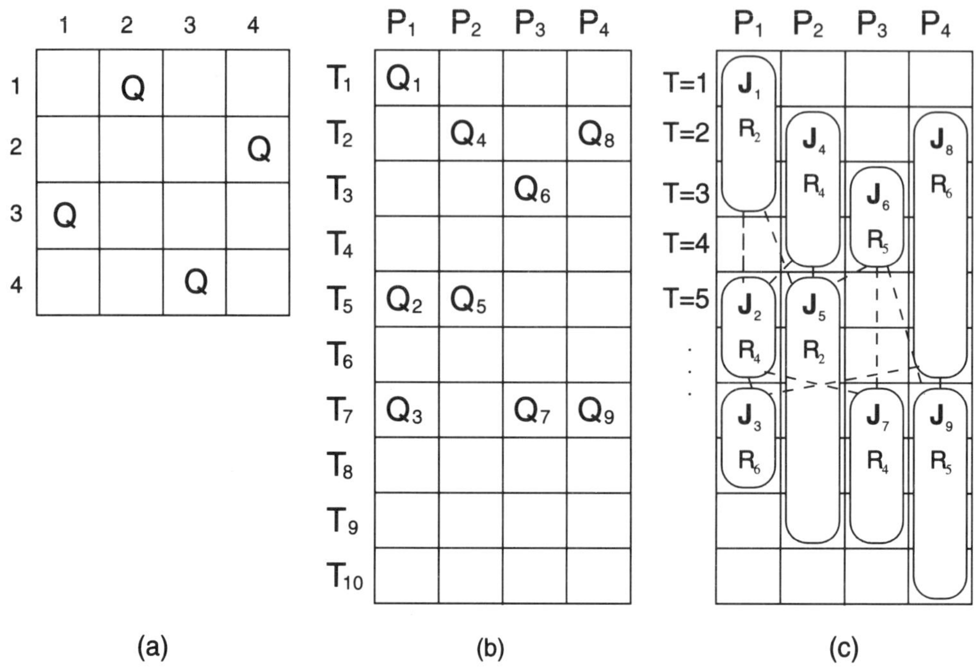

FIGURE 4. The n-queen scheduling model. Instead of minimizing conflicts among queens on diagonal lines, here we reduce conflicts among jobs and minimize the longest execution path following the given topological constraints.

first local search algorithm developed for solving the large size n-queen problem [**34, 117, 112, 113**]. Three improved local search algorithms for the n-queen problem were developed during 1988 to 1990 [**76, 77, 67, 100**]. $QS2$ is a near linear-time local search algorithm with an efficient random variable selection strategy [**114**]. $QS3$ is a near linear-time local search algorithm with efficient pre- and random variable selection and assignment [**114**]. $QS4$ is a linear time local search algorithm with efficient partial and random variable selection and assignment techniques [**115, 116**]. Compared to the first local search algorithm [**34**], partial and random variable selection/assignment heuristics have significantly improved search efficiency by orders of magnitude. $QS4$, for example, was able to solve 3,000,000 queens in a few seconds.

Three years after releasing the $QS1$ algorithm, Minton *et al.* independently reported a similar local search algorithm for the n-queen problem [**84, 85**]. A major difference between Minton's algorithms and Sosic and Gu's algorithms was that Minton's algorithm was a one dimensional local search without using random heuristics.

Early local search solutions for scheduling applications were developed during the late eighties. Since then more than one hundred industrial companies worldwide have developed similar scheduling software systems for various industrial scheduling applications.

3.4. A Simulated Annealing Algorithm for *Max-SAT*. Motivated by the method of simulated annealing, Hansen and Jaumard [**56**] proposed a steepest ascent, mildest descent algorithm for the maximum satisfiability (Max-SAT) problem. In this approach, Hansen and Jaumard focused on *a* local change and defined an objective function based on *a* switching variable and its related clauses. The objective function maximizes local compensation for each variable which can be used for solving the Max-SAT problem. The objective function can not be used for the SAT problem unless another objective function whose global minimum corresponds to a solution of the SAT problem is given. Furthermore, Hansen and Jaumard used local optima checking to handle the local optimum and found it by providing additional guidance to the search direction.

3.5. Theoretical Study for *2-SAT* and *SAT1*. During the study of the complexity of a certain natural generalization of SAT, Papadimitriou gave a randomized algorithm for the 2-SAT problem [**89**]. Furthermore Papadimitriou showed that such a randomized algorithm finds assignments for 2-SAT instances in $O(n^2)$ steps with probability approaching one, where n is the number of variables. With further extensions [**90**], the algorithm can be applied to solve the random 3-SAT problems.

During the early nineties, researchers started to work on the theoretical analysis of local search algorithms for CSP and SAT problems. In 1991 two theoretical studies that focused on the SAT problem were reported. Gu and Gu took three algorithms (i.e., $SAT1.1$, $SAT1.2$, and $SAT1.3$) from the *SAT1* algorithm family and made average time complexity study for the SAT problem [**50**].

Early on, local search methods for the large size SAT, n-queen and scheduling problems attracted great attention in the AI area. This was partly due to the close relationship among SAT, n-queen and scheduling that all three problems are special cases in CSP. If one can find an efficient (non-analytical) search algorithm for one of them, then the algorithm can be directly translated to an efficient algorithm for the other problems.

4. Local Search

Local search is a major class of discrete, unconstrained optimization procedures that can be applied to a discrete search space. In this section, we briefly summarize the basic framework of local search, describe a three-level search space model and several essential components of local search.

4.1. Framework. *Local search*, or *local optimization*, is one of the primitive forms of continuous optimization applied to a discrete search space. It was one of the early techniques proposed to cope with the overwhelming computational intractability of NP-hard combinatorial optimization problems [**91**].

There have been two major periods for the development of local search. Early local search method was able to solve small size, unconstrained problems such as small TSP instances [**80**]. During the middle and late eighties, more powerful techniques (see Section 4.3) for randomized local search were developed which were capable of solving large size, highly constrained CSP, SAT and scheduling problems efficiently [**34, 39, 115, 116**].

Given a minimization (maximization) problem with objective function f and feasible region R, a typical local search procedure requires that, with each solution point $\mathbf{x}_k \in R$, there is a predefined *neighborhood* $N(\mathbf{x}_k) \subset R$. Given a current solution point $\mathbf{x}_k \in R$, the set $N(\mathbf{x}_k)$ is searched for a point $\mathbf{x}_{k+1}$ with $f(\mathbf{x}_{k+1}) < f(\mathbf{x}_k)$ ($f(\mathbf{x}_{k+1}) > f(\mathbf{x}_k)$). If such a point exists, it becomes the new current solution point, and the process is iterated. Otherwise, $\mathbf{x}_k$ is retained as a *local optimum* with respect to $N(\mathbf{x}_k)$. Then, a set of feasible solution points is generated, and each of them is "locally" improved within its neighborhood. To apply local search to a particular problem, one needs only to specify the neighborhood and the procedure for obtaining a feasible starting solution.

Local search can be efficient for two reasons. First, at the beginning of search, a full assignment is assigned to all the variables in the search space. Search efforts are focused on a single path in the search space. Second, local search refines for improvement within its local neighborhood using a testing for improvement and, if there is any improvement, takes an action for improvement. Since the objective function has a polynomial number of input numbers, both *testing* and *action* can be done efficiently. Little effort is needed to generate the next solution point. A major weakness of local search is that the algorithm has a tendency to get stuck at a locally optimum configuration, i.e., a local minimum.

Many search techniques, such as statistical optimization [**17, 18, 103**], simulated annealing [**74**], stochastic evolution [**101**], and conflict minimization [**34, 85, 112, 116**], are either local search or variations of local search. For most search problems encountered, in terms of computing time and memory space, local search often achieves many orders of magnitude of performance improvement over conventional techniques such as branch-and-bound [**39, 41, 97, 114, 116**].

4.2. A Three-Level Search Space Model. A large number of real experimental data suggest that there are several typical local minimum structures when solving NP-hard optimization problems (see Figure 5). A *valley* and a *basin* are ideal cases that one can find a global minimum quickly. Local search and the related heuristics can handle a *terrace* and a *plateau* without much difficulty. The most difficult situation is a *trap* where a group of local minima is confined in a "well." The search process walks around the set of local minima periodically and cannot get away without special mechanism. In general there may be many traps in a search problem. The characteristics of a trap are closely related to the objective function used, the search space structure, and the search algorithm structure.

Further observations suggest that a search space may be roughly divided into several different levels, depending on the problem structures (Figure 6). In the model, a search space is roughly viewed in three levels: top level, middle level, and bottom level. The *top level* is the upper open portion of the search space with smoothing edges. For a minimization problem, most optimization algorithms can descend quickly in the top level and thus perform quite well. The *middle level* is the middle portion of the search space where there are relatively big "mountain peaks." During the descent, the search process may encounter problems and it may have to use some tunneling and random heuristics (see Section 4.3) to proceed. The *bottom level* is the bottom portion of the valleys (particular the lowest valley) where there are many traps. When a local search process falls into a trap it may become locked into a loop of local minima. Most algorithms do not succeed in this stage and have difficulty continuing.

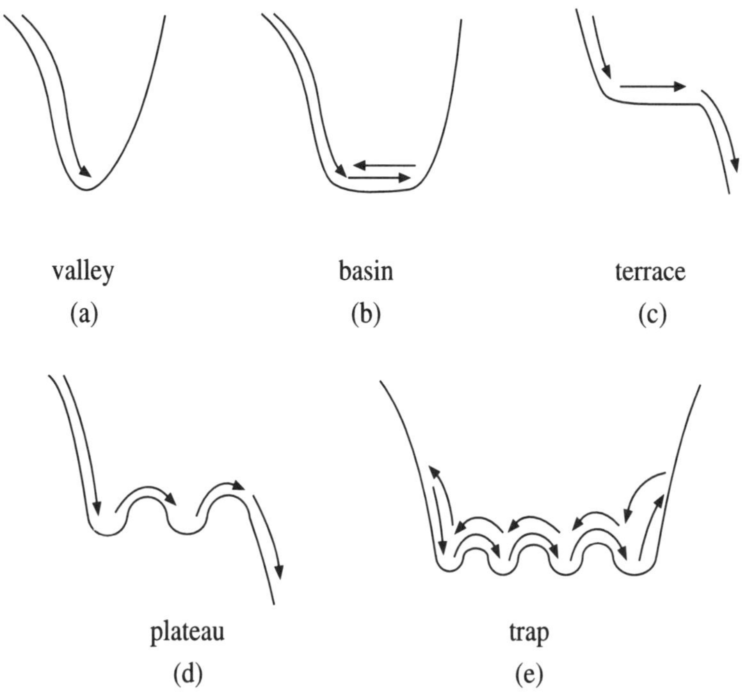

FIGURE 5. There are a number of local minimum structures. A trap is a "well" of local minima and is difficult to deal with.

The above observations suggest to use multiphase search to handle the NP-hard problems [**39, 42, 51**]. That is, we may use an *open search* in the top level, a *peak search* for searching "coarse" peak structures in the middle level, and a *trap search* for tracking "fine" rugged trap surface structures in the valleys. Following this, a variety of heuristics and techniques have been proposed to improve the performance of local search.

4.3. Four Components in Local Search. A number of efficient local search algorithms for SAT, CSP and scheduling problems have been developed since 1987. Past lessons showed that the following **four components** are crucial to the development of an efficient local search algorithm for the satisfiability and NP-hard problems. They are: (1) the min-conflict heuristics, (2) the best-neighbor heuristics, (3) the random value/variable selection heuristics, and (4) the trap handling heuristics.

1. The Min-Conflicts Heuristics.

Different forms of min-conflict heuristics were proposed during 1985 and 1987 for solving SAT, CSP and scheduling problems [**34**].[2] The min-conflict heuristics

[2]In the early days *min-conflict* was variously called inconsistency removing, inconsistency resolution, conflict resolution, enforce local consistency, and local conflict minimization [**34**]. Later, Minton shortened these words into a concise term: min-conflicts.

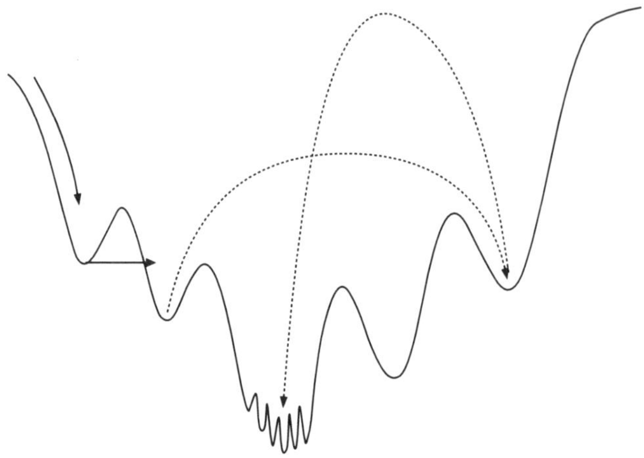

FIGURE 6. An informal example of the three-level search space model. A search process would go through an *open search* in the upper open portion of the search space, a *peak search* in the middle portion of the search space, and a *trap search* in the valley portion of the search space.

aim at performing local conflict minimization in Boolean, discrete, and real spaces which are important to handle constraints in a constrained optimization problem. [**100**]: [3]

Min-Conflict Heuristic (Boolean Space) [34]. Multiple values to be assigned to a variable are represented by a vector of Boolean labels. Each Boolean label, either "1" or "0," indicates the variable's instantiation to a specific value. Two labels are *conflicting* if their values do not satisfy the given constraint. The *conflicts* (due to an assignment) are formulated as a set of objective functions. The objective functions are minimized by changing values assigned to the labels.

Min-Conflict Heuristic (Discrete Space) [34]. Interrelated objects are chosen as variables. Two variables are *conflicting* if their values do not satisfy the given constraint. The *number of conflicts* (due to an assignment) is formulated in an objective function. The objective function is iteratively minimized by changing values assigned to the variables.

Min-Conflict for SAT [35, 50, 39, 40, 41]. Using *conflicts* as objective [**34**], the objective function for the SAT problem gives the number of unsatisfied clauses. A *CNF* is true *if and only if* the objective function takes the global minimum value 0 on the corresponding solution point.

This objective function is the basis for the design of the *SAT1*, *SAT2*, *SAT3*, and *GSAT* algorithms [**39, 40, 50, 41, 107**].

---

[3]The min-conflict heuristics also work in real space (see examples in [**40, 38, 44**]).

Performance. The min-conflict heuristics have been applied to solve the CSP and SAT problems since 1985 [**34, 35, 112, 117, 50, 39**]. They showed significant performance improvements when compared to traditional backtracking search algorithms. The effectiveness of min-conflicts heuristic was further observed by Russel and Norvig [**100**], Kumar [**76, 77**], Johnson [**67**], Minton *et al.* [**85**], and Selman *et al.* [**107**].

Early local search algorithms can handle small size path-finding problem such as small TSP instances. By grouping conflicts into an objective function, local search with min-conflicts heuristics can now solve large size, highly constrained SAT, CSP and scheduling problems.

2. The Best-Neighbor Heuristics.

Local search proceeds by taking *any* feasible solution point that reduces the objective function. Among many neighboring feasible solution points, local search does not take into account its neighbors' relative performance with respect to the objective function.

Best-Neighbor Heuristic [**39, 41, 44**]. A *greedy* local search algorithm selects the *best neighbor* that yields the minimum value to the objective function and takes this *best neighbor direction* as the descent direction of the objective function.

In a real search space, continuous optimization algorithms can find the best neighbor feasible solution efficiently. A number of local and global optimization algorithms have been developed to solve the SAT problem [**39, 41, 44**]. The first version of the *GSAT* algorithm was proposed as a *greedy* local search algorithm [**107**].

Performance. Greedy local search pursues only paths where every step leads to an improvement, but this leads to a procedure that becomes stuck much more often than the randomized local search. Greedy local search procedure gets stuck in flat places as well as at local minima depending on the application cases. In practice, the best neighbor heuristic should be used in conjunction with random value/variable selection and trap handling heuristics described next.

3. The Random Value/Variable Heuristics.

Random value assignment and random variable selection techniques are fundamental to the design of an effective local search algorithm for NP-hard problems [**48, 33, 49**].

Random Flip Heuristic [**34, 39, 40, 50, 41**]: Randomly flip the truth values of $1 \leq k \leq n$ variables in the SAT formula.

This simple heuristic was first implemented in several *SAT1* algorithms as *local handler*(s) in 1987. It has been proven to be effective in improving the performance of local search algorithms [**39, 50, 40, 41**].

During 1988 to 1990 a similar heuristic, **random swap**, was used to develop local search algorithms for the CSP (e.g., n-queen) problems. It showed significant performance improvement for solving large size n-queen problems [**34, 113, 114, 115, 116**].

Random Value (Assignment) Heuristics [34, 39, 50, 41, 80, 92, 114, 115, 116]. These include: randomly select a value that generates the minimum number of conflicts; randomly select a value if there is a symmetry (i.e., more than one value producing the same performance); and randomly select a value for conflict minimization when local minima are encountered.

A simple random value assignment heuristic, random disturbance, was used early in solving the TSP problem.

Random Variable (Selection) Heuristics [34, 39, 50, 41]. There are two important heuristics:

1. **Any Variable Heuristic:** select any variable randomly.
2. **Bad Variable Heuristic:** select a variable from the conflicting variables randomly.

The random variable selection heuristic is one of the most important heuristics in the design of local search algorithms for NP-hard problems. It was first used in the local search solution for the SAT problem [**35**] and then used for the local search solution for the CSP (e.g., n-queen) problems [**114**].

Conflicting variables in the SAT problem contribute to the unsatisfied clauses. Accordingly we have:

Bad Variable Heuristic for the SAT problem [39, 41, 114, 115, 116, 113, 117]: randomly select a variable in the unsatisfied clauses for conflict minimization.

The bad variable heuristic was first implemented to solve the large size n-queen problems during 1988 to 1990 [**114, 115, 116, 113, 117**] and was implemented in the *SAT2* algorithm in 1990 [**39, 41**]. The bad variable heuristic was independently developed by Papadimitriou for solving the 2-SAT problem in 1991 [**89**] and was recently used in the development of *GSAT* and *WSAT* algorithms by Selman *et al.* [**105**].

Partial/Pre- Random Variable Selection Heuristics [34, 39, 50, 41]. Partial variable random selection makes use of partial or alternating variable selection techniques. Variants of partial random selection include partial and alternating selection of conflicting and consistent variables, a combination of partial deterministic and partial random variable selection, partial interleaved selection of the different search phases, and partial random selection with meta-heuristic control. The simplest selection strategies include: select a variable deterministically (randomly) and select another variable randomly for conflict minimization; select a variable deterministically (randomly) from the conflicting variables and select another variable randomly for conflict minimization; select a variable deterministically and select another variable randomly from the conflicting variables for conflict minimization; during certain periods of search, select a variable deterministically (randomly) and select another variable randomly for conflict minimization; during certain periods of search, select a variable deterministically (randomly) from the conflicting variables and select another variable randomly for conflict minimization; during certain periods of search, select a variable deterministically and select another variable randomly from the conflicting variables for conflict minimization.

Partial Random Variable Selection Heuristics for the SAT problem: a variable may be selected from the unsatisfied clauses in a random, partially alternating, partially periodic, or partially interleaving order.

The partial and pre- random variable selection heuristics were implemented in the *SAT3* algorithm in 1990 [**39, 41**] and were used to solve the large size n-queen problems around 1990 [**114, 115, 116, 113, 117**]. A similar heuristic to the partial random variable selection, *random walk*, was developed by Selman, Kautz, and Cohen independently in 1994 [**105**].

Performance. Random and partial variable selection heuristics were introduced in the design of $SAT1$, $QS2$, $QS3$, and $QS4$ algorithms [**35, 50, 39, 41, 112, 114, 115, 116**]. They can overcome the weakness of the greedy local search algorithms. Compared to greedy local search, they can offer many orders of magnitude of performance improvements in terms of computing time, solving hard, large size SAT, CSP and scheduling problems with multi-million variables in seconds [**39, 41, 114, 116**]. They were used in the design of $SAT1.5$, $SAT2$, and $SAT3$ algorithms [**39, 41**].

Selman *et al.* have recently applied a number of random variable selection heuristics to improve the performance of the $GSAT$ algorithm and found them effective [**105**].

4. The Trap Handling Heuristics.

Search is a process of combating local minima. When the search process is approaching the final search stage, trap handling heuristics are needed to cope with local minima and traps (see Sections 4.2 and 5.3).

Tunneling Heuristic [**50, 41, 123**]: Change the value of a variable if it does not change the value of the objective function.

Tunneling Heuristic for the SAT Problem [**50, 41, 107**]: Flip the truth value of a variable if it does not change the value of the objective function (see Section 5).

Local Tracking Heuristics [**39, 51**]. Local tracking heuristics are used to track and break local loops (a periodic occurrence of a set of local minima). Several frequently used heuristics include: track local loop(s) when falling into a trap; give low priority to flip to the variables in a local minimum loop; give high priority to flip to variables that lead to a *new* descending direction; lock and release trapping variables periodically, adaptively, or statistically; move gently in a trap to handle fine local structures; move strongly in a trap to handle coarse local structures; jump out of a trap if walking inside it sufficiently long.

Multiphase Search Heuristics [**35, 39, 115, 116, 41, 97, 123, 51**]. Multiphase heuristics are a part of multispace search heuristics [**36, 48, 49**]. They have been developed to adapt to the different phases of a search process: perform a poor initial search and then a fine local search for conflict minimization; perform a good initial search and then a fine local search for conflict minimization; perform a good initial search, then a rough local search, and a fine local search for conflict minimization; perform an initial search, and then a rough local search and a fine local search alternatively for conflict minimization; perform a rough initial search, then a coarse local search, and finally, a fine local search for conflict minimization.

Multispace Search Heuristics [**36, 48, 49**]. Structural multispace operations have been developed that empower a search process with an information flux which is derived from a sequence of stepwise structural transformations. These include multispace scrambling, extradimension transition, search space smoothing, multiphase search, local to global passage, tabu search, and perturbations. They can disturb the environment of forming local minima and facilitate efficient local search when there are many local minima.

Performance. Trap handling heuristics have significantly improved the search efficiency of the *SAT1.5* algorithm [**39, 41, 51**] (see Section 5.3). Multiphase and multispace search heuristics have been applied to a variety of practical applications and found to be effective [**36, 39, 41, 51, 97, 123, 115, 116**].

In the next two sections, we show how to use these techniques to design local search algorithms for SAT and scheduling problems.

5. Local Search for Satisfiability (SAT) Problem

An instance of the satisfiability (SAT) problem is a Boolean formula that has three components [**15, 27**]:

- A set of n variables: x_1, x_2, ..., x_n.
- A set of literals. A literal is a variable ($Q = x$) or a negation of a variable ($Q = \bar{x}$).
- A set of m distinct clauses: C_1, C_2, ..., C_m. Each clause consists of only literals combined by just logical *or* ($\vee$) connectives.

The *goal* of the satisfiability problem is to determine whether there exists an assignment of truth values to variables that makes the following Conjunctive Normal Form (*CNF*) formula satisfiable:

$$C_1 \wedge C_2 \wedge \cdots \wedge C_m, \tag{5.1}$$

where $\wedge$ is a logical *and* connective.

Local search is a major class of discrete, unconstrained optimization procedures that can be applied to a discrete search space. Such procedures can be used to solve SAT by introducing an objective function that counts the number of unsatisfiable (*CNF*) or satisfiable (*DNF*) clauses and solving to minimize the value of this function [**35, 39, 41, 50, 89, 107**].

In this section, we will briefly review the existing SAT algorithms and then we will describe local search algorithms for SAT problem.

5.1. SAT Algorithms. The existing SAT algorithm are discussed in a recent survey [**33**]. Most existing SAT algorithms can be grouped into the following categories.

Discrete, constrained algorithms. Algorithms in this category treat a SAT formula as an instance of a constrained decision problem, applying discrete search and inference procedures to determine a solution. One straightforward way to solve an instance of SAT is to enumerate all possible truth assignments and check to see if one satisfies the formula. Many improved techniques, such as consistency algorithms [**48, 82**], backtracking algorithms [**6, 8, 10, 79, 93**], term-rewriting [**22, 62**], production system [**110**], multi-valued logic [**108**], Binary Decision Diagrams [**9, 4**], chip and conquer [**28**], resolution and regular resolution [**29, 81, 88, 99, 121, 125**], independent set algorithm [**65**], and matrix inequality system [**120**], have been proposed.

Other specific algorithms using these principles include simplified DP algorithms [**26, 31, 94**], and a simplified DP algorithm with strict ordering of variables [**64**]. The DP algorithm improved in certain aspects over Gilmore's proof method [**30**]. Analyses of SAT algorithms often concentrates on algorithms that are simple because it is difficult to do a correct analysis of the best algorithms. Under those conditions where simple algorithms are fast, related practical algorithms are also fast.

Discrete, unconstrained algorithms. In this approach, the number of unsatisfiable *CNF* (or satisfiable *DNF*) clauses is formulated as the value of the objective function, transforming the SAT formula into a discrete, unconstrained minimization problem to the objective function. Local search is a major class of discrete, unconstrained search methods [**39, 41, 48, 50, 89, 107**]. It can be used to solve the transformed formula.

Early work in constraint satisfaction and complexity study contributed to the development of local search algorithms for the SAT problem (see Sections 3, 4.3, and Figure 7). There were two major approaches in this area: randomized local search (*SAT1*) and greedy local search (*GSAT*). The *SAT1* algorithm was the first local search algorithm developed from the VLSI engineering and scheduling applications. The *GSAT* algorithm was derived from the early local search algorithms for the n-queen problem.

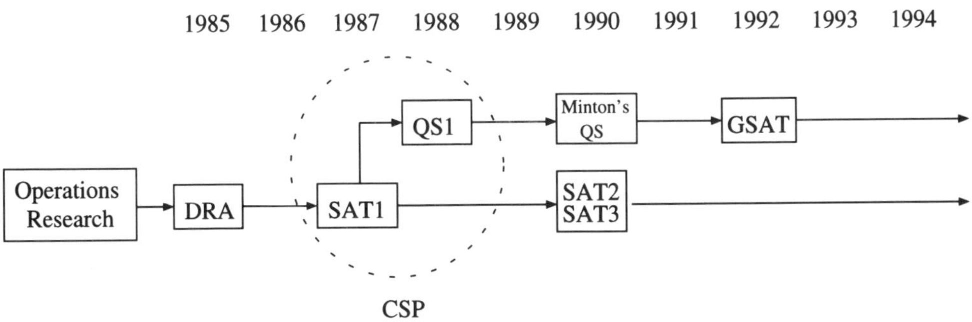

FIGURE 7. Early development of local search algorithms for SAT problem. There were two major approaches: randomized local search (*SAT1*) and greedy local search (*GSAT*). The *SAT1* was the first local search algorithm developed for the VLSI engineering and scheduling applications. The *GSAT* algorithm was derived from the early local search algorithms for the n-queen problem.

Constrained programming algorithms. Methods in this class were developed based on the fact that *CNF* or *DNF* formulas can be transformed to instances of Integer Programming, and possibly solved using Linear Programming relaxations [**7, 59, 60, 66, 70, 69, 92, 124**]. Many approaches, including branch-and-bound [**7**], cutting-plane [**60, 58**], branch-and-cut [**61**], interior-point [**70, 69**], and improved interior-point [**109**], have been proposed to solve the integer program representing the inference problem. Researchers found integer programming methods faster than resolution for certain classes of problems, although these methods are not robust and often fail to solve hard instances of satisfiability [**7, 59, 60, 66, 70, 69**].

Unconstrained, global optimization algorithms. Special models have been formulated to transform a discrete formula on Boolean space $\{0,1\}^n$ (a decision problem) into an unconstrained *UniSAT* problem on real space E^n (an unconstrained global optimization problem). The transformed formulas can be solved by many existing global optimization methods [**35, 39, 52, 44, 48**].

Some SAT algorithms are *complete* (they definitely determine whether an input has a solution or does not have one) [**9, 20, 19, 81, 99**], while others are *incomplete* (they sometimes determine whether or not the input has a solution, but in other cases they cannot find one) [**41, 44, 69, 89**].

The techniques used in complete SAT algorithms can usually be adapted to provide *exact solutions* to optimization problems. The techniques used in incomplete SAT algorithms can usually be adapted to provide *approximate solutions* to optimizations problems. They normally lead to algorithms that produce low (but not necessarily the lowest) value of the function.

For hard random 3-SAT problem instances, a complete algorithm can solve a SAT problem instance with few hundred variables while an incomplete local search algorithm can solve large size SAT formulas with thousand variables. This gives some advantages of using incomplete search method to solve large size SAT problem instances.

Presently two best local search algorithms for satisfiability testing are *GSAT* (*WSAT*) and *SAT1.5* algorithms. A *GSAT* algorithm can comfortably solve a SAT problem with 2,000 variables on an SGI Challenge with a 70 MHz MIPS R4400 processor [**105, 106**]. A typical *SAT1.5* algorithm can solve large size SAT problem instances with over 10,000 variables efficiently.

5.2. Basic Local Search Algorithms. In the following discussion, let:

- $F(\mathbf{x})$ be a function from $\{0,1\}^n$ to integer N,
- x_j be the jth variable,
- $\mathbf{x}$ be a vector of n variables,
- $C_i(\mathbf{x})$ be the ith clause function, and
- $Q_{i,j}(\mathbf{x})$ be the jth literal function of the ith clause function,

where $i = 1, ..., m$ and $j = 1, ..., n$.

Model. Most discrete local search procedures were developed based on a discrete, unconstrained optimization model, the *SAT1 model* [**35, 39, 41, 50**]. In

```
procedure SAT1.0 ()
begin
    /* initialization */
    get_a_SAT_instance();
    x_0 := select_a_random_initial_point();
    F(x_0) := evaluate_objective_function(x_0);

    /* search */
    k := 0;
    while F(x_k) ≠ 0 do
    begin
        for each variable i := 1 to n do
            /* if flip(x_i, x̄_i) does not increase F */
            if test_flip(x_i, x̄_i) then
            begin
                x_{k+1} := perform_flip(x_i, x̄_i);
                F(x_{k+1}) := evaluate_objective_function(x_{k+1});
            end;
        /* random flips */
        if local then local_handler();
        k := k + 1;
    end;
end;
```

FIGURE 8. **SAT1.0:** a randomized local search procedure for the SAT problem [**39, 41, 50**]. Random flips are introduced (1) to disorder the sequence with which the variables are selected for local optimization, and (2) to perturb local search with randomized downhill, tunneling, or uphill moves [**50, 41**].

this model, the truth values assigned to the variables are defined as:

$$x_i = \begin{cases} 1 & \text{if the variable has value } \textit{true} \\ -1 & \text{if the variable has value } \textit{false} \end{cases} \tag{5.2}$$

The objective function, $F(\mathbf{x})$, in the *SAT1* model counts the number of unsatisfied clauses as its objective value. A *CNF* is true *if and only if* $F(\mathbf{x})$ takes the global minimum value 0 on the corresponding $\mathbf{x}$.

Basic Local Search. The *SAT1.0* algorithm for the SAT problem is shown in Figure 8. It consists of an initialization stage and a search stage. At the beginning of search, a SAT formula is generated. An initial random solution is chosen. The number of unsatisfiable clauses is computed and is assigned as the value of the objective function. During each iteration, function *test_flip*() performs a test to see if the objective function would not increase after a flip. If *test_flip*() returns *true*, a flip operation is performed by procedure *perform_flip*(). Then function *evaluate_objective_function*() updates the objective function.

The procedure terminates when the objective function is reduced to zero, i.e., a solution to the given SAT instance is found. In practice, before the objective function reduces to zero, the procedure may become stuck at local minima. In the *SAT1.0* algorithm [**39, 50**], a simple randomized local handler performing random flips was used (Figure 9). It combines local descent (reducing objective function)

> **procedure** Local_handler ()
> **begin**
> random select *some* variable x_t's;
> $\mathbf{x}_{k+1}$:= perform_flip(x_t's,$\overline{x_t}$'s);
> $F(\mathbf{x}_{k+1})$:= evaluate_objective_function($\mathbf{x}_{k+1}$);
> **end**;

FIGURE 9. A simple randomized local handler used in the *SAT1* algorithms [**39, 41, 50**]. Random flips or a new random solution were applied to the algorithm if (1) $F \neq 0$ (*SAT1.0* algorithm), (2) $F > 0$ (*SAT1.1* algorithm), (3) $F(\mathbf{x}_{k+1}) = F(\mathbf{x}_k)$ (*SAT1.2* algorithm), and (4) $F > 0$ and $F(\mathbf{x}_{k+1}) = F(\mathbf{x}_k)$ (*SAT1.3* algorithm).

with the random uphill moves (increasing objective function), improving *SAT1*'s convergence performance effectively. In the *SAT1* algorithm family, one or more local handlers were implemented [**50, 39, 41**]. If the algorithms have difficulty to proceed, the algorithms will call the local handlers and use other heuristics (see Section 4.3) to improve algorithms' convergence performance.

The random flips used in the *SAT1* algorithms make the *order* of selecting which variable for local examination (i.e., the *for* loop) trivial [**50, 39, 41**]. One can essentially select any variable randomly for examination.

Tunneling Heuristic. A local handler and its activating condition(s) have significantly effect on the performance (e.g., running time and average running time) of a local search algorithm for the SAT problem. The conditions for activating local handlers differ from algorithm to algorithm (see Figure 9). In *SAT*1.1 algorithm, the local handler is called if the objective function is not zero (an aggressive strategy) [**50, 41**]. In *SAT*1.2 algorithm, the local handler is called if the objective function does not increase [**50, 41**]. In *SAT*1.3 algorithm, the local handler is called if the objective function does not increase or the objective function is greater than zero for some iterations [**50, 41**]. In the last two algorithms, the condition *"objective function does not increase"* means that the objective value is either reduced (*local descent*) or remained unchanged (*tunneling heuristic*).

Instead of making a random swing in the vertical direction in the search space, whenever a local minimum is encountered, one can *tunnel* through the rugged terrain structure in a horizontal direction, moving from one local basin to another local basin in an attempt to locate a better locally optimal solution (Figure 10). A tunnel can be thought of as a short-cut passing through a mountain separating points of equal elevation. Whenever a local minimum is encountered, a tunnel is made through a mountain to a neighboring basin as long as this does not change/increase the objective function. Tunneling can be used to search a region with local minima effectively. The behavior of local search with tunneling illustrates the fact

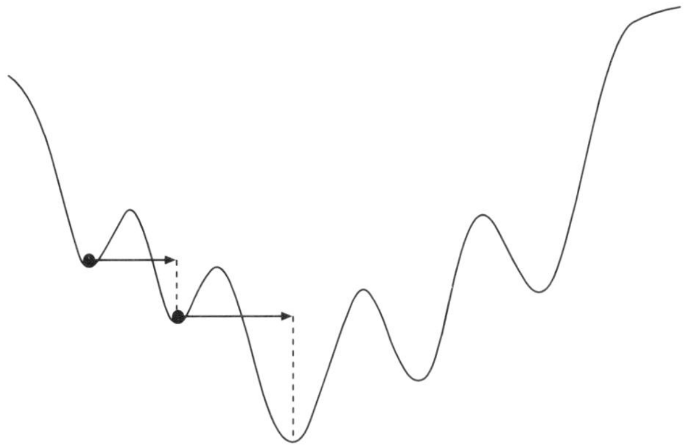

FIGURE 10. Tunneling

that seemingly innocuous changes in an optimization routine can have a surprisingly large effect on its performance. When tunneling was first implemented in the $SAT1$ algorithm in the late eighties, it was proven to be effective for solving SAT problem instances.

Random Flips (Noise). If the local search procedure becomes stuck at a local minimum, it may proceed by using a noise perturbation to change its current location in the search space. The effectiveness with which local minima are handled significantly affects the performance of a local search algorithm. Researchers have proposed a number of techniques such as *jumping*, *climbing*, *annealing*, and *indexing* to handle local minima [**36**]. In simulated annealing, a search process occasionally moves up rather than down in the search space, with large uphill moves being less likely than small ones. The probability of large uphill moves is gradually reduced as the search progresses.

A variety of local handlers have been designed for use in the local search algorithms [**50, 41**]. $SAT1.0$ [**35, 50, 39, 41**] uses a local handler that may randomly negate the truth values of one or up to n variables (Figure 9). The basic idea is to generate *random exchanges* in some current solution points when the search is stuck at a local minimum. The search accepts a modified point as a new current solution *not only when the value of the objective function is better but also when it is worse* [**50, 35, 39, 41**] This differs from the traditional local search such as $GSAT$ that used greedy local descent [**107**].

Parallel Local Search. Several parallel algorithms and VLSI architectures have been developed to accelerate CSP and SAT computations [**53, 52, 41, 118**]. Depending on implementations, there are several ways of grouping variables or clauses together in parallel so they can be evaluated simultaneously. In the $SAT1$ algorithms, the most frequently used part of computation is the function *evaluate_objective_function*(). It takes $O(ml)$ time to update the objective function. The execution of *evaluate_objective_function* can be done in a simple bit-parallel manner in $O(m)$ time on a sequential computer.

A computer word has 32 or 64 bits. The number of literals in a clause of most practical *CNF* formulas is much less than 32. In a local search algorithm, therefore, one can pack all the literals in a clause into the bits of a computer word and then evaluate all the literals in one clause in parallel. For m clauses, instead of $O(ml)$, it will take procedure *evaluate_objective_function* $O(m)$ time to evaluate and update the objective function. Occasionally, a clause may have more than 32 literals which can be packed in several computer words and be evaluated simultaneously. This general bit-parallel evaluation method was implemented in the $SAT1.7$ algorithm [**35, 41**].

Complete Local Search. Local search algorithms are incomplete, *i.e.*, they can find some solutions for certain *CNF* formulas and give no answer if the *CNF* formula is not satisfiable. To overcome this problem, researchers developed complete local search algorithms to test satisfiability as well as unsatisfiability. The basic idea in the *SAT1.11* and *SAT1.13* algorithms [**35, 41**] was to combine local search with a systematic search procedure, keeping local search's efficiency while maintaining search completeness by a systematic search method. If at a node of the search tree a solution point is found unsatisfiable, then the algorithm backtracks and continues searching until a solution is found or unsatisfiability is proven.

The *SAT1.11* and *SAT1.13* algorithms were two early experiments of complete local search algorithms. Probe order backtracking is a simplified version of complete local search [**96, 95**]. Recently Crawford studied a complete local search algorithm [**16**]. He used weights assigned to clauses to help choose branch variables. Variables occurring in heavily weighted clauses were given precedence.

5.3. Randomized Local Search with Trap Handling. For the SAT problem, with high probability, a greedy local search will fall into a trap much more easily. In this case some variables are updated very quickly. The related clauses oscillate between the *sat* and *unsat* states. The search is limited to these trapping states. Without any help, there is little chance of getting out to explore other solution states.

Based on early observation of trap phenomenon and the development of a three-level search space model (Section 4.2), Gu *et al.* developed a *SAT1.5* algorithm with trap handling ability [**39, 51**]. The *SAT1.5* algorithm can monitor and break local minimum loops and can handle multiple traps during the search. The current version of the *SAT1.5* algorithm contains advanced data structures and complicated trap detection/handling methods [**39, 41, 51**]. For the sake of simplicity, Figure 11 gives a brief outline of the algorithm.

The *SAT1.5* starts with an initial random solution and a set of limiting parameters. Parameter *Max_Time* specifies the maximum number of times allowed to restart a new search. The number of unsatisfiable clauses is computed and is assigned as the value of the objective function. The first *while* loop is limited by the *Max_Time*. Procedure *complete_flip*() flips all the variables that can reduce the value of the objective function. *Evaluate_objective_function*() updates the objective function.

The second *while* loop is a randomized local search with trap tracking and handling. Trap detection facilities are installed several places in the *while* loop to record trap statistics. They are essential to figure out trap "height," "width," and other parameters for subsequent decision making. A trap may contain a global minimum solution and it must be searched with reasonable effort. Leaving a trap too early

```
procedure SAT1.5 ()
begin
    /* initialization */
    get_a_SAT_instance();
    x_0 := select_a_random_initial_point();
    F(x_0) := evaluate_objective_function(x_0);

    /* search */
    k := 0; Restart_Times := 0;
    while F > 0 and Restart_Times < Max_Time do
    begin
        /* Open Search: flip all variables that reduce F */
        x_{k+1} := complete_flip(x_k);
        F := evaluate_objective_function(x_{k+1});
        /* parameters for trap tracking */
        Clean_trap_records(); Trapping_Times := 0;
        /* Peak Search: randomized local search */
        while F > 0 and Trapping_Times < Max_Trapping_Times do
        begin
            /* randomly select one var for randomized local search */
            x_i := select_one_var_to_flip(x_{k+1});
            x_{k+1} := randomized_local_search(x_i,x̄_i);
            F := evaluate_objective_function(x_{k+1});
            /* Trap Search */
            if a trap is detected then
            begin
                    Trapping_Times + +;
                /* random flip vars in conflicting clauses */
                    x_{k+1} := weak_flip(x_{k+1});
                    F := evaluate_objective_function(x_{k+1});
                /* random flip a few percent (pct) of variables */
                    x_{k+1} := gentle_flip(x_{k+1},pct);
                    F := evaluate_objective_function(x_{k+1});
                /* random flip a small set of variables */
                    if Trapping_Times is large then
                        x_{k+1} := strong_flip(x_{k+1},set);
                        F := evaluate_objective_function(x_{k+1});
                /* initialization for a new trap */
                    Clean_trap_records();
            end;
        end;
        if F > 0 then
            x_{k+1} := restart_a_new_random_point();
            Restart_Times + +;
        k := k + 1;
    end;
end;
```

FIGURE 11. **SAT1.5:** a randomized local search procedure with trap handling ability.

or too late could result in either losing solutions or wasting computing time. The time to jump out of a trap is determined by parameter *Max_Trapping_Times*.

When the search algorithm jumps out of a trap, there are several alternatives to pursue. One is to start a new search. In the *while* loop, several randomized local search procedures deploying random value and random variable heuristics are grouped together with partial random selection heuristics (see Section 4.3). They together select a variable for randomized local search (the objective function F may increase during the search).

If a trap is detected, a number of strategies can be used to perform a trap search [**41, 51**]. In one approach proposed by Gu *et al.*, a sequence of random flip operations is performed (see Figure 11). The intensity of the flip operations evolves from weak to strong, according to the trap structure and search times. That is, a variable in each *unsat clause* is flipped to a *true* value (procedure *weak_flip*()), followed by a random flip of a few percent of the variables (procedure *gentle_flip*()). And finally, if the *Trapping_Times* is sufficiently large, procedure *strong_flip*() would flip a small set of variables randomly. Additional facilities for hill climbing, tabu search, and variable locking/unlocking were developed. The *SAT1.5* algorithm can walk on the rugged surface of a trap adaptively.

The real execution performance of the *SAT1.5* algorithm (Section 5.5) suggests that it is presently one of the fastest local search algorithms for the SAT problem.

5.4. Greedy Local Search. Traditional local search proceeds by taking a feasible solution point that reduces the value of the objective function. Among many neighboring solution points, local search does not evaluate neighbors' relative performance with respect to the objective function. A *greedy* algorithm selects the *best neighbor* that yields the minimum value of the objective function and takes this *best neighbor direction* as the descent direction of the objective function.

Selman *et al.* proposed a *greedy* local search procedure, i.e., *GSAT*, for the SAT problem [**107**]. During each search step, the algorithm evaluates all the moves and selects the best move that gives the *greatest* decrease in the total number of unsatisfied clauses. If the algorithm becomes stuck at a local minimum, *GSAT* uses *side-walk* (a form of tunneling heuristic) to move aside. In *GSAT* procedure, two parameters, *MAX-TRIES* and *MAX-FLIPS*, were used to control the algorithm's maximum running state. Based on greedy local search, *GSAT* pursues only paths where every step leads to an improvement, this leads to a procedure that becomes stuck much more often than the randomized local search.

Satisfiability problem is an important problem in VLSI circuit design. Late eighties VLSI researchers experimented a large number of practical SAT formulas and found that greedy local search became stuck much more easily at local minima than randomized local search procedures. Accordingly, Gu proposed a method of combining local descent with randomized multiphase search and trap handling heuristics (see Section 4.3 and Section 5.3). These ideas were used in the subsequent *SAT1* algorithm family design [**50, 35, 39, 41**].

In order to overcome the weekness of greedy local search, recently Selman *et al.* combined bad variable and partial random variable selection heuristics (Section 4.3) in *random walk* for the *GSAT* algorithm [**105, 106**]. They found that these random heuristics (such as random flips, selecting a variable in unsat clause, and partial random variable selection) improved the performance of the *GSAT* algorithm significantly [**105**].

procedure GSAT ()
begin
  for i := 1 to MAX-TRIES
    T := a randomly generated truth assignment
      for j := 1 to MAX-FLIPS
        if T satisfies α **then return** T
        p := a propositional variable such that a change
            in its truth assignment gives the **largest**
            increase in the total number of clauses
            of α that are satisfied by T
        T := T with the truth assignment of p reversed
      end for
  end for
  return "no satisfying assignment found"
end

FIGURE 12. **GSAT:** a **G**reedy local search procedure for the **SAT** problem [**107**]. MAX-FLIPS, MAX-TRIES are constants, and α is a set of clauses. During each search step, *GSAT* takes the best neighbor that gives the maximum descent to the objective function.

5.5. Performance Study. In this section, we compare the performance of the following representative SAT algorithms in solving random l-SAT formulas and random average l-SAT formulas:

1. *SAT1.3*: a sequential *CNF* local search algorithm [**35, 50, 39, 41**],
2. *SAT1.7*: a parallel *CNF* local search algorithm [**35, 41**],
3. *SAT1.13*: a complete *CNF* local search algorithm [**35, 41**],
4. *SAT1.4*: a sequential *DNF* local search algorithm [**35**],
5. *SAT1.8*: a parallel *DNF* local search algorithm [**35**],
6. *SAT1.18*: a complete *DNF* local search algorithm [**35**],
7. *SAT14.6*: an optimized, discrete global optimization algorithm [**35, 39, 44**],
8. *SAT14.16*: a complete global optimization algorithm [**35, 39, 44**],
9. *SAT14.7*: a continuous, global optimization [**35, 39, 46**],
10. *SAT14.11*: a complete, continuous global optimization [**35, 39, 38**],
11. *DPL*: a Davis-Putnam algorithm in Loveland form [**19**],
12. *GSAT*: a sequential, greedy local search algorithm [**107**],
13. *WSAT*: a *GSAT* algorithm with random walk [**105, 106**],
14. *TSAT*: a Tabu search algorithm for SAT [**83**],
15. *SAT1.5*: a randomized local search algorithm with trap handling ability [**39, 51**], and
16. *IP*: a parallel interior point zero-one integer programming algorithm [**70, 69**].

Real Execution Times. In Table 1 we give real execution times of some local search and global optimization algorithms for solving l-SAT instances. All the results were run on a SUN SPARC 2 workstation. The number of clauses (m), the number of variables (n), and the number of literals per clause (l), are given in the first three columns. Symbol "G/L" in Column 4 stands for the number of times that all the algorithms hit *global/local* minimum points. From these results we can

TABLE 1. Real execution performance averaged over ten runs of some **local and global optimization algorithms** on a SUN SPARC 2 Workstation. Time Units: seconds. Symbol "G/L" stands for the number of times that all the algorithms hit *global/local* minimum points.

Problems			Execution Time						
m	n	l	G/L	SAT1.3	SAT1.7	SAT1.13	SAT1.8	SAT14.6	SAT14.16
100	100	3	10/0	0.003	0.001	0.001	0.002	0.002	0.003
200	100	3	10/0	0.007	0.004	0.010	0.004	0.005	0.008
300	100	3	10/0	0.035	0.008	0.015	0.004	0.014	0.012
400	100	3	10/0	0.464	0.027	0.145	0.030	0.040	0.215
1000	1000	3	10/0	0.036	0.030	0.048	0.029	0.033	0.051
1500	1000	3	10/0	0.087	0.055	0.078	0.049	0.058	0.081
2000	1000	3	10/0	0.192	0.084	0.113	0.080	0.093	0.115
2500	1000	3	10/0	0.371	0.124	0.158	0.114	0.133	0.180
3000	1000	3	10/0	0.872	0.179	0.310	0.164	0.241	0.359
3500	1000	3	10/0	6.878	0.636	1.008	0.588	0.919	1.357
1000	1000	4	10/0	0.026	0.022	0.045	0.027	0.029	0.040
2000	1000	4	10/0	0.094	0.061	0.103	0.061	0.057	0.092
3000	1000	4	10/0	0.239	0.094	0.160	0.091	0.109	0.166
4000	1000	4	10/0	0.483	0.144	0.230	0.135	0.162	0.234
5000	1000	4	10/0	1.004	0.227	0.338	0.210	0.267	0.330
6000	1000	4	10/0	2.410	0.383	0.465	0.359	0.388	0.478
7000	1000	4	10/0	5.999	0.865	0.729	0.852	0.756	0.840
8000	1000	4	10/0	36.17	1.896	2.088	1.821	2.595	2.641
8500	1000	4	10/0	140.3	10.79	7.974	10.51	12.79	12.12
10000	1000	5	10/0	2.899	0.451	0.610	0.393	0.464	0.567
11000	1000	5	10/0	3.799	0.489	0.800	0.426	0.580	0.750
12000	1000	5	10/0	6.729	0.593	0.839	0.505	0.649	0.844
13000	1000	5	10/0	9.541	0.761	1.154	0.681	0.978	1.064
14000	1000	5	10/0	21.41	1.107	1.308	0.969	1.282	1.652
15000	1000	5	10/0	60.80	1.671	2.207	1.429	2.047	2.166
10000	400	6	10/0	12.58	0.497	0.625	0.463	0.514	0.771
10000	500	6	10/0	4.353	0.377	0.640	0.342	0.345	0.553
10000	600	6	10/0	2.571	0.328	0.439	0.280	0.331	0.534
10000	700	6	10/0	1.989	0.284	0.550	0.248	0.289	0.491
10000	800	6	10/0	1.776	0.277	0.494	0.256	0.287	0.452
10000	900	6	10/0	1.305	0.289	0.523	0.248	0.278	0.476
10000	1000	6	10/0	1.140	0.264	0.488	0.227	0.269	0.473
20000	1000	7	10/0	3.238	0.500	1.124	0.421	0.496	1.004
30000	2000	7	10/0	4.110	0.882	1.733	0.722	0.910	1.460
40000	3000	7	10/0	5.557	1.289	2.382	1.114	1.250	2.196
50000	4000	7	10/0	6.793	1.666	3.036	1.386	1.632	2.730
60000	5000	7	10/0	7.942	1.260	3.719	1.833	1.971	3.402
10000	1000	10	10/0	0.143	0.050	0.377	0.034	0.048	0.312
20000	2000	10	10/0	0.408	0.124	0.821	0.090	0.099	0.664
30000	3000	10	10/0	0.726	0.258	1.311	0.197	0.179	1.076
40000	4000	10	10/0	0.963	0.305	1.826	0.241	0.328	1.511
50000	5000	10	10/0	1.262	0.441	2.372	0.357	0.395	1.887

observe that, in terms of global convergence and local convergent rate, these local search and global optimization algorithms exhibit desirable convergent properties and fast computing speed for instances in the table.

TABLE 2. Performance comparison averaged over ten runs between a **DPL** and some optimization algorithms on a SUN SPARC 2 workstation for solving 3-SAT problem instances. Time units: seconds. Symbol "S/F" in Column 4 stands for DPL's *success/failure* to give an answer within a time limit of $120 \times m/n$ seconds, whereas symbol "G/L" stands for the number of times that all the remaining SAT optimization algorithms hit *global/local* minimum points.

Problems			Execution Time							
m	n	l	S/F	DPL	G/L	SAT1.7	SAT1.13	SAT1.8	SAT14.6	SAT14.16
500	500	3	10/0	2.159	10/0	0.013	0.021	0.011	0.013	0.027
750	500	3	10/0	2.916	10/0	0.015	0.030	0.013	0.021	0.033
1000	500	3	10/0	3.657	10/0	0.035	0.048	0.032	0.031	0.047
1250	500	3	9/1	5.797	10/0	0.049	0.064	0.044	0.067	0.072
1500	500	3	6/4	9.147	10/0	0.108	0.117	0.089	0.115	0.108
1000	500	4	10/0	4.684	10/0	0.026	0.040	0.024	0.024	0.038
1500	500	4	10/0	7.960	10/0	0.040	0.075	0.043	0.042	0.066
2000	500	4	8/2	10.27	10/0	0.066	0.105	0.062	0.069	0.088
2500	500	4	2/8	15.96	10/0	0.074	0.130	0.085	0.109	0.152
3000	500	4	1/9	46.33	10/0	0.118	0.201	0.115	0.153	0.234
3000	500	5	10/0	16.90	10/0	0.094	0.137	0.082	0.074	0.139
4000	500	5	5/5	28.39	10/0	0.119	0.204	0.103	0.144	0.188
5000	500	5	0/10	>1200	10/0	0.180	0.253	0.170	0.196	0.288
6000	500	5	0/10	>1440	10/0	0.313	0.370	0.267	0.313	0.471
7000	500	5	0/10	>1680	10/0	0.591	0.575	0.478	0.604	0.623
10000	1000	10	10/0	101.8	10/0	0.047	0.382	0.038	0.049	0.315
12000	1000	10	10/0	124.3	10/0	0.073	0.458	0.053	0.052	0.382
14000	1000	10	10/0	145.2	10/0	0.077	0.562	0.058	0.063	0.430
16000	1000	10	10/0	167.1	10/0	0.098	0.596	0.073	0.078	0.517
18000	1000	10	10/0	188.6	10/0	0.136	0.716	0.102	0.095	0.583

Among optimization algorithms, the parallel *CNF* local search ($SAT1.7$) algorithm was much faster than the sequential local search ($SAT1.3$) algorithm. The $SAT1.7$ algorithm had comparable computing performance with the *DNF* parallel local search ($SAT1.8$) algorithm. Discrete global optimization ($SAT14.6$) algorithm was slightly slower than parallel local search algorithms. Complete local search ($SAT1.13$) algorithm and complete global optimization ($SAT14.16$) algorithm, due to a systematic bookkeeping, were slightly slower than parallel local search but significantly faster than the sequential local search algorithm.

As discussed in [**50, 41**], beyond a certain range of hardness, for example, for $m = 8500, n = 1000$, and $l = 4$, the computing time of these optimization algorithms started to increase.

The experimental results shown in Table 1 were collected from early reports in [**35, 41, 44**]. The present local and global optimization algorithms are much more faster than their previous versions [**39, 41, 51**].

Performance Comparison with the DP Algorithm. The execution results of the DPL algorithm and some optimization algorithms for solving l-SAT instances are given in Table 2. We executed each algorithm ten times and report the average execution times. Because DPL was slow for large size instances, we set a maximum execution time of $120 \times m/n$ seconds as the time limit of its execution. Symbol "S/F" in Column 4 stands for DPL's *success/failure* in giving an answer within such a time limit. For DPL, the average execution time does not include the maximum execution time limit if some of the ten executions were successful; the average execution time was taken as the maximum execution time limit only if all

TABLE 3. Performance comparison between some optimization algorithms running on a SUN SPARC 2 workstation and the **GSAT algorithm** running on a MIPS computer with comparable computing power for the 3-SAT problem instances. Time units: seconds.

Problems			Execution Time					
m	n	l	GSAT	SAT1.7	SAT1.13	SAT1.8	SAT14.6	SAT14.16
215	50	3	0.400	0.019	0.100	0.006	0.020	0.030
301	70	3	0.900	0.054	0.010	0.020	0.020	0.020
430	100	3	6.000	0.336	0.040	0.420	0.050	0.370
516	120	3	14.00	0.596	0.810	1.136	0.410	0.040
602	140	3	14.00	0.260	8.060	0.750	1.990	0.170
645	150	3	45.00	0.102	0.190	0.120	0.040	0.870
860	200	3	168.0	1.776	0.970	0.070	6.710	0.490
1062	250	3	246.0	3.106	20.71	0.070	12.43	0.090
1275	300	3	720.0	8.822	19.66	3.750	19.14	4.510

ten executions failed. Symbol "G/L" in Column 6 stands for the number of times that all the remaining SAT optimization algorithms hit the *global/local* minimum points.

From numerous algorithm executions, we observe that, for random l-SAT instances listed in Table 2, DPL was slower than the rest of the SAT optimization algorithms. As the input size increases, the number of failures, F, increased quickly. For some slightly large inputs, such as $m = 5000, n = 500$, and $l = 5$, all ten algorithm executions failed after a reasonably long time limit. Due to its $O(m^{O(n/l)})$ average run-time complexity, even for some fairly easy instances, such as $m = 10000, n = 1000$, and $l = 10$, DPL took an excessive amount of time to find a solution. In comparison, local search and global optimization algorithms were successful for all ten executions. They were able to find a solution to the given instances efficiently.

Table 2 suggests that DPL may not be a suitable candidate for large size random l-SAT instances. This observation should not be generalized to other application cases. In many applications, as observed by others [**12, 11, 21**], DPL performed very well.

Performance Comparison with *GSAT*. Table 3 compares the performance between some local search and global optimization algorithms running on a SUN SPARC 2 workstation and *GSAT* [**107**] running on a MIPS computer with comparable computing power [**104**]. Since *GSAT* is essentially a version of sequential local search (*i.e.*, *SAT*1) algorithm, for solving 3-SAT instances generated from the same input model used in [**86**], local search and global optimization algorithms performed approximately tens to hundreds times faster than *GSAT*. Among them, parallel *DNF* local search (*SAT*1.8) algorithm and complete global optimization (*SAT*14.16) were the best.

TABLE 4. Performance comparison between some optimization algorithms running on a SUN SPARC 2 workstation and **an interior point zero-one integer programming algorithm** running on a *KORBX(R)* parallel/vector computer for solving average 3-SAT problem instances. Time units: seconds. Symbol "S/F" stands for the number of times that IP hits the *global/local* minimum points, whereas symbol "G/L" stands for the number of times that the remaining SAT algorithms hit the *global/local* minimum points.

Problems			Execution Time							
m	n	l	S/F	IP	G/L	SAT1.7	SAT1.13	SAT1.8	SAT14.6	SAT14.16
100	50	5	52/0	0.7	10/0	0.001	0.004	0.001	0.001	0.004
200	100	5	70/0	1.1	10/0	0.006	0.010	0.006	0.005	0.007
400	200	7	69/0	3.5	10/0	0.007	0.014	0.007	0.007	0.018
800	400	10	31/0	5.6	10/0	0.009	0.034	0.009	0.003	0.030
800	400	7	20/0	7.8	10/0	0.014	0.032	0.014	0.009	0.026
1000	500	10	49/0	7.4	10/0	0.012	0.037	0.012	0.006	0.039
2000	1000	10	10/0	18.5	10/0	0.032	0.091	0.032	0.019	0.083
2000	1000	7	50/0	21.5	10/0	0.056	0.099	0.056	0.055	0.055
2000	1000	3	49/1	50.4	10/0	2.657	0.162	2.657	3.917	27.19
4000	1000	4	1/1	1085.4	10/0	10.63	11.07	10.63	6.826	9.555
4000	1000	10	10/0	25.1	10/0	0.055	0.189	0.055	0.044	0.163
8000	1000	10	10/0	38.0	10/0	0.219	0.456	0.219	0.254	0.353
16000	1000	10	10/0	66.4	10/0	0.603	1.042	0.603	0.625	1.052
32000	1000	10	10/0	232.4	10/0	1.701	2.720	1.701	1.611	2.434

Performance Comparison with Interior Point Zero-One Integer Programming Algorithm. Recently, Kamath *et al.* used an interior point zero-one integer programming algorithm to solve SAT [**70, 69**]. They implemented their algorithm in FORTRAN and C languages and ran the algorithm on a *KORBX(R)* parallel/vector computer with instances generated from the average 3-SAT input model. The *KORBX(R)* parallel computer operates in scalar mode at approximately 1 MFlops and at 32 MFlops with full vector concurrent mode. Their execution results are given in Columns 4 and 5 of Table 4.

We ran local search and global optimization algorithms for the same instances (listed in [**70, 69**]) on a SUN SPARC 2 workstation. The results are given in Table 4. Apparently, as compared to the interior point zero-one integer programming algorithm running on a parallel computer, in addition to improved global convergence, local search and global optimization algorithms were much simpler and achieved several orders of magnitude of performance improvements in terms of computing time.

Experiments on Hard Random Formulas. We compare the performance of two local search algorithms and a tabu search algorithm for the hard random 3-SAT problem instances generated from the **mwff** generator. All three programs were written in *C*. Table 5 give the real execution performance of the *WSAT* (*GSAT* with random walk) on an SGI Challenge with a 70 MHz MIPS R4400 processor [**105, 106**].

Tables 6 and 7 show the experimental results of *WSAT* and *TSAT* (Tabu search for SAT) programs written in *C* under Linux 1.1.59 for PC [**83**]. On the same machine, Mazure, Sais, and Gregoire compared the *GSAT* with *TSAT* and found that *TSAT* was more efficient in most cases. In addition, *TSAT* was able to solve more problem instances compared to the *GSAT*. The testing for *TSAT* for large size

TABLE 5. **WSAT** (*GSAT* with random walk)'s real execution performance for hard random 3-SAT problem instances on an SGI Challenge with a 70 MHz MIPS R4400 processor. Time unit: seconds [**105**].

Problems		*GSAT*			Walk		
n	m	time	flips	R	time	flips	R
100	430	.4	7554	8.3	.2	2385	1.0
200	860	22	284693	143	4	27654	1.0
400	1700	122	2.6×10^6	67	7	59744	1.1
600	2550	1471	30×10^6	500	35	241651	1.0
800	3400	*	*	*	286	1.8×10^6	1.1
1000	4250	*	*	*	1095	5.8×10^6	1.2
2000	8480	*	*	*	3255	23×10^6	1.1

TABLE 6. **WSAT** (*GSAT* with random walk)'s real execution performance for hard random 3-SAT problem instances on a PC. Time unit: seconds [**83**].

n	m	inst.	time	flips	solved	ratio
100	430	500	0.18	2803	88%	31,85
200	860	500	1.99	18626	73%	255,85
400	1700	500	15.03	204670	100%	2046,70
600	2550	500	19.59	250464	62%	4013,85
800	3400	500	140.61	1809986	67%	26854,39
1000	4250	500	369.88	4633763	57%	81009,84
2000	8240	50	3147.26	26542387	16%	1658899,19

TABLE 7. **TSAT**'s real execution performance for hard random 3-SAT problem instances on a PC. Time unit: seconds [**83**].

n	m	inst.	time	flips	solved	ratio
100	430	500	0.11	1633	93%	17,60
200	860	500	0.73	9678	74%	130,78
400	1700	500	11.51	145710	100%	1457,10
600	2550	500	13.92	167236	65%	2580,80
800	3400	500	99.45	1143444	71%	16150,34
1000	4250	500	292.10	3232463	62%	51802,29
2000	8240	50	3269.15	29415465	40%	735386,63

example with $n = 2000$ and $m = 8240$, however, was terminated at $n/m = 4.12$, before entering into the hard region of the random 3-SAT instances.

The performance of the *SAT1.5* algorithm [**39, 51**] is shown in Table 8. For hard problem instances in the transition region [**86**], *SAT1.5* can solve large-size SAT problem instances efficiently. It took *WSAT* on average 3,255 seconds to solve the $n = 2,000$ and $m = 8,480$ instances on an SGI Challenge with a 70 MHz MIPS

TABLE 8. Real execution performance of the **SAT1.5 algorithm** for hard random 3-SAT problem instances on a SUN SPARC 20 workstation. For each problem, 30 random instances were tested. The minimum (T_{min}), maximum (T_{max}), and average (T_{mean}) execution times were recorded. "S" indicates the number of success cases of finding solutions within the time limit (*T-limit*). Time unit: second.

n	m	m/n	S/30	T_{min}	T_{mean}	T_{max}	T-limit
1000	4230	4.2300	20/30	6.04	206.90	956.15	1000
1000	4240	4.2400	15/30	0.55	223.26	891.11	1000
1000	4250	4.2500	14/30	1.69	88.370	454.79	1000
1000	4260	4.2600	10/30	24.4	243.25	914.35	1000
2000	8460	4.2300	12/30	115.8	779.44	2069.9	3000
2000	8480	4.2400	14/30	17.64	530.32	1360.9	3000
2000	8500	4.2500	7/30	58.09	789.35	1677.6	3000
2000	8510	4.2550	9/30	59.08	840.33	2322.8	3000
2000	8460	4.2300	15/30	58.95	684.32	4508.2	5000
2000	8480	4.2400	15/30	15.31	1273.5	4057.9	5000
2000	8500	4.2500	15/30	112.7	1527.8	3644.5	5000
2000	8520	4.2600	9/30	123.2	1522.9	4338.5	5000
3000	12690	4.2300	12/30	430.6	1787.2	2876.4	5000
3000	12700	4.2333	18/30	122.5	2101.5	4479.4	5000
3000	12720	4.2400	11/30	270.6	1503.6	3840.7	5000
3000	12740	4.2467	12/30	229.1	2062.9	4807.4	5000
3000	12680	4.2267	15/30	356.1	2788.3	9510.2	10000
3000	12700	4.2333	11/30	503.3	3681.1	8247.9	10000
3000	12720	4.2400	15/30	30.09	2300.3	7002.3	10000
3000	12740	4.2467	8/30	563.3	2620.5	5330.9	10000
4000	16920	4.2300	11/30	739.83	4064.5	11498.2	12000
4000	16930	4.2325	10/30	1733.5	5472.0	10187.8	12000
4000	16940	4.2350	7/30	571.20	1948.9	4768.92	12000
4000	16960	4.2400	10/30	294.80	3709.0	9921.77	12000
5000	21150	4.2300	8/30	2024.7	3867.9	8134.81	9000
5000	21175	4.2350	6/30	1640.1	2982.7	4193.68	9000
5000	21200	4.2400	3/30	2935.8	4435.7	6357.65	9000
5000	21225	4.2450	4/30	3883.5	6025.6	10980.9	15000
10000	41000	4.1000	30/30	294.44	1315.6	3849.38	20000
10000	41800	4.1800	8/18	4294.5	8387.9	16654.8	20000
10000	42000	4.2000	4/30	963.53	5877.8	12020.3	20000
10000	42200	4.2200	2/30	9270.6	14241.9	19213.4	20000

R4400 processor. On a SUN SPARC 20 workstation, the *SAT1.5* algorithm was able to solve the same problem instance in some 530 seconds on average [**51, 47**]. For hard, large size SAT problem instances with $n > 5,000$, *SAT1.5* algorithm was able to handle the problems effectively.

6. Local Search for Scheduling and Task Assignment

Scheduling computational tasks to multiprocessors is one of the key issues in high-performance computing today. Scheduling can be performed at compile-time or runtime. Scheduling performed at compile-time is called *static scheduling.* Scheduling performed at runtime is called *dynamic scheduling.* The flexibility inherent in dynamic scheduling allows for adaptation to the unforeseen applications requirements at runtime. However, load balancing suffers from run-time overhead due to load information transfer among processors, load balancing decision making process, and communication delay due to task relocation. Furthermore, most runtime scheduling algorithms utilize neither the characteristics information of application problems, nor global load information for load balancing decision. The major advantage of static scheduling is that the overhead of the scheduling process is incurred at compile time, resulting in a more efficient execution time environment compared to dynamic scheduling. Static scheduling can utilize the knowledge of problem characteristics to reach a well-balanced load.

We consider static scheduling algorithms that schedule an edge-weighted *directed acyclic graph* (DAG), or called *task graph,* to a set of homogeneous processors to minimize the completion time. Since the static scheduling problem is NP-hard in its general forms [**27**], and optimal solutions are known in restricted cases [**14, 24, 32**], there has been considerable research effort in this area resulting in many heuristic algorithms [**13, 23, 73, 102, 111, 126, 128**].

In this section, we present a fast, deterministic local search algorithm, namely *TASK* (Topological Assignment and Scheduling Kernel), that can effectively schedule DAGs onto multiprocessors [**37, 54, 127**]. Randomized local search was efficient but was unable to produce the best scheduling results. Based on the dynamic reconstruction of directed acyclic graph in each move, *TASK* can quickly determine the optimal search direction and systematically minimize a given schedule in a *topological order.* Tested through real multiprocessor machines, *TASK* is faster than presently existing scheduling algorithms. It can effectively reduce the scheduling length generated by other existing scheduling algorithms by 20% to 30%.

6.1. DAG Scheduling. A directed acyclic graph (DAG) consists of a set of nodes $\{n_1, n_2, ..., n_n\}$ connected by a set of edges, each of which is denoted by $e_{i,j}$. Each node represents a task, and the weight of node n_i, $w(n_i)$, is the execution time of the task. Each edge represents a message transferred from one node to another node, and the weight of edge $e_{i,j}$, $w(e_{i,j})$ is equal to the transmission time of the message. The communication-to-computation ratio (CCR) of a parallel program is defined as its average communication cost divided by its average computation cost on a given system. In a DAG, a node which does not have any parent is called an entry node whereas a node which does not have any child is called an exit node. A node cannot start execution before it gathers all of the messages from its parent nodes. In static scheduling, the number of nodes, the number of edges, the node weight, and the edge weight are assumed to be known before program execution. The weight among two nodes assigned to the same processing element (PE) is assumed to be zero.

The objective in static scheduling is to assign nodes of a DAG to PEs such that the schedule length or makespan is minimized without violating the precedence constraints. There are many approaches that can be employed in static scheduling. In the classical approach [**63**], which is also called list scheduling, the basic idea is to make a priority list of nodes, and then assign these nodes one by one to PEs. In the scheduling process, the node with the highest priority is chosen for scheduling. The PE that allows the earliest start time, is selected to accommodate this node. Most of the reported scheduling algorithms are based on this concept employing variations in the priority assignment methods such as HLF (Highest level First), LP (Longest Path), LPT (Longest Processing Time) and CP (Critical Path) [**2, 126, 75**]. In the following we review some of contemporary static scheduling algorithms including MCP, DSC, DLS, and CPN methods.

The Modified Critical Path (MCP) algorithm is based on the *as-late-as-possible (ALAP)* time of a node [**126**]. The ALAP time is defined as $T_L(n_i) = T_{critical} - level(n_i)$, where $T_{critical}$ is the length of the critical path, and $level(n_i)$ is the length of the longest path from node n_i to an exit node, including node n_i [**24**]. The MCP algorithm was designed to schedule a DAG on a bounded number of PEs. It sorts the node list in the increasing ALAP order. The first node in the list is scheduled to the PE that allows the earliest start time, considering idle time slots. Then the node is deleted from the list and this operation repeats until the list is empty.

The Dominant Sequence Clustering (DSC) algorithm is designed based on an attribute for a task graph called the dominant sequence (DS) [**128**]. A DS is defined for a partially scheduled task graph as the path with the maximum sum of communication costs and computation costs in the graph. Nodes on the DS are considered to be relatively more important than others. The ready nodes with the highest priority in terms of $blevel + tlevel$ will be scheduled first. Then the priorities of the child nodes of the scheduled node will be updated and this operation repeats until all nodes are scheduled. The dynamic cost (i.e., *tlevel*) is used to quickly determine the critical path length. This idea has been incorporated into our *TASK* algorithm to reduce its complexity.

The Dynamic Level Scheduling (DLS) algorithm determines node priorities by assigning an attribute called dynamic level (DL) to each node at every scheduling step [**111**]. DL is the difference between the static level and message ready time. DLS computes DL for each ready node on all available processors. Suppose $DL(n_i, J)$ is the largest among all pairs of ready node and available processor. Schedule n_i to processor J. Repeat this process until all nodes are scheduled.

Recently, a new algorithm has been proposed by using the Critical Path Node (CPN) [**78**]. This algorithm is based on the CPN-dominate priority. If the next CPN is a ready node, it is put in the CPN-dominate list. For a non-ready CPN, its parent node n_y with the largest *blevel* is put in the list if all the parents of n_y are already in the list; otherwise, all the ancestor nodes of n_y are recursively included in the list before the CPN node is in the list. The first node in the list is scheduled to the PE that allows the earliest start time. Then the scheduled node is removed from the list and this operation repeats until the list is empty. The CPN algorithm utilizes the two important properties of DAG: the critical path and topological order. It potentially generates a good schedule.

Although these algorithms produce relatively good schedules, they are usually not optimal. Sometimes, the generated schedule is far from optimal. In this section, we propose a fast, deterministic local search algorithm, *TASK*, to improve the quality of schedules generated by an initial scheduling algorithm.

6.2. Local Search for Scheduling and Task Assignment. There were several significant local search solutions to the scheduling and task assignment problems.

Early local search method was able to solve small size, unconstrained problems such as small TSP instances [**80**]. During the middle and late eighties, more powerful techniques for randomized local search were developed (see Section 4.3). These include conflict minimization, random variable selection, and pre-, partial, and stochastic variable selection [**34, 39, 41, 112, 113, 115, 114, 116**]. They were capable of solving large size, highly constrained CSP, SAT and scheduling problems efficiently [**34, 39, 115, 116**].

The scheduling and task assignment problems are well-known as CSP/satisfiability problems. In fact, SAT, n-queen and scheduling are three typical problems in constraint satisfaction. SAT model formulates scheduling and task assignment problems precisely in algebraic expressions. CSP model expressively characterizes scheduling and task assignment operations in simple geometric patterns (e.g., n-queen and *hyper*-queen models). Following this, several local search solutions to SAT and the n-queen problems were applied to solve large-scale industrial scheduling and task assignment problems. A detailed discussion of the early work for SAT, n-queen and scheduling was given in Section 3.3.

There are two basic strategies in local search. The first one is random search, in which the local search direction is randomly selected. If the initial solution point is improved, it moves to the refined solution point. Otherwise, another search direction is randomly selected. The random strategy is simple and effective for some problems, such as the n-queens problem [**112**]. However, it may not be efficient for other problems such as the microword length minimization [**97**] and DAG scheduling problem.

The second strategy utilizes certain criteria to find out a search direction that most likely leads to a better solution point. In the computer microword length minimization [**97**], a compatibility class will be considered only when moving some nodes from the class may reduce the cost function. This strategy effectively reduces the search space by guiding the search towards a more promising direction. With carefully selected criteria, local search for DAG scheduling becomes much more efficient. The scheduling quality can be improved significantly.

6.3. Random Local Search Algorithm. A number of local search algorithms for scheduling have been developed based on n-queen and *hyper*-queen scheduling models [**37, 43, 116, 54, 127, 55, 78**]. A typical random local search algorithm, *RAND*, was designed and implemented [**55, 78**]. The algorithm makes a direct use of the random variable selection heuristics developed from the *QS* algorithms [**115**].

In *RAND* algorithm, a node is randomly picked from the blocking node list, where a blocking node is defined as a node that has the potential to block critical path nodes. Then the node is moved to a randomly selected PE. If the schedule length is reduced, the move is accepted. Otherwise, the node is moved back the its original PE. Each move, successful or not, takes $O(e)$ time to compute the schedule length, where e is the number of edges in the graph. To reduce its complexity, a constant $Max\_iteration$ is defined to limit the number of steps so that only $Max\_iteration$ nodes are inspected. The time taken for the algorithm is proportional to $e \times Max\_iteration$. Moreover, randomly selected nodes and PEs

```
procedure RAND (G,DAG_Schedule)
begin
    construct_a_scheduled_DAG(G);
    construct_the_blocking_node_list();
    k := 0;
    while (k++ < Max_iteration) do
    begin
        pick a node n_i randomly from the blocking node list;
        pick a PE P randomly;
        assign n_i to PE P;
        if schedule length increases then;
            move n_i back to its original PE;
        if local then random_local_handler();
    end;
end;
```

FIGURE 13. A random local search algorithm for DAG scheduling.

may not be able to significantly reduce the length of a given schedule. Even if the *Max_iteration* is equal to the number of nodes, leading to a complexity of $O(en)$, the random search algorithm still cannot provide a satisfied performance.

6.4. Local Search with Topological Ordering for Scheduling and Task Assignment. We propose a fast, deterministic local search algorithm utilizing topological ordering for effective DAG scheduling. The algorithm is called *TASK* (Topological Assignment and Scheduling Kernel) [**37, 54, 127**]. In this algorithm, the nodes in the DAG are inspected in a *topological order*. In this order, it is not required to visit every edge to determine whether the schedule length is reduced. Time spent on each move can be drastically reduced so that inspecting every node in a large graph becomes feasible. Also, in this order, we can compact the given schedule systematically.

For a given graph, in order to describe the *TASK* algorithm succinctly, several terms are defined as follows:

- $tlevel(n_i)$, the largest sum of communication and computation costs at the top level of node n_i, i.e., *from an entry node to* n_i, excluding its own weight $w(n_i)$.
- $blevel(n_i)$, the largest sum of communication and computation costs at the bottom level of node n_i, i.e., *from* n_i *to an exit node.*
- The critical path, *CP*, is the longest path in a DAG. The length of the critical path of a DAG is

$$L_{\text{CP}} = \max_{n_i \in V}\{L(n_i)\},$$

where $L(n_i) = tlevel(n_i) + blevel(n_i)$ and V is the node set of the graph.

If the given graph has been previously scheduled, more terms are defined:

- Node n_i has been scheduled on *PE* $pe(n_i)$.

- Let $p(n_i)$ be the *predecessor node* that has been scheduled immediately before node n_i on *PE* $pe(n_i)$. If node n_i is the first node scheduled on the *PE*, $p(n_i)$ is *null.*
- Let $s(n_i)$ be the *successor node* that has been scheduled immediately after node n_i on *PE* $pe(n_i)$. If node n_i is the last node scheduled on the *PE*, $s(n_i)$ is *null.*

```
procedure TASK (DAG_Schedule)
begin
    /* initialization */
    Construct a scheduled DAG;
    for node i := 0 to n − 1 do
        L(n_i) := tlevel(n_i) + blevel(n_i);
    L_CP := max_{0≤i<n} L(n_i), the longest path in DAG;

    /* search */
    while there are nodes in DAG to be scheduled do
    begin
        i := pick_a_node_with_Max_L(n_i);
        for each PE k := 0 to p − 1 do
            L^k(n_i) := move_n_i_to_PE_k;
        t := pick_a_PE_with_Min_L^k, where k = 0, ..., p − 1;
        /* if no improvement */
        if t == pe(n_i) then
            let node n_i stay at PE pe(n_i);
        /* if there are improvements */
        else begin
            move node n_i from PE pe(n_i) to PE t;
            modify_pseudo_edges_in_DAG;
            propagate_tlevel_of_n_i_to_its_children;
        end;
        mark n_i as being scheduled;
    end;
end;
```

FIGURE 14. **TASK:** Topological Assignment and Scheduling Kernel, a fast, deterministic local search algorithm based on topological ordering for fast scheduling.

One of characteristics of the *TASK* algorithm is its independence of the algorithm that was used to generate the initial schedule. As long as the initial schedule is correct and every node n_i has available $pe(n_i)$, $p(n_i)$, and $s(n_i)$ nodes, application of the local compaction algorithm guarantees that the new schedule of the graph is better than or equal to the initial one. The *TASK* algorithm is shown in Figure 14.

The input of the algorithm is a given DAG schedule generated by any heuristic DAG scheduling algorithm. In the first initialization step, a *scheduled DAG* is constructed, which contains scheduling and execution order information. To enforce the execution order in each *PE*, some *pseudo edges* (with zero weights) are inserted to incorporate the initial schedule into the graph. The *for* loop in the initialization computes the value of *blevel* of each node in the scheduled DAG and initializes *tlevel* for entry nodes. Every edges are marked *unvisited*. The variable $next_k$ points to the next node that has not been inspected in PE k. Initially, none of nodes is inspected so $next_k$ points to the first node in PE k.

In the *while* loop, a ready node n_i with the maximum value $L(n_i) = tlevel(n_i) + blevel(n_i)$ is selected for inspection. Ties are broken by $tlevel(n_i)$. A node is ready when all its parents have been inspected. In this way, the nodes are inspected in a topological order. Although other topological orders, such as *blevel*, *tlevel*, or *CPN-dominate* can be used, L has been shown to be a good indicator to determine the node order of inspection [**128**].

To inspect node n_i, in the *for* loop, the value $L(n_i) = tlevel(n_i) + blevel(n_i)$ is re-calculated for each PE. To conduct the recalculation at *PE* k, node n_i is pretended to be inserted right in front of $next_k$. Here, $tlevel(n_i)$ can be varied if any of its parent nodes was scheduled to either *PE* k or *PE* $pe(n_i)$. Similarly, $blevel(n_i)$ can be varied if any of its child nodes was initially scheduled to either *PE* k or *PE* $pe(n_i)$. Because the *tlevels* of its parent nodes are available and the *blevels* of its child nodes are unchanged, the value of $L(n_i)$ in every PE can be easily computed. The values indicate the degree of improvement by local search. With the new $L(n_i)$'s recalculated for every *PE*, node n_i is then moved to the PE that allows the minimum value of $L(n_i)$. If node n_i has been moved to PE t, the corresponding pseudo edges are to be modified by the *modify_pseudo_edges_in_DAG*(). The *tlevel* of n_i is propagated to its children so that when a node becomes ready, its *tlevel* can be computed. This process continues until every node is inspected.

The *TASK* algorithm satisfies the following properties [**54, 127**]:

1. The critical path length L_{CP} will not increase after each step of the *TASK* algorithm.

2. If the nodes in a DAG are inspected in a topological order and each ready node is appended to the previous node list in the PE, the *blevel* of a node is invariant *before* it is inspected and the *tlevel* of a node is invariant *after* it is inspected.

3. The time complexity of the *TASK* algorithm is $O(e + nlogn + np)$, where e is the number of edges, n is the number of nodes, and p is the number of PEs.

In the following, we use an example to illustrate the operation of the *TASK* algorithm.

Example: Assume the DAG shown in Figure 15 has been scheduled to three PEs by a DAG scheduling algorithm. The schedule is shown in Figure 16(a), in which three pseudo (dashed) edges have been added to construct a scheduled DAG, one from node n_6 to node n_8, one from node n_3 to node n_9, and one from node n_4 to node n_5 (not shown in Figure 16(a)). The schedule length is 14. The *blevel* of each node is computed as shown in Table 9. Tables 10 and 11 trace the $tlevel + blevel = L$ values for each step. In Table 10, "√" indicates the node with the largest L value and is to be inspected in the current step. In Table 11, "*" indicates the original PE and "√" the PE where the node is moved to.

First, there is only one ready node, n_1, which is a CP node. Its L value on PE 0 is $L^0(n_1) = 0 + 14 = 14$. Then the L values on other PEs are computed: $L^1(n_1) = 0 + 14 = 14$, $L^2(n_1) = 0 + 12 = 12$, as shown in Table 11. Thus, node n_1 is moved from PE 0 to PE 2, as shown in Figure 16(b). The L_{CP} of the DAG is

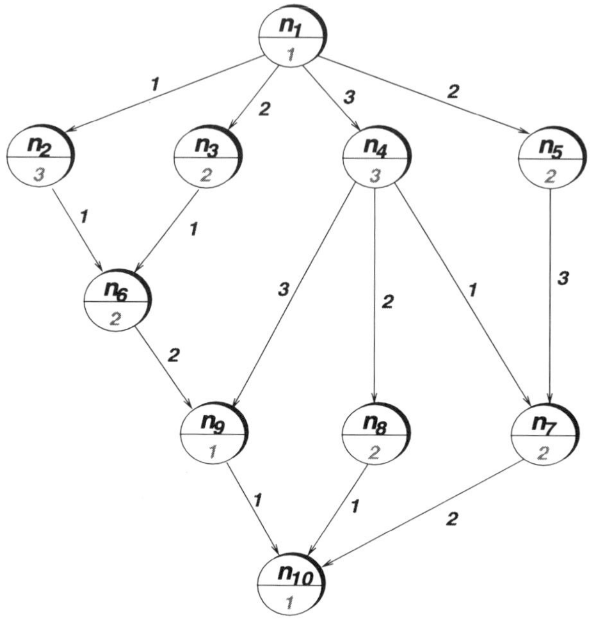

FIGURE 15. A DAG example for scheduling.

reduced to 12. In iterations 2, 3, and 4, moving nodes n_3, n_4, and n_2 do not reduce any L value. In iteration 5, node n_6 is moved from PE 0 to PE 1 as the L value is reduced from 12 to 11, as shown in Figure 16(c). In the rest five iterations, nodes n_5, n_7, n_8, n_9 and n_{10} do not move.

TABLE 9. The initial *blevel* value of each nodes for Example 1.

Node	n_1	n_2	n_3	n_4	n_5	n_6	n_7	n_8	n_9	n_{10}
blevel	14	9	9	10	7	6	5	4	2	1

6.5. Performance Study. The *TASK* algorithm is faster than presently existing scheduling algorithms for DAG scheduling. It can effectively reduce the scheduling length generated by other existing scheduling algorithms by 15% to 20%.

Table 12 compares the complexities of several important scheduling algorithms (where n is the number of nodes, e the number of edges, and p the number of PEs). The *TASK* algorithm has the lowest complexity among the five.

In the following, we present the performance results of the *TASK* algorithm and compare its performance with those of the random local search algorithm for scheduling, which was one of the best scheduling algorithms. We performed experiments using synthetic DAGs as well as real workload generated from the Gaussian elimination program.

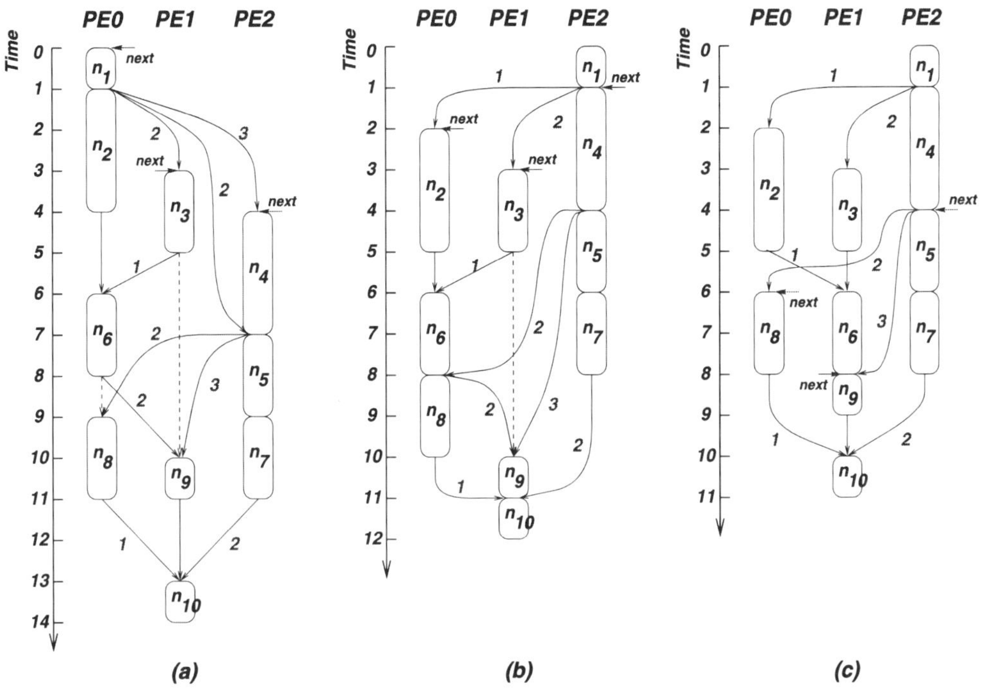

FIGURE 16. An example of *TASK*'s operations.

TABLE 10. The L values of ready nodes for selecting a node to be inspected.

Iteration			
1	n_1 (0+14=14) $\surd$		
2	n_2 (2+9=11),	n_3 (3+9=12) $\surd$,	n_4 (1+10=11)
3	n_2 (2+9=11),	n_4 (1+10=11) $\surd$	
4	n_2 (2+9=11) $\surd$,	n_5 (4+7=11)	
5	n_5 (4+7=11),	n_6 (6+6=12) $\surd$	
6	n_5 (4+7=11) $\surd$,	n_8 (6+4=10),	n_9 (8+2=10)
7	n_7 (6+5=11) $\surd$,	n_8 (6+4=10),	n_9 (8+2=10)
8	n_8 (6+4=10) $\surd$,	n_9 (8+2=10)	
9	n_9 (8+2=10) $\surd$		
10	n_{10} (10+1=11) $\surd$		

We use the same random graph generator for performance comparison. The synthetic DAGs are randomly generated graphs consisting of thousands of nodes. These large DAGs are used to test the scalability and robustness of the local search algorithms. These DAGs were synthetically generated in the following manner. Given N, the number of nodes in the DAG, we first randomly generated the height of the DAG from a uniform distribution with mean roughly equal to $\sqrt{N}$. For

TABLE 11. The L values of node n_i on each PE to select a PE.

Iteration	Node	PE 0	PE 1	PE 2
1	n_1	0+14=14*	0+14=14	0+12=12 $\surd$
2	n_3	3+11=14	3+9=12*	1+12=13
3	n_4	4+12=16	5+9=14	1+10=11*
4	n_2	2+9=11*	5+10=15	4+10 =14
5	n_6	6+6 =12*	6+4=10$\surd$	6+9=15
6	n_5	5+10 =15	8+10 =18	4+7 =11*
7	n_7	9+6 =15	9+4 =13	6+5=11*
8	n_8	6+4 =10*	8+4=12	8+4 =12
9	n_9	10+3=13	8+2=10*	8+4=12
10	n_10	10+1=11	10+1=11*	10+1=11

TABLE 12. Complexities of different scheduling algorithms.

MCP [**126**]	$n^2 log\, n$
DSC [**128**]	$(e+n) log\, n$
DLS [**111**]	pn^3
$RAND$ [**78**]	$MAXSTEP \times e$
$TASK$	$e + nlogn + pn$

each level, we generated a random number of nodes which was also selected from a uniform distribution with mean roughly equal to $\sqrt{N}$. Then, we connected the nodes from the higher level to the lower level randomly. The edge weights were also randomly generated. The sizes of the random DAGs were varied from 1000 to 4000 with an increment of 1000. Three values of the communication-computation-ratio (CCR) were selected to be 0.1, 1, and 10. The weights on the nodes and edges were generated randomly so that the average value of CCR corresponded to 0.1, 1, or 10.

We evaluate performance of these algorithms in two aspects: the schedule length generated by the algorithm and the running time of the algorithm.

Tables 13 and 14 show the comparison of the random local search algorithm and the *TASK* algorithm on 4 PEs and 16 PEs, respectively, where "*RAND*" is the random local search algorithm and "*TASK*" is the *TASK* algorithm. $MAXSTEP$ is set to be 64 according to [**78**]. The comparison is conducted on 12 DAGs of different sizes and different CCRs. The CPN-Dominate algorithm [**78**] generates the initial schedules. The value in the column "Initial" is the length of the initial schedule; that in the column "+*RAND*" is after the random local search algorithm conducted on the initial schedule; and that in the column "+*TASK*" is after the *TASK* algorithm conducted on the initial schedule. The columns "%" following "+*RAND*" and "+*TASK*" are the percentage of improvement of the initial schedule. The running times of the random local search algorithm and the *TASK* algorithm are also shown in the table, as well as their ratio. From the tables, it can be seen that *TASK* is much more effective and faster than random local search for scheduling and task assignment problem. Increasing the value of $MAXSTEP$ in

TABLE 13. Comparison for synthetic DAGs (4 PEs).

# of nodes	CCR	Schedule length					Running time		
		Initial	+*RAND*	%	+*TASK*	%	*RAND*	*TASK*	ratio
1000	0.1	2536	2536	0.0	2531	0.2	16.4	0.76	21.6
	1	2818	2813	0.2	2669	5.3	5.43	0.25	21.7
	10	5095	5083	0.2	4462	12.4	9.13	0.43	21.2
2000	0.1	5011	5011	0.0	4994	0.3	41.8	1.91	21.9
	1	5506	5498	0.1	5226	5.1	18.4	0.82	22.4
	10	11002	10980	0.2	8873	13.4	21.7	1.02	21.3
3000	0.1	7728	7728	0.0	7582	1.4	28.1	1.27	22.1
	1	7707	7698	0.1	7475	2.1	81.2	3.57	22.7
	10	15619	15586	0.2	11953	15.6	38.7	1.76	22.0
4000	0.1	10088	10087	0.01	9937	1.0	57.3	2.54	22.6
	1	10665	10648	0.2	10145	4.1	88.2	3.90	22.6
	10	21436	21419	0.1	17390	12.5	55.5	2.58	21.5

TABLE 14. Comparison for synthetic DAGs (16 PEs).

# of nodes	CCR	Schedule Length					Running time		
		Initial	+*RAND*	%	+*TASK*	%	*RAND*	*TASK*	ratio
1000	0.1	663	660	0.5	653	1.5	16.1	0.82	19.6
	1	960	960	0.0	910	5.2	5.18	0.29	17.9
	10	3201	3187	0.4	3092	4.4	7.86	0.50	15.7
2000	0.1	1348	1348	0.0	1317	2.3	40.1	2.09	19.2
	1	1831	1830	0.05	1738	5.1	17.6	0.95	18.5
	10	6784	6784	0.0	6606	2.6	19.8	1.13	17.5
3000	0.1	2233	2233	0.0	2123	4.7	26.6	1.43	18.6
	1	2038	2038	0.0	1976	3.0	74.1	3.81	19.4
	10	8770	8770	0.0	8471	3.4	34.4	1.90	18.1
4000	0.1	2931	2931	0.0	2775	5.3	54.6	2.80	19.5
	1	2992	2992	0.0	2865	4.2	79.9	4.14	19.3
	10	12924	12920	0.03	12757	1.3	51.6	2.76	18.7

the random local search algorithm can improve its performance. However, even if the $MAXSTEP$ is equal to the number of nodes, the random search algorithm does not perform as well as the *TASK* algorithm. The search order is important. The order with the L value is superior than the random search order. Furthermore, if $MAXSTEP$ is equal to the number of nodes, the random search algorithm is extremely slow because the running time is proportional to $MAXSTEP$.

We also tested the local search algorithms with the DAGs generated from a real application, Gaussian elimination with partial pivoting. The Gaussian elimination program operates on matrices. The matrix is partitioned by columns. The finest grain size of this column partitioning scheme is a single column. However, this fine-grain partition generates too many nodes in the graph. For example, the fine-grain partition of a 1k×1k matrix generates a DAG of 525,822 nodes. To reduce the number of nodes, a medium-grain partition is used. Table 15 lists the number of nodes in different matrix sizes and grain sizes (number of columns). These graphs are generated by the Hypertool from an annotated sequential Gaussian elimination program [**126**]. Table 16 shows the comparison of the random local search algorithm and the *TASK* algorithm on different DAGs and different number of PEs. Random local search does not improve quality on these schedules except the first one, whereas, *TASK* can always reduce the schedule length. Again, *TASK* is much faster than random local search.

TABLE 15. The number of nodes in different matrix sizes and grain sizes for Gaussian elimination.

Matrix size	1k×1k				2k×2k			
Grain size	64	32	16	8	64	32	16	8
# of nodes	138	530	2082	8258	530	2082	8258	32898

TABLE 16. Comparison for Gaussian elimination.

Matrix size	Grain size	# of PEs	Length					Running time		
			Initial	+*RAND*	%	+*TASK*	%	*RAND*	*TASK*	ratio
1k	64	4	209	189	9.7	191	9.0	0.45	0.02	22.5
×	32	8	110	110	0.0	97.0	11.7	0.51	0.03	17.0
1k	16	16	56.6	56.6	0.0	49.8	12.0	1.80	0.09	20.0
	8	32	28.9	28.9	0.0	25.4	12.1	7.60	0.35	21.7
2k	64	8	868	868	0.0	786	9.5	0.45	0.03	15.0
×	32	16	449	449	0.0	394	12.3	1.84	0.11	16.7
2k	16	32	386	386	0.0	353	8.6	7.43	0.45	16.5
	8	64	116	116	0.0	102	12.1	32.0	1.88	17.0

From the real algorithm executions, *TASK* can improve the scheduling lengths produced by the *RAND* algorithm by a factor between 26 to 238. In addition, it can speed up the execution time of *RAND* by a factor between 15 to 22.

7. Conclusions

Random techniques are simple and effective for a wide range of algorithm design. The key issue lies to know when and how to use random techniques for what type of algorithm design. In this research, we have studied randomized and deterministic local search algorithms for solving very large size satisfiability and scheduling problems. Experimental results indicate that randomized local search algorithms are particularly efficient for large size NP-complete problems that have many solutions. For an NP-hard optimization problem with fewer optimum solutions, randomized local search algorithms perform satisfactorily well and, in some cases, a deterministic local search algorithm can find higher quality solutions much more efficiently.

Acknowledgments

Panos Pardalos suggested that this article be written. Reviewers provided helpful comments that improved the quality of this paper.

[1] B. Abramson and M. Yung. Divide and conquer under global constraints: A solution to the n-queens problem. *Journal of Parallel and Distributed Computing*, 6:649–662, 1989.
[2] T.L. Adam, K.M. Chandy, and J.R. Dickson. A comparison of list scheduling for parallel processing systems. *Communications of ACM*, 17(12):685–690, Dec. 1974.
[3] W. Ahrens. *Mathematische Unterhaltungen und Spiele (in German).* B.G. Teubner (Publishing Company), Leipzig, 1918-1921.
[4] P. Ashar, A. Ghosh, and S. Devadas. Boolean satisfiability and equivalence checking using general Binary Decision Diagrams. *Integration, the VLSI journal*, 13:1–16, 1992.
[5] B. Bernhardsson. Explicit solutions to the n-queens problems for all n. *ACM SIGART Bulletin*, 2(2):7, Apr. 1991, ACM Press.
[6] J. R. Bitner and E. M. Reingold. Backtrack programming techniques. *Comm. of ACM*, 18(11):651–656, Nov. 1975.
[7] C.E. Blair, R.G. Jeroslow, and J.K. Lowe. Some results and experiments in programming techniques for propositional logic. *Computers and Operations Research*, 5:633–645, 1986.
[8] C.A. Brown and P.W. Purdom. An average time analysis of backtracking. *SIAM J. on Computing*, 10(3):583–593, Aug. 1981.
[9] R.E. Bryant. Graph-based algorithms for Boolean function manipulation. *IEEE Trans. on Computers*, C-35(8):677–691, Aug. 1986.
[10] K. Bugrara and P. Purdom. Clause order backtracking. Technical Report 311, 1990.
[11] M. Buro and H.K. Buning. Report on a SAT competition. *Bulletin of the European Association for Theoretical Computer Science*, 49:143–151, Feb. 1993.
[12] M. Buro and H.K. Buning. Report on a SAT competition. Technical report, FB-17 — Mathematik/Informatik, Universitat Paderborn, Nov. 1992.
[13] Y.C. Chung and S. Ranka. Applications and performance analysis of a compile-time optimization approach for list scheduling algorithms on distributed memory multiprocessors. *Supercomputer'92*, Nov. 1992.
[14] E. Coffman, editor. *Computer and Job-Shop Scheduling Theory.* John Wiley & Sons, 1976.
[15] S.A. Cook. The complexity of theorem-proving procedures. In *Proceedings of the Third ACM Symposium on Theory of Computing*, pages 151–158, 1971.
[16] J.M. Crawford. Solving satisfiability problems using a combination of systematic and local search. Submitted to the DIMACS Challenge II Workshop, 1994.
[17] V. Cěrny. A thermodynamical approach to the travelling salesman problem: An efficient simulation algorithm. Technical report, Institute of Physics and Biophysics, Comenius University, Bratislava, 1982.
[18] V. Cěrny. Thermodynamical approach to the traveling salesman problem: An efficient simulation algorithm. *Journal of Optimization Theory and Applications*, 45:41–51, 1985.
[19] M. Davis, G. Logemann, and D. Loveland. A machine program for theorem proving. *Communications of the ACM*, 5:394–397, 1962.
[20] M. Davis and H. Putnam. A computing procedure for quantification theory. *J. of ACM*, 7:201–215, 1960.
[21] R. Dechter. Directional resolution: The davis-putnam procedure revisited. Presented at the 3rd International Symposium on AI & Mathematics, Jan. 1994.
[22] N. Dershowitz, J. Hsiang, N. Josephson, and D. Plaisted. Associative-commutative rewriting. In *Proceedings of IJCAI*, pages 940–944, 1983.
[23] H. El-Rewini and T.G. Lewis. Scheduling parallel program tasks onto arbitrary target machines. *Journal of Parallel and Distributed Computing*, Jun. 1990.
[24] H. El-Rewini, T.G. Lewis, and H. H. Ali. *Task Scheduling in Parallel and Distributed Systems.* Prentice Hall, 1994.
[25] B.-J. Falkowski and L. Schmitz. A note on the queens' problem. *Information Processing Letters*, Vol. 23:39–46, July 1986.
[26] Z. Galil. On the complexity of regular resolution and the Davis-Putnam procedure. *Theoretical Computer Science*, pages 23–46, 1977.
[27] M.R. Garey and D.S. Johnson. *Computers and Intractability: A Guide to the Theory of NP-Completeness.* W.H. Freeman and Company, San Francisco, 1979.
[28] A.V. Gelder. A satisfiability tester for non-clausal propositional calculus. *Information and Computation*, 79(1):1–21, Oct. 1988.

[29] M.R. Genesereth and N. J. Nilsson. *Logical Foundations of Artificial Intelligence.* Morgan Kaufmann Publishers, Los Altos, California, 1987.
[30] P.C. Gilmore. A proof method for quantification theory. *IBM J. Res. Develop.*, 4:28–35, 1960.
[31] A. Goldberg, P.W. Purdom, and C.A. Brown. Average time analysis of simplified davis-putnam procedures.[4] *Information Processing Letters*, 15(2):72–75, 6 Sept. 1982.
[32] R.L. Graham, E.L. Lawler, J.K. Lenstra, and A.H.G. Rinnoy Kan. Optimization and approximation in deterministic sequencing and scheduling: A survey. *Annals of Discrete Mathematics*, (5):287–326, 1979.
[33] J. Gu, P.W. Purdom, J. Franco, and B.W. Wah. Algorithms for satisfiability (SAT) problem: A survey. *DIMACS Volume Series on Discrete Mathematics and Theoretical Computer Science, Vol. 35: The Satisfiability (SAT) Problem*, pages 19–151, American Mathematical Society, 1997.
[34] J. Gu. Parallel algorithms and architectures for very fast search (PhD Thesis). Technical Report UUCS-TR-88-005, Jul. 1988.
[35] J. Gu. How to solve Very Large-Scale Satisfiability problems. Technical Report UUCS-TR-88-032. 1988, and UCECE-TR-90-002, 1990.
[36] J. Gu. Optimization by multispace search. Technical Report UCECE-TR-90-001, Jan. 1990.
[37] J. Gu. Local search for large-scale scheduling and task assignment problems. Lecture Notes in Algorithm and Optimization, 1991–1996.
[38] J. Gu. The *UniSAT* problem models (appendix). *IEEE Trans. on Pattern Analysis and Machine Intelligence*, 14(8):865, Aug. 1992.
[39] J. Gu. Efficient local search for very large-scale satisfiability problem. *SIGART Bulletin*, 3(1):8–12, Jan. 1992, ACM Press.
[40] J. Gu. *On Optimizing the SAT Problem. In* Advanced Series on Artificial Intelligence, Vol. 1, chapter 2, pages 63–105. World Scientific, New Jersey, Jan. 1992.
[41] J. Gu. Local search for satisfiability (SAT) problem. *IEEE Trans. on Systems, Man, and Cybernetics*, 23(4):1108–1129, Jul. 1993, and 24(4):709, Apr. 1994.
[42] J. Gu. *Multispace Search: A New Optimization Approach (Summary).* In *Lecture Notes in Computer Science, Vol. 834*, pages 252–260. 1994.
[43] J. Gu. *Optimization Algorithms for the Satisfiability (SAT) Problem. In* Advances in Optimization and Approximation., pages 72–154. Kluwer Academic Publishers, 1994.
[44] J. Gu. Global optimization for satisfiability (SAT) problem. *IEEE Trans. on Knowledge and Data Engineering*, 6(3):361–381, Jun. 1994, and 7(1):192, Feb. 1995.
[45] J. Gu. Parallel algorithms for satisfiability (SAT) problem. *DIMACS Volume Series on Discrete Mathematics and Theoretical Computer Science,* Vol. 22, pages 105–161, American Mathematical Society, 1995.
[46] J. Gu, Q.P. Gu, and D.-Z. Du. Convergence properties of optimization algorithms for the satisfiability (SAT)problem. *IEEE Trans. on Computers*, 45(2):209–219, Feb. 1996.
[47] J. Gu. The *Multi-SAT* algorithm. *Presented at* Siena SAT Workshop *and* Workshop on Semidefinite and Interior-Point Methods, *May 1996. To appear in Discrete Applied Mathematics.*
[48] J. Gu. *Constraint-Based Search.* Cambridge University Press, New York, 1997.
[49] J. Gu. Multispace search for satisfiability and NP-hard problems. *DIMACS Volume Series on Discrete Mathematics and Theoretical Computer Science, Vol. 35*, pages 407–517, American Mathematical Society, 1997.
[50] J. Gu and Q.P. Gu. Average time complexities of several local search algorithms for the satisfiability problem. Technical Report UCECE-TR-91-004, 1991. In *Lecture Notes in Computer Science*, Vol. 834, pp. 146-154, 1994, and to appear in *IEEE Trans. on Knowledge and Data Engineering.*
[51] J. Gu, H. Li, Z. Zhou, and B. Du. An efficient implementation of SAT1.5 algorithm. Technical report, Dept. of Computer Science, USTC, Sept. 1995.
[52] J. Gu and W. Wang. A novel discrete relaxation architecture. *IEEE Trans. on Pattern Analysis and Machine Intelligence*, 14(8):857–865, Aug. 1992.

[4]Some printer errors in this paper were corrected in *Information Processing Letters*, 16:213, 1983.

[53] J. Gu, W. Wang, and T. C. Henderson. A parallel architecture for discrete relaxation algorithm. *IEEE Trans. on Pattern Analysis and Machine Intelligence*, PAMI-9(6):816–831, Nov. 1987.

[54] J. Gu, M.Y. Wu, and W. Shu. Fast local search algorithms for DAG scheduling. Research Collaborations, Summer 1995.

[55] J. Gu, B. Du, and Y.K. Kwok. Design an efficient local search algorithm for DAG scheduling. COMP680 Applied Optimization Course Project Meeting, HKUST, Fall 1995.

[56] P. Hansen and B. Jaumard. Algorithms for the maximum satisfiability problem. *Computing*, 44:279–303, 1990.

[57] E. J. Hoffman, J. C. Loessi, and R. C. Moore. Constructions for the solution of the m queens problem. *Mathematics Magazine*, pages 66–72, 1969.

[58] J.N. Hooker. Generalized resolution and cutting planes. *Annals of Operations Research*, 12:217–239, 1988.

[59] J.N. Hooker. A quantitative approach to logical inference. *Decision Support Systems*, 4:45–69, 1988.

[60] J.N. Hooker. Resolution vs. cutting plane solution of inference problems: Some computational experience. *Operations Research Letter*, 7(1):1–7, 1988.

[61] J.N. Hooker and C. Fedjki. Branch-and-cut solution of inference problems in propositional logic. Technical Report 77-88-89, GSIA, Carnegie Mellon University, Aug. 1989.

[62] J. Hsiang. Refutational theorem proving using term-rewriting systems. *Artificial Intelligence*, pages 255–300, 1985.

[63] T.C. Hu. Parallel sequencing and assembly line problems. *Operations Research*, 9(6):841–848, 1961.

[64] T.H. Hu, C.Y. Tang, and R.C.T. Lee. An average case analysis of a resolution principle algorithm in mechanical theorem proving. *Annals of Mathematics and Artificial Intelligence*, 6:235–252, 1992.

[65] K. Iwama. CNF satisfiability test by counting and polynomial average time. *SIAM J. on Computing*, pages 385–391, 1989.

[66] R.G. Jeroslow. Computation-oriented reductions of predicate to propositional logic. *Decision Support Systems*, 4:183–197, 1988.

[67] W. Lewis Johnson. Letter from the editor. *SIGART Bulletin*, 2(2):1, April 1991, ACM Press.

[68] M.D. Johnston. Scheduling with neural networks — the case of the hubble space telescope. NASA Memo, 1989.

[69] A.P. Kamath, N.K. Karmarkar, K.G. Ramakrishnan, and M.G.C. Resende. Computational experience with an interior point algorithm on the satisfiability problem. *Annals of Operations Research*, 25:43–58, 1990.

[70] A.P. Kamath, N.K. Karmarkar, K.G. Ramakrishnan, and M.G.C. Resende. Computational experience with an interior point algorithm on the satisfiability problem. *Mathematical Sciences Research Center, AT&T Bell Laboratories*, Oct. 1989.

[71] R.M. Karp. *Probabilistic Algorithms. In* Algorithms and Complexity: New Directions and Recent Results, pages 21–39. Academic Press, New York, 1976.

[72] R.M. Karp. *The Probabilistic Analysis of Some Combinatorial Search Algorithms. In* Algorithms and Complexity: New Directions and Recent Results, pages 1–19. Academic Press, New York, 1976.

[73] A.A. Khan, C.L. McCreary, and M.S. Jones. A comparison of multiprocessor scheduling heuristics. *Int'l Conf. on Parallel Processing*, II:243–250, Aug. 1994.

[74] S. Kirkpatrick, C.D. Gelat, and M.P. Vecchi. Optimization by simulated annealing. *Science*, 220:671–680, 1983.

[75] B. Kruatrachue and T.G. Lewis. Grain size determination for parallel processing. *IEEE Software*, pages 23–32, Jan. 1988.

[76] V. Kumar. Algorithms for constraint satisfaction problems: A survey. Technical Report TR-91-28, Dept. of Computer Science, Univ. of Minnesota, 1991.

[77] V. Kumar. Algorithms for constraint satisfaction problems: A survey. *The AI Magazine*, 13(1):32–44, 1992.

[78] Y.K. Kwok, I. Ahmad, and J. Gu. FAST: A low-complexity algorithm for efficient scheduling of DAGs on parallel processors. In *Proc. of Int'l Conference on Parallel Processing*, pages (II) 150 — 157, Aug. 1996.

[79] T. Larrabee. Test pattern generation using Boolean satisfiability. *IEEE Trans. on Computer-Aided Design*, 11(1):4–15, Jan. 1992.
[80] S. Lin. Computer solutions of the traveling salesman problem. *Bell Sys. Tech. Journal*, 44(10):2245–2269, Dec. 1965.
[81] D.W. Loveland. *Automated Theorem Proving: A Logical Basis.* North-Holland, 1978.
[82] A. K. Mackworth. Consistency in networks of relations. *Artificial Intelligence*, 8:99–119, 1977.
[83] B. Mazure, L. Sais, and E. Gregoire. Tabu search for SAT. In *Proceedings of CP'95 Workshop on Solving Really Hard Problems*, pages 127–130, 1995.
[84] S. Minton, M.D. Johnston, A.B. Philips, and P. Laird. Solving large-scale constraint satisfaction and scheduling problems using a heuristic repair method. In *Proceedings of AAAI'90*, pages 17–24, Aug. 1990.
[85] S. Minton, M.D. Johnston, A.B. Philips, and P. Laird. A heuristic repair method for constraint satisfaction and scheduling problems. *Artificial Intelligence*, 58:161–205, 1992.
[86] D. Mitchell, B. Selman, and H. Levesque. Hard and easy distributions of SAT problems. In *Proceedings of AAAI'92*, pages 459–465, Jul. 1992.
[87] R. Motwani and P. Raghavan. *Randomized Algorithms.* Cambridge University Press, New York, 1995.
[88] N.J. Nilsson. *Principles of Artificial Intelligence.* Tioga Publishing Company, Palo Alto, California, 1980.
[89] C.H. Papadimitriou. On selecting a satisfying truth assignment. In *Proceedings of the 32nd Annual Symposium of the Foundations of Computer Science*, pages 163–169, 1991.
[90] C.H. Papadimitriou. Private Communications, 1996.
[91] C.H. Papadimitriou and K. Steiglitz. On the complexity of local search for the traveling salesman problem. *SIAM J. on Computing*, 6(1):76–83, 1977.
[92] C.H. Papadimitriou and K. Steiglitz. *Combinatorial Optimization: Algorithms and Complexity.* Prentice Hall, Englewood Cliffs, 1982.
[93] P.W. Purdom. Search rearrangement backtracking and polynomial average time. *Artificial Intelligence*, 21:117–133, 1983.
[94] P.W. Purdom and C.A. Brown. The pure literal rule and polynomial average time. *SIAM J. on Computing*, 14:943–953, 1985.
[95] P.W. Purdom and G.N. Haven. Probe order backtracking. *SIAM J. on Computing*, 1997.
[96] P.W. Purdom and G.N. Haven. Backtracking and probing. Technical Report No. 387, Dept. of Computer Science, Indiana University, Aug. 1993.
[97] R. Puri and J. Gu. An efficient algorithm for computer microword length minimization. *IEEE Transactions on CAD*, 12(10):1449–1457, Oct. 1993.
[98] M. Reichling. A simplified solution of the n queens' problem. *Information Processing Letters*, Vol. 25:253–255, June 1987.
[99] J.A. Robinson. A machine–oriented logic based on the resolution principle. *Journal of the ACM*, pages 23–41, 1965.
[100] S. Russell and P. Norvig. *Artificial Intelligence: A Modern Approach.* Prentice-Hall, Englewood Cliffs, 1995.
[101] Y.G. Saab and V. B. Rao. Combinatorial optimization by stochastic evolution. *IEEE Transactions on CAD*, CAD-10(4):525–535, Apr. 1991.
[102] V. Sarkar. *Partitioning and Scheduling Parallel Programs for Multiprocessors.* The MIT Press, 1989.
[103] B.M. Schwartzschild. Statistical mechanics algorithm for Monte Carlo optimization. *Physics Today*, 35:17–19, 1982.
[104] B. Selman. Private Communications, Aug. 1992.
[105] B. Selman, H. Kautz, and B. Cohen. Noise strategies for improving local search. In *Proc. of the 12th National Conference on Artificial Intelligence*, pages 337–343, Jul. 1994.
[106] B. Selman, H.A. Kautz, and B. Cohen. Local search strategies for satisfiability testing. *DIMACS Series Volume: Clique, Graph Coloring, and Satisfiability — Second DIMACS Implementation Challenge.* Editors: D.S. Johnson and M.A. Trick, American Mathematical Society, 1996.
[107] B. Selman, H. Levesque, and D. Mitchell. A new method for solving hard satisfiability problems. In *Proceedings of AAAI'92*, pages 440–446, Jul. 1992.

[108] M.J. Shensa. A computational structure for the propositional calculus. In *Proceedings of IJCAI*, pages 384–388, 1989.
[109] R.C.-J. Shi, A. Vannelli, and J. Vlach. An improvement on Karmarkar's algorithm for integer programming. *COAL Bulletin of the Mathematical Programming Society*, 1992.
[110] P. Siegel. *Representation et Utilization de la Connaissances en Calcul Propositionel.* PhD thesis, University Aix-Marseille II, 1987.
[111] G.C. Sih and E.A. Lee. A compile-time scheduling heuristic for interconnection-constrained heterogeneous processor architectures. *IEEE Transactions on Parallel and Distributed Systems*, 4(2):175–187, Feb. 1993.
[112] R. Sosič and J. Gu. How to search for million queens. Technical Report UUCS-TR-88-008, Dept. of Computer Science, Univ. of Utah, Feb. 1988.
[113] R. Sosič and J. Gu. A polynomial time algorithm for the n-queens problem. *SIGART Bulletin*, 1(3):7–11, Oct. 1990, ACM Press.
[114] R. Sosič and J. Gu. Fast search algorithms for the n-queens problem. *IEEE Trans. on Systems, Man, and Cybernetics*, SMC-21(6):1572–1576, Nov./Dec. 1991.
[115] R. Sosič and J. Gu. 3,000,000 queens in less than one minute. *SIGART Bulletin*, 2(2):22–24, Apr. 1991.
[116] R. Sosič and J. Gu. Efficient local search with conflict minimization. *IEEE Trans. on Knowledge and Data Engineering*, 6(5):661–668, Oct. 1994.
[117] R. Sosič and J. Gu. Quick n-queen search on VAX and Bobcat machines. CS 547 AI Class Project, Winter Quarter Feb. 1988.
[118] R. Sosič, J. Gu, and R. Johnson. The Unison algorithm: Fast evaluation of Boolean expressions. *ACM Transactions on Design Automation of Electronic Systems*, 1(4):pp. 456–477, Oct. 1996.
[119] H.S. Stone and J.M. Stone. Efficient search techniques – an empirical study of the n-queens problem. *IBM J. Res. Develop.*, 31(4):464–474, July 1987.
[120] K. Truemper. Polynomial algorithms for problems over d-systems. Presented at the 3rd International Symposium on AI & Mathematics, Jan. 1994.
[121] G.S. Tseitin. *On the Complexity of Derivations in Propositional Calculus. In* Structures in Constructive Mathematics and Mathematical Logic, Part II*, A.O. Slisenko, ed.*, pages 115–125. 1968.
[122] W. Wang, J. Gu, and K.F. Smith. A regular layout for concurrent discrete relaxation computation. In *Proceedings of 1987 IEEE International CAS Symposium*, pages 927–933, May 1987.
[123] W. Wang and C.K. Rushforth. An adaptive local search algorithm for channel assignment problem. *IEEE Trans. on Vehicular Technology*, 45(3):459–466, Aug. 1996.
[124] H.P. Williams. Linear and integer programming applied to the propositional calculus. *Systems Research and Information Sciences*, 2:81–100, 1987.
[125] P.H. Winston. *Artificial Intelligence.* Addison-Wesley, Reading, 1984.
[126] M.Y. Wu and D.D. Gajski. Hypertool: A programming aid for message-passing systems. *IEEE Transactions on Parallel and Distributed Systems*, 1(3):330–343, July 1990.
[127] M.Y. Wu, W. Shu, and J. Gu. Efficient local search for DAG scheduling. In *Proc. of Int'l Conference on Parallel Processing*, pages 174 — 180, 1997.
[128] T. Yang and A. Gerasoulis. DSC: Scheduling parallel tasks on an unbounded number of processors. *IEEE Transactions on Parallel and Distributed Systems*, 5(9):951–967, Sept. 1994.

Dept. of Electrical and Computer Engineering, Univ. of Calgary, Calgary, Alberta T2N 1N4, Canada

E-mail address: `gu@enel.ucalgary.ca, jungu@dimacs.rutgers.edu`

DIMACS Series in Discrete Mathematics
and Theoretical Computer Science
Volume **43**, 1999

An Approximation Scheme for Scheduling of Malleable Parallel Tasks

Klaus Jansen

ABSTRACT. We study the problem of scheduling a set of n independent malleable parallel tasks on a fixed number of processors. We propose a polynomial time approximation scheme (based on a linear program) that finds a schedule of length within a factor of $(1+\epsilon)$ of optimal in the non-preemptive model.

The approximation scheme can be extended directly to special interconnection networks (e.g. hypercubes, meshes) where an assigned processor set must have a special structure (e.g. subcubes, submeshes).

1. Introduction

In this paper, we study scheduling of parallel malleable tasks. The tasks can be executed on an arbitrary number of available processors, and the running time of a task is given by a function that depends on the number of processors allotted to it. We consider the problem of generating a schedule for a set of n independent parallelizable tasks on a multiprocessor system consisting of m identical processors. The goal is to construct a non-preemptive schedule that minimizes the schedule length, or makespan. This problem is called malleable parallel task scheduling (MPTS). Several authors have studied this problem (see for example [**3, 11, 22, 25**]).

We consider also an additional variant of MPTS. In several applications, a network topology is specified for the m processors (see [**10, 12, 13, 19, 22**]). The tasks are scheduled not on an arbitrary subset of processors, but on a subset of processors with a particular interconnection network (e.g. lines, hypercubes, meshes) that depends on the computer architecture. For example on a hypercube the tasks would require subcubes, and on a mesh the tasks would require submeshes. If the underlying architecture is not taken into account, we refer to this type of scheduling as scheduling on a PRAM.

The problem of non-malleable parallel task scheduling (NPTS) is a restriction of MPTS in which the allotments of processors for the tasks are known a priori. In this case, the chosen number of processors and the execution time of each task are given as part of the input [**7**]. Closely related to NPTS are the resource constrained scheduling (see e.g. [**6, 14**]) and rectangle packing problem (see e.g. [**5, 8, 18, 24**]).

1991 *Mathematics Subject Classification.* Primary 68R10, 68Q25; Secondary 90C05,05B40.

This work was supported in part by DIMACS and by EU ESPRIT LTR Project No. 20244 (ALCOM-IT).

Du and Leung [**11**] have proved that NPTS is strongly NP-complete even for a fixed number $m \geq 5$ of processors. For $m = 2, 3$ they have given a pseudo polynomial algorithm for MPTS under the condition that the execution time of a task does not increase with the number of processors. If a further condition holds that the work (product of execution time and number of processors) does not decrease with the number of processors, we obtain the restricted version called monotonic MPTS (or short MMPTS). Belkhale and Banerjee [**3**] proposed a $(2 - \frac{2}{m+1})$ - approximation algorithm for MMPTS restricted to PRAMs and lines.

Turek, Wolf and Yu [**25**] showed that some approximation algorithms for NPTS on a PRAM or line can be used to obtain algorithms for MPTS with the same approximation factor. Furthermore, Ludwig and Tiwari [**22**] presented an approximation algorithm with improved running time that works also for other interconnection networks. The best algorithm for MPTS achieves a 2 - approximation for PRAMs, lines and hypercubes. For meshes, an algorithm for NPTS and MPTS with approximation factor $\frac{46}{7}$ is known [**21, 22**]. Interestingly, the problem NPTS on a mesh is equivalent to a 3 - dimensional packing problem [**21**].

Let $A(I)$ denote the schedule length used by algorithm A on input I, and let $OPT(I)$ denote the minimum schedule length. A polynomial time approximation scheme (PAS) for an optimization problem is an algorithm A, which takes as input both the input I and a constant $\epsilon > 0$, runs in time polynomial in the length of I and has the performance guarantee $R_A(I, \epsilon) = \frac{A(I)}{OPT(I)} \leq 1+\epsilon$. We can view such an algorithm A as a family of algorithms $\{A_\epsilon | \epsilon > 0\}$ such that $A_\epsilon(I) \leq (1+\epsilon)OPT(I)$. A fully polynomial approximation scheme is an approximation scheme $\{A_\epsilon\}$ where each algorithm A_ϵ runs in time polynomial in the length of I and $\frac{1}{\epsilon}$.

Recently, Kopidakis and Zissimopoulos [**19**] have given a linear time approximation scheme for NPTS restricted to hypercubes of fixed dimension. The running time depends on the minimum task processing time, and the algorithm does not work for arbitrary small processing times. Recently, Kopidakis [**20**] has proposed an approximation scheme that works also for arbitrary small processing times and runs in $O(n \cdot \log n)$ time. List scheduling approximation algorithms for NPTS on hypercubes have been presented by Chen and Lai [**9**] and by Zhu and Ahuja [**26**]. The first authors have obtained an approximation using LDLPT (largest dimension largest processing time) with factor $2 - \frac{1}{2^{m-1}}$, and the second authors have got an algorithm with factor $2 - \frac{1}{2^m}$ using LDF (largest dimension first).

In this paper, we use a linear program to get a polynomial time approximation scheme for scheduling of malleable parallel tasks for a fixed number m of processors in the non-preemptive model. Our work was inspired by the approximation scheme of Amoura et al. [**1**] for scheduling of independent tasks on dedicated processors, where each task requires a prespecified subset of processors. For NPTS our algorithm runs in time $O(n)$ and improves the complexity of the $O(n \cdot \log n)$ approximation scheme of Kopidakis [**20**]. Here, we consider only the simpler case where all possible subsets of processors are feasible. We note that our algorithm can be directly extended to arbitrary network topologies (e.g. lines, hypecubes and meshes).

1.1. Definitions. Formally, we are given a set of tasks $T = \{T_0, \ldots, T_{n-1}\}$ and a set of identical processors $\{1, \ldots, m\}$. Each task T_j has an associated function $t_j : \{1, \ldots, m\} \to \mathbb{Q}^+$ that gives the execution time $t_j(\beta_j)$ of task T_j in terms of the number of processors $\beta_j \in \{1, \ldots, m\}$ that are assigned to T_j. For interconnection networks (e.g. hypercubes, meshes), not all subsets of processors are feasible. In this case, we have to use a set $D \subset 2^{\{1,\dots,m\}}$ of feasible processor sets (e.g. subcubes, submeshes). In the following we have no restrictions for the processor sets and use $D = 2^{\{1,\dots,m\}}$.

The β_j processors allotted to task T_j are required to execute task T_j in union. That means that these processors are all required to start task T_j at some starting time τ_j, and they will complete T_j at the completion time $\tau_j + t_j(\beta_j)$. A schedule consists of a processor allotment $\beta_j \in \{1,\dots,m\}$ and a starting time $\tau_j \geq 0$ for each task T_j without preemptions. If the number of active processors does not exceed the total number of processors:

$$\sum_{\{T_j | \tau \in [\tau_j, \tau_j + t_j(\beta_j))\}} \beta_j \leq m$$

for each time step τ, then the schedule is called feasible. Our goal is to find an optimum feasible schedule such that the overall makespan given by

$$max_{0 \leq j \leq n-1}\{\tau_j + t_j(\beta_j)\}$$

is minimized.

1.2. Main Ideas. The main ideas of the approximation scheme are the following. First, we compute take $d_j = min_{1 \leq \ell \leq m} t_j(\ell)$ for each task T_j and take a constant number of tasks $T_{j_1},\dots,T_{j_k}$ with the largest times d_j. Next, we build all relative schedules for the set $L = \{T_{j_1},\dots,T_{j_k}\}$ consisting of a processor assignment for the tasks in L and of an execution order of the tasks in L. For each relative schedule, we construct a linear program to get a solution for all tasks in T respecting the relative schedule of L. In the solution of the linear program some of the variables $y_{j,\ell}$ (corresponding to the processor allotments) can be fractional. If there are at least two variables $y_{j,\ell}, y_{j,\ell'} \neq 0$ for a task T_j then T_j has a non-unique processor allotment. We delete all these tasks and generate a schedule for the other tasks according to the LP solution.

In the resulting schedules, some of the tasks are scheduled incorrectly. A task (with ℓ processors alloted to the task) is scheduled incorrectly, if the task is preempted or processed on more than ℓ processors in parallel. Another possibility is a change of the processor assignment during the execution. We remove the incorrect scheduled tasks, and schedule these tasks and the tasks with non - unique processor allotments sequentially at the end of the schedule. Fortunately, the number of removed tasks can be bounded by a constant and the total execution time of these tasks by $\epsilon \cdot OPT$ where OPT is the minimum schedule length. Furthermore, the minimum objective value among all linear programs (or relative schedules) gives a lower bound for the minimum schedule length OPT. The running time of our algorithm is polynomial in n, but super exponential in $\frac{1}{\epsilon}$.

2. Linear Program

2.1. Non-malleable Tasks. In this subsection, we consider the non-malleable parallel task scheduling NPTS problem and describe a linear program to solve this problem. In the next subsection, we extend the linear program for the malleable problem. In section 3, we describe our approximation scheme that is based on this linear program. For each task T_j, the number of allotted processors is denoted by β_j and the execution time by p_j.

In the following, we consider a subset $L \subset T$ of long tasks. A processor assignment for L is a mapping $f : L \to 2^{\{1,\dots,m\}}$ such that $|f(T_j)| = \beta_j$ for each task $T_j \in L$. Two tasks T_j and $T_{j'}$ are compatible, if the assigned processor sets have an empty intersection: $f(T_j) \cap f(T_{j'}) = \emptyset$. Given a set of tasks L and a processor assignment f, a snapshot of L is a subset of compatible tasks $M \subset L$. A relative schedule of L is a processor assignment $f : L \to 2^{\{1,\dots,m\}}$ together with a sequence $M(1), \dots, M(g)$ of snapshots such that

(1): each task $T_j \in L$ occurs in a subsequence $M(s_j), \dots, M(t_j)$ of consecutive snapshots (with $1 \leq s_j \leq t_j < g$). Here, $M(s_j)$ is the first and $M(t_j)$ is the last snapshot that contains T_j,

(2): consecutive snapshots are different: $M(t) \neq M(t')$ for $1 \leq t \neq t' \leq g$.

A relative schedule corresponds to an order of executing the long tasks in L. One can associate a relative schedule for each non-preemptive schedule of L by looking at the schedule at every time where a task of L starts or ends and creating a snapshot right after that time step. Creating snapshots this way, $M(1) \neq \emptyset$, $M(g) = \emptyset$ and the number of snapshots can be bounded by $2|L|$ (since we have at most $2|L|$ such events). Given a relative schedule $(f, M(1), \dots, M(g))$, the processor set used in snapshot $M(i)$ is given by $P(i) = \cup_{T_j \in M(i)} f(T_j)$. If $P(i) = \{1, \dots, m\}$ then all processors are used in snapshot $M(i)$.

For the other tasks in $T \setminus L$, we define τ-configurations. Given a set τ (the free processors), a τ-configuration is a collection $C_\tau^{(j)}$ of processor sets $\tau_i \subset \tau$ such that

(1): the processor sets of $C_\tau^{(j)}$ are pairwise disjoint,

(2): $\cup_{\tau_i \in C_\tau^{(j)}} \tau_i = \tau$.

Clearly, a τ-configuration forms a partition of τ (the set of free processors). For each relative schedule $(f, M(1), \dots, M(g))$ of L, we build a linear program. The variables of the LP have the following interpretation:

t_i: the time step where we change from snapshot $M(i)$ to $M(i+1)$, for $1 \leq i \leq g-1$. The starting time of the schedule or snapshot $M(1)$ is denoted by $t_0 = 0$ and the finishing time by t_g.

e_τ: the total processing time where processors of $\{1, \dots, m\} \setminus \tau$ are executing long tasks of L and processors of τ are not executing tasks of L (these processors are either idle or executing short tasks of $T \setminus L$).

$x(C_\tau^{(j)})$: the total processing time where only processors of τ are executing short tasks and where each subset of processors $\tau_i \in C_\tau^{(j)}$ executes at most one short task at each time step in parallel.

Given a τ-configuration $C_\tau^{(j)}$, a subset $\tau_i \in C_\tau^{(j)}$ of processors of size ℓ is reserved for ℓ-processor tasks (tasks that use ℓ processors). We allow (for our relaxed solution) that

(1): short tasks are preemptable,

(2): ℓ-processor short tasks can be executed on multiple subsets $\tau_i \in C_\tau^{(j)}$ of cardinality ℓ,

(3): the processor assignments of short tasks can be changed during the execution.

Let D^ℓ be the total processing time for all short tasks in $T' = T \setminus L$ executed on ℓ processors. For each configuration $C_\tau^{(j)}$, the number of processor sets $\tau_i \in C_\tau^{(j)}$ with cardinality ℓ is denoted by $a_\ell(C_\tau^{(j)})$. This number gives the parallelization factor for ℓ - processor tasks in the configuration $C_\tau^{(j)}$. Notice that $a_\ell(C_\tau^{(j)})$ can be zero.

The constraints of the linear program are the following:

(1): $t_i > t_{i-1}$ for each $i \in \{1, \dots, g\}$,
(2): $t_j - t_i = p_r$ if t_j is the finishing time of task T_r started at t_i,
(3): $\sum_{i|P(i)=\{1,\dots,m\}\setminus\tau} t_i - t_{i-1} = e_\tau$ for each non-empty set $\tau \subset \{1, \dots, m\}$,
(4): $\sum_{C_\tau^{(j)}} x(C_\tau^{(j)}) \leq e_\tau$ for each non-empty set $\tau \subset \{1, \dots, m\}$,
(5): $\sum_{\tau, C_\tau^{(j)}} a_\ell(C_\tau^{(j)}) \cdot x(C_\tau^{(j)}) \geq D^\ell$ for each number $\ell \in \{1, \dots, m\}$,
(6): $x(C_\tau^{(j)}) \geq 0$ for each τ-configuration $C_\tau^{(j)}$.

The objective function of the linear program is to minimize t_g. The relative schedule $(f, M(1), \dots, M(g))$ together with the constraints in (1) and (2) defines a feasible schedule of L. In (3), we determine the total processing times e_τ. Clearly, these equalities can be inserted directly into (4). The inequalities in (4) mean that the total length of the τ-configurations must be at most e_τ. Furthermore, the inequalities (5) guarantee that there is enough time for the execution of all ℓ-processor tasks in T'.

For our approximation algorithm, we replace the strict inequalities (1) by

(1'): $t_i \geq t_{i-1}$ for each $i \in \{1, \dots, g\}$.

In this case, we enlarge the search space of the linear program and can get as a solution a modified relative schedule for L (if we have to delete some snapshots).

2.2. Malleable Tasks. In this section, we show how we can extend the linear program for the malleable parallel task scheduling (MPTS) problem. Let L be a set of k tasks (later k will be a fixed constant). Furthermore, we assume that we have a processor allotment β_j for each task $T_j \in L$ and a relative schedule for L. In this case, the running time p_j of task T_j is given by $t_j(\beta_j)$. The relative schedule consists of a processor assignment $f : L \to 2^{\{1,\dots,m\}}$ and a sequence of snapshots $M(1), \dots, M(g)$ with $g \leq 2k$.

We use additional 0/1-variables $y_{j,\ell}$ for each task $T_j \in T' = T \setminus L$ and each number $\ell \in \{1, \dots, m\}$ with the interpretation that

$$y_{j,\ell} = \begin{cases} 1 & \text{if } T_j \text{ is executed on } \ell \text{ processors,} \\ 0 & \text{otherwise.} \end{cases}$$

For the extended linear program we use the following constraints:

(7): $\sum_{\ell=1}^{m} y_{j,\ell} = 1$ for each task $T_j \in T'$,
(8): $\sum_{T_j \in T'} t_j(\ell) \cdot y_{j,\ell} = D^\ell$ for each number $\ell \in \{1, \dots, m\}$,
(9): $y_{j,\ell} \geq 0$ for each $T_j \in T'$ and $\ell \in \{1, \dots, m\}$.

Again, the equalities (8) can be inserted directly into the inequalities (5). The (in-)equalities (7) and (9) determine the processor allotment for the tasks $T_j \in T'$. Furthermore, using the equalities (8) we can calculate the total processing times D^ℓ of all tasks in T' that are executed on ℓ processors. In general, the variables $y_{j,\ell}$ are fractional. Note that the objective value of this linear program gives a lower bound for the schedule length (respecting the chosen relative schedule).

3. Algorithm

In this section, we describe our approximation scheme for the MPTS problem for the three-processor case. This approach can be directly generalized to the m-processor case for any fixed m. First, we give the main algorithm and, later, we describe the construction of the schedules from the linear program solutions. In the following we assume that the number of tasks $n > 8^{\lceil \frac{1}{\epsilon'} \rceil + 1}$. If the number of tasks $n \leq 8^{\lceil \frac{1}{\epsilon'} \rceil + 1}$, then we take all tasks into L and the algorithm finds the optimum solution in constant time (by computing all possible schedules for L).

3.1. The Main Algorithm.

(1): compute the minimum execution time $d_j = min_{1 \leq \ell \leq 3} t_j(\ell)$ achieved on α_j processors for each task T_j. Set $D = \sum_{j=0}^{n-1} d_j$ and $\epsilon' = \frac{\epsilon}{3}$,

(2): sort the $8^{\lceil \frac{1}{\epsilon'} \rceil + 2}$ longest tasks with respect to the times d_j such that $d_0 \geq d_1 \geq \ldots \geq d_{8^{\lceil \frac{1}{\epsilon'} \rceil + 2} - 1}$. Find $i \in \{2, \ldots, \lceil \frac{1}{\epsilon'} \rceil + 1\}$ such that $d_{8^i} + \ldots + d_{8^{i+1}-1} < \epsilon' \cdot D$. Let $k = 8^i$. Partition the set of tasks T into two sets T' and L where L is the set of k tasks with the longest execution times d_j,

(3): construct all possible feasible relative schedules consisting of processor allotments β_j, processor assignments $f(T_j) \subset \{1, 2, 3\}$ of cardinality β_j for each task $T_j \in L$ and sequences of snapshots $M(1), \ldots, M(g)$ for L,

(4): for each relative schedule (f, M) of L,

- **(4.1):** build the corresponding linear program and compute a fractional solution with objective value $t_g(f, M)$,
- **(4.2):** delete all tasks $T_j \in T'$ that have a non-unique processor allotment (that means: $y_{j,\ell} \neq \emptyset$ and $y_{j,\ell'} \neq \emptyset$ for $1 \leq \ell \neq \ell' \leq 3$), let T'' be the set of tasks with unique processor allotment,
- **(4.3):** construct a non-feasible schedule $Schedule(f, M)$ for the tasks in T'' (using the algorithm described in section 3.2).
- **(4.4):** compute the set $T''' \subset T''$ of tasks which are scheduled incorrectly in $Schedule(f, M)$. Then, modify the solution $Schedule(f, M)$ by executing the tasks T_j in $T''' \cup (T' \setminus T'')$ at the end of the schedule sequentially using α_j processors.

(5): among all schedules produced in stage (4.4), take the one that has the smallest makespan.

3.2. Generating a schedule. In step (4.3), we have to compute a schedule $Schedule(f, M)$ for the tasks in T'' with unique processor allotments. In this schedule, we allow that some tasks are preempted and that some ℓ-processor tasks may be executed parallel on multiple blocks of ℓ processors. In this subsection, we describe an algorithm to compute such a schedule but with only few incorrect scheduled tasks. These tasks and the tasks with non-unique processor allotment are processed at the end of the schedule (see step (4.4)). Furthermore, we give some upper bounds for the number of incorrect scheduled tasks and the number of tasks with non-unique processor allotments.

Let $\bar{D}^\ell$ be the total processing time for all tasks in $T'' \subset T'$ (with unique processor allotment) assigned to ℓ processors. Then, using equality (8) we obtain

$$\bar{D}^\ell \leq D^\ell = \sum_{T_j \in T'} t_j(\ell) \cdot y^*_{j,\ell}$$

for the optimum solution $(t_i^*, x^*(C_\tau^{(h)}), y_{j,\ell}^*)$ of the linear program. First, we order all ℓ-processor tasks in T'': $T_{\ell,1}, \ldots, T_{\ell,n_\ell}$ for $1 \leq \ell \leq 3$.

Then, we analyse all intervals $[t_i^*, t_{i+1}^*)$ with $t_i^* < t_{i+1}^*$ one after another starting with $[t_0^* = 0, t_1^*)$. Each interval $[t_i^*, t_{i+1}^*)$ has an associated processor set $P(i+1)$ used by the tasks in L. Let L_{i+1} be the length of the interval. For this interval, we consider all $\tau = \{1, \ldots, 3\} \setminus P(i+1)$ configurations. We choose the first τ-configuration with $x^*(C_\tau^{(h)}) > 0$ (possibly a part of $x^*(C_\tau^{(h)})$ has been considered before). Then, we select for each type $\tau_a \in C_\tau^{(h)}$ the first not-completely scheduled task $T_{\ell,j}$ that needs $\ell = |\tau_a|$ processors. The processor set τ_a is filled with the tasks $T_{\ell,j}, T_{\ell,j+1}, \ldots$ until the total processing time of the selected tasks becomes greater than or equal to

$$L = min(x^*(C_\tau^{(h)}), L_{i+1})$$

or until all ℓ - processor tasks have been scheduled. If the total processing time is strictly greater than L, then the last selected task is preempted such that the total time is exactly L.

After the assignments of the tasks for the types $\tau_a \in C_\tau^{(h)}$, we compare the length of the configuration $x^*(C_\tau^{(h)})$ with the length of the interval L_{i+1}. If $x^*(C_\tau^{(h)}) < L_{i+1}$, then we set $L_{i+1} = L_{i+1} - x^*(C_\tau^{(h)})$, $x^*(C_\tau^{(h)}) = 0$ and consider the next τ-configuration $x^*(C_\tau^{(h+1)})$. If $x^*(C_\tau^{(h)}) \geq L_{i+1}$, we reduce the length $x^*(C_\tau^{(h)})$ by L_{i+1} and go over to the next interval $[t_{i+1}^*, t_{i+2}^*)$.

Using the inequality $\sum_{\tau, C_\tau^{(h)}} a_\ell(C_\tau^{(h)}) \cdot x^*(C_\tau^{(h)}) \geq \bar{D}^\ell$, all ℓ - processor tasks are scheduled completely by this algorithm. But it is possible that some tasks are preempted or that some ℓ - processor tasks are assigned to several blocks of ℓ processors. Furthermore, some tasks can get different processor assignments in different intervals. In what follows, we give an upper bound for the number of incorrect scheduled tasks.

LEMMA 3.1. *The number of incorrect scheduled tasks in T'' can be bounded by* $4k + 10$.

PROOF. First, we may assume that there is at most one $\tau = \{1, 2, 3\}$ - configuration in the last interval $[t_{g-1}, t_g)$; otherwise we can shift all such intervals at the end of the schedule. There are two different cases of incorrect scheduled tasks.

Case 1: Inside of an interval $[t_i, t_{i+1})$ caused by a change from a configuration $C_\tau^{(h)}$ to $C_\tau^{(h+1)}$.

Case 2: At the end of an interval $[t_i, t_{i+1})$ or inside of an interval without a change of configurations.

In the second case, we have at most two new incorrect scheduled tasks (not counted before) for each interval $[t_i, t_{i+1})$. If we have a $\tau \neq \{1, 2, 3\}$ configuration, there are at most two free processors. Moreover, for a $\tau = \{1, 2, 3\}$ configuration in the interval $[t_{g-1}, t_g)$, the last selected task is either not preempted or counted before. In total, we obtain at most $2g \leq 4k$ incorrect scheduled tasks in case 1.

In the first case, we can order the τ - configurations for each τ such that the total number of incorrect scheduled tasks is bounded by 10. For example, consider the configurations for $\tau = \{2, 3\}$. We have two configurations for $\tau = \{2, 3\}$ and take the order $C_{\{2,3\}}^{(1)} = (\{2, 3\})$ and $C_{\{2,3\}}^{(2)} = (\{2\}, \{3\})$. Then, we get at most

variables	t_i	$x(C_\tau^{(h)})$	$y_{j,\ell}$
numbers	k	14	$2\|T'\|$

TABLE 1. Upper bounds for the numbers of variables

inequalities	(1′)	(4)	(5)	(6)	(9)
numbers	$2k$	7	3	14	$3\|T'\|$

TABLE 2. Upper bounds for the numbers of inequalities

one incorrect scheduled task (processed parallel on the processors 2 and 3) caused by a change from $C^{(1)}_{\{2,3\}}$ to $C^{(2)}_{\{2,3\}}$. □

In the next part of this section, we give an upper bound for the number of tasks in $T' \setminus T''$.

LEMMA 3.2. *The number of tasks with a non-unique processor allotment can be bounded by* $2k+19$.

PROOF. First, we replace all variables in the linear program that are given by equality constraints by other variables (e.g. $t_j = t_i + p_r$, $y_{j,3} = 1 - y_{j,1} - y_{j,2}$). Then, the number of remaining variables is at most $k + 2|T'| + 14$. Upper bounds for the numbers are given in Table 1.

If we insert one further inequality $t_g \leq B := \sum_{j=1}^k p_j + \sum_{j \in T'} \bar{d}_j + 1$ with $\bar{d}_j = max_{1 \leq \ell \leq 3} t_j(\ell)$ as an upper bound for the schedule length, then all variables of the linear program are bounded and we obtain a convex polytope. The dimension d of the polytope is at least $2|T'| + 5$ (counting the variables $y_{j,\ell}$ and the variables for the $\tau = \{1,2,3\}$ - configurations). Furthermore, the number of remaining inequalities (given in Table 2) is at most $3|T'| + 2k + 24$.

Each vertex of the polytope (and the corresponding optimum solution) is determined by d linearly independent rows of the constraint matrix such that each of these constraints is satisfied with equality [**23**]. Since we have $2k + 24$ inequalities of type (1) – (6) (in the LP for the non-malleable case), we have to select at least $2|T'| - (2k + 19)$ further inequalities of type (9) for each vertex. Furthermore, for each task $T_j \in T'$, at least one of the three variables $y_{j,1}$, $y_{j,2}$ or $y_{j,3} = 1 - y_{j,1} - y_{j,2}$ must be unequal zero. This implies that at least one of the inequalities $y_{j,1} \geq 0$, $y_{j,2} \geq 0$, $y_{j,1} + y_{j,2} \leq 1$ is not satisfied with equality. Then, there are at most $2k+19$ further inequalities of type (9) that are not satisfied with equality. This means, that there are at least $|T'| - (2k+19)$ tasks with an unique processor allotment (for these tasks exactly two of the inequalities are satisfied with equality). □

In total, we get the following bound for the number of tasks in $T''' \cup (T' \setminus T'')$ that are scheduled at the end of the schedule sequentially.

LEMMA 3.3. *The number of tasks that have a non-unique processor allotment or that are scheduled incorrectly can be bounded by* $6k + 29$.

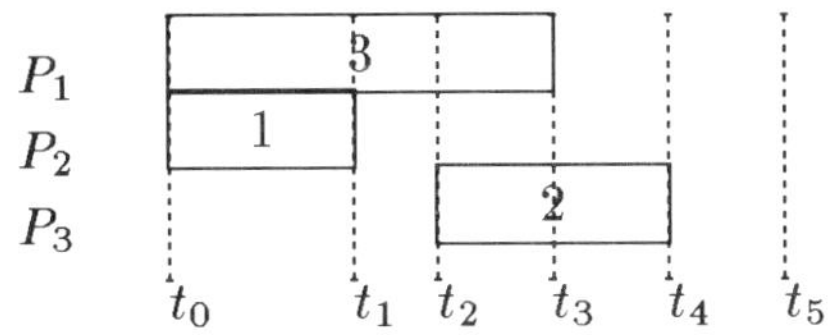

FIGURE 1. A relative schedule for three tasks 1, 2, 3.

task T_j	4	5	6	7	8	9	10
$t_j(1)$	5	6	6	4	10	10	10
$t_j(2)$	3	2	2	2	5	6	7
$t_j(3)$	2	1	1	1	2	4	5

TABLE 3. The execution times for the tasks in $T \setminus L$.

4. Example for MPTS

In this section, we give an example for the MPTS problem. Consider a set of tasks $L = \{1, 2, 3\}$ with processor assignments $f(1) = \{2\}$, $f(2) = \{3\}$, $f(3) = \{1\}$ and execution times $p_1 = 12$, $p_2 = 15$ and $p_3 = 25$. A relative schedule for L is given in Figure 1. In this schedule the sequence of snapshots is

$$(\{1, 3\}, \{3\}, \{2, 3\}, \{2\}, \emptyset)$$

with processor sets $P(1) = \{1, 2\}$, $P(2) = \{1\}$, $P(3) = \{1, 3\}$, $P(4) = \{3\}$ and $P(5) = \emptyset$. The execution times of the other tasks in $T \setminus L$ are given in Table 3.

For the total processing times of the types (the sets of free processors) we use the variables $e_{\{3\}}, e_{\{2,3\}}, e_{\{2\}}, e_{\{1,2\}}$ and $e_{\{1,2,3\}}$. Therefore, we have to consider the following τ-configurations: $C^{(1)}_{\{3\}} = (\{3\})$, $C^{(1)}_{\{2,3\}} = (\{2, 3\})$, $C^{(2)}_{\{2,3\}} = (\{2\}, \{3\})$, $C^{(1)}_{\{2\}} = (\{2\})$, $C^{(1)}_{\{1,2\}} = (\{1, 2\})$, $C^{(2)}_{\{1,2\}} = (\{1\}, \{2\})$, $C^{(1)}_{\{1,2,3\}} = (\{1, 2, 3\})$, $C^{(2)}_{\{1,2,3\}} = (\{1\}, \{2, 3\})$, $C^{(3)}_{\{1,2,3\}} = (\{2\}, \{1, 3\})$, $C^{(4)}_{\{1,2,3\}} = (\{3\}, \{1, 2\})$, $C^{(5)}_{\{1,2,3\}} = (\{1\}, \{2\}, \{3\})$.

For the relative schedule above, we construct the following linear program:

(0): $min\ t_5$ s.t.
(1): $t_1 = 12,\ t_2 \geq t_1,\ t_3 = 25,\ t_3 \geq t_2,\ t_4 = t_2 + 15,\ t_4 \geq t_3,\ t_5 \geq t_4,$
(2): $e_{\{3\}} = 12,\ e_{\{2,3\}} = t_2 - t_1,\ e_{\{2\}} = t_3 - t_2,\ e_{\{1,2\}} = t_4 - t_3,\ e_{\{1,2,3\}} = t_5 - t_4,$
(3): $x(C^{(1)}_{\{3\}}) \leq e_{\{3\}}, \ldots, x(C^{(1)}_{\{1,2,3\}}) + \ldots + x(C^{(5)}_{\{1,2,3\}}) \leq e_{\{1,2,3\}},$
(4): $x(C^{(1)}_{\{3\}}) + 2x(C^{(2)}_{\{2,3\}}) + \ldots + 3x(C^{(5)}_{\{1,2,3\}}) \geq D^1, \ldots, x(C^{(1)}_{\{1,2,3\}}) \geq D^3,$
(5): $x(C^{(1)}_{\{3\}}), \ldots, x(C^{(5)}_{\{1,2,3\}}) \geq 0,$
(6): $y_{4,1} + y_{4,2} + y_{4,3} = 1, \ldots, y_{10,1} + y_{10,2} + y_{10,3} = 1,$
(7): $5y_{4,1} + 6y_{5,1} + \ldots + 10y_{10,1} = D^1, \ldots, 2y_{4,3} + y_{5,3} + \ldots + 5y_{10,3} = D^3,$
(8): $y_{4,1}, \ldots, y_{10,3} \geq 0.$

In the solution of the linear program, all tasks have a unique processor allotment: $\beta_4 = 1$, $\beta_5 = 3$, $\beta_6 = 3$, $\beta_7 = 1$, $\beta_8 = 3$, $\beta_9 = 1$ and $\beta_{10} = 1$ (this is not true in general). The assignment of the tasks in $T \setminus L$ to the free processors

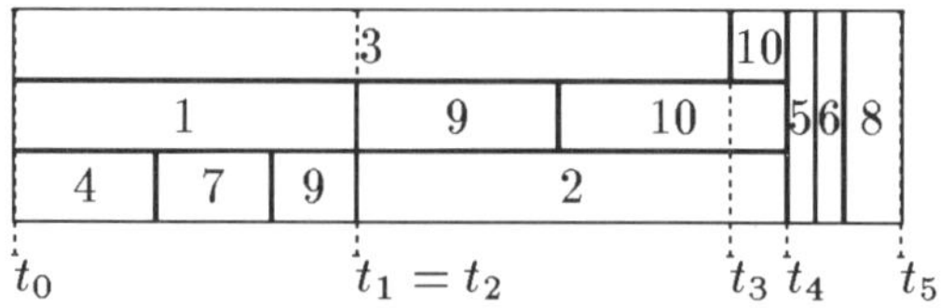

FIGURE 2. A non-feasible schedule for all tasks in T.

is illustrated in Figure 2. We note that $t_2 = t_1$ and that the tasks 9 and 10 are scheduled incorrectly. For task 9 we have a change of the processor assignments at time t_1, and task 10 with allotment $\beta_{10} = 1$ is executed on two processors during the interval $[t_3, t_4)$. At step (4.4) of our algorithm, both tasks are removed and scheduled sequentially at the end of the schedule after time t_5 on three processors.

5. Analysis of the algorithm

First, we prove that we have only a constant number of relative schedules, if k is a fixed constant.

LEMMA 5.1. *If $|L| = k$, then the total number of relative schedules of L is at most $(28k^2)^k$.*

PROOF. We have to create a snapshot when a task of L starts or ends its execution. Since $|L| = k$, there are at most $2k$ such events and, consequently, at most $2k$ snapshots. Since we have only 7 different non-empty subsets of $\{1, 2, 3\}$, there are at most 7^k possible processor assignments for the tasks in L.

Furthermore, we have to choose for each task $T_j \in L$ the first and last snapshot that contains T_j. This gives at most $4k^2$ possible choices for each task. Therefore, the number of different sequences of snapshots is at most $(4k^2)^k$. In total, we obtain at most $(28k^2)^k$ different relative schedules. □

The next Lemma shows us a lower bound for the minimum schedule length OPT.

LEMMA 5.2. *Let $T = \{T_0, \ldots, T_{n-1}\}$ be a set of tasks with execution times $t_j(\ell) > 0$ for $0 \leq j \leq n-1$ and $1 \leq \ell \leq 3$. If $d_j = min_{1 \leq \ell \leq 3} t_j(\ell)$ and $D = \sum_{T_j \in T} d_j$, then the minimum schedule length $OPT \geq \frac{1}{3} \cdot D$.*

PROOF. Consider an optimum solution for the tasks with schedule length OPT, processor allotments $\beta_j \geq 1$ and execution times $t_j(\beta_j)$. Then, we have directly $d_j \leq t_j(\beta_j)$. Furthermore (using an average argument),

$$OPT \geq \sum_{j=0}^{n-1} \frac{\beta_j \cdot t_j(\beta_j)}{3} \geq \frac{1}{3} \cdot \sum_{j=0}^{n-1} d_j = \frac{1}{3} \cdot D.$$

□

Since at step (4) of the algorithm all possible relative schedules of the tasks in L are considered, we get the following result.

LEMMA 5.3. *The minimum objective value obtained in step (4.1) among all relative schedules gives a lower bound for the minimum schedule length OPT.*

In the next Lemma, we describe how the constant k can be found.

LEMMA 5.4. *Let* $d_0 \geq d_1 \geq \ldots \geq d_{n-1} > 0$ *be a sequence of numbers such that* $\sum_{i=0}^{n-1} d_i = D$ *and* $n > 8^{\lceil \frac{1}{\epsilon'} \rceil + 1}$, *and let* $\epsilon > 0$ *and* $\epsilon' = \frac{\epsilon}{3}$. *Then, there exists a* $k \geq 64$ *and* $k \leq 8^{\lceil \frac{1}{\epsilon'} \rceil + 1}$ *such that*

$$d_k + d_{k+1} + \ldots + d_{8k-1} < \epsilon' \cdot D.$$

PROOF. Decompose the sum $d_0 + \ldots + d_{n-1}$ into blocks $B_0 = d_0 + \ldots d_7$, $B_1 = d_8 + \ldots d_{8^2-1}$, $B_2 = d_{8^2} + \ldots + d_{8^3-1}$, ..., $B_i = d_{8^i} + \ldots + d_{8^{i+1}-1}$. The sum of the blocks $B_2, B_3, \ldots$ is smaller than D. Then, at most $\lceil \frac{1}{\epsilon'} \rceil - 1$ of these blocks are greater than $\epsilon' \cdot D$; otherwise the sum is at least D. Take the first block B_i with $i \geq 2$ which is less than $\epsilon' \cdot D$. We have $i \leq \lceil \frac{1}{\epsilon'} \rceil + 1$ and for $k = 8^i$ we obtain $B_i = d_k + \ldots + d_{8k-1} < \epsilon' \cdot D$. □

If the number of tasks $n \leq 8^{\lceil \frac{1}{\epsilon'} \rceil + 1}$, then we take all tasks into L and the algorithm finds the optimum solution in constant time (by computing all possible schedules for L). If $n > 8^{\lceil \frac{1}{\epsilon'} \rceil + 1}$, we take the first k longest tasks (with respect to the execution times $d_j = min_{1 \leq \ell \leq 3} t_j(\ell)$) into L and apply our algorithm from Section 3. Notice that we can obtain also other execution times rather than $d_0, \ldots, d_{k-1}$ in the relative schedules for L. We have at most $6k + 29 \leq 7k$ (for $k \geq 29$) tasks in $T''' \cup (T' \setminus T'')$ scheduled sequentially on α_j processors at the end of the schedule. Since $k \geq 64$, the total length of these tasks can be bounded as follows:

$$\sum_{T_j \in T''' \cup (T' \setminus T'')} d_j \leq \sum_{i=k}^{8k-1} d_i < \epsilon' \cdot D \leq \epsilon \cdot OPT.$$

Since we can find an optimum vertex of the linear programs in polynomial time [**17, 16**], we get our main result:

THEOREM 5.1. *There is an algorithm A which given a set of n independent malleable parallel tasks, a fixed number m of processors, execution times* $t_j(\ell)$ *and a constant* $\epsilon > 0$, *produces in polynomial time a schedule of T whose makespan is at most* $(1 + \epsilon)OPT$.

6. Conclusions

In this paper, we have proposed an approximation scheme for the malleable parallel task scheduling MPTS problem for any fixed number of machines. The algorithm can be directly extended to special interconnection networks (e.g. hypercubes, meshes) where each parallel task requires a special subset of processors (subcubes, submeshes). For 1-dimensional meshes, we get an approximation scheme for a rectangular packing problem. Furthermore, using the algorithm for NPTS on meshes we obtain an approximation scheme for a 3 - dimensional packing problem.

We notice that Amoura et al. [**2**] have found independently also an approximation scheme for the non-malleable problem NPTS and fixed number m of processors. Our approximation scheme for NPTS runs in time linear in n (the number of tasks), since the linear programs for NPTS have only a constant number of variables and constraints. Furthermore, using our approach we obtain also a polynomial time approximation scheme for the problem $Pm|set_j|C_{max}$ where we have different execution times $t_j(S)$ in dependence to the chosen processor sets $S \subset \{1, \ldots, m\}$. The best previous algorithm for this problem had an approximation ratio of m [**4**].

The algorithm given is polynomial in the size of the input, but super exponential in $\frac{1}{\epsilon}$. Since no fully polynomial approximation scheme can be expected (as the problem is strongly NP-hard for $m \geq 5$) **[15]**, this result is the best performance guarantee one might hope. It remains interesting whether there is an approximation scheme with a lower time complexity in m and $\frac{1}{\epsilon}$. The main open problem is to find an algorithm for an arbitrary number m of processors and with an approximation factor better than 2. Furthermore, it would be interesting whether randomized algorithms can be applied to get better and / or faster approximation algorithms.

References

[1] A.K. Amoura, E. Bampis, C. Kenyon and Y. Manoussakis, Scheduling independent multiprocessor tasks, *European Symposium on Algorithms* ESA 97, Graz, LNCS (1997).

[2] A.K. Amoura, E. Bampis and C. Kenyon, private communication.

[3] K. Belkhale and P. Banerjee, Approximate algorithms for the partitionable independent task scheduling problem, *International Conference on Parallel Processing*, Vol. 1 (1990), 72-75.

[4] L. Bianco, J. Blazewicz, P. Dell Olmo and M. Drozdowski, Scheduling multiprocessor tasks on a dynamic configuration of dedicated processors, *Annals of Operations Research* 58 (1995), 493-517.

[5] B. Baker, E. Coffman and R. Rivest, Orthogonal packings in two dimensions, *SIAM Journal on Computing* 9 (1980), 846-855.

[6] J. Blazewicz, W. Cellary, R. Slowinski and J. Weglarz, Scheduling under resource constraints - deterministic models, *Annals of Operations Research* 7 (1986).

[7] J. Blazewicz, M. Drabowski and J. Weglarz, Scheduling multiprocessor tasks to minimize schedule length, *IEEE Transactions on Computers*, C-35-5 (1986), 389-393.

[8] E. Coffman, M. Garey, D.S. Johnson and R. Tarjan, Performance bounds for level-oriented two-dimensional packing problems, *SIAM Journal on Computing*, 9 (1980), 808-826.

[9] G.I. Chen and T.H. Lai, Scheduling independent jobs on hypercubes, *Symposium on Theoretical Aspects of Computer Science* STACS 88, Bordeaux, LNCS 294 (1988), 273-280.

[10] M. Drozdowski, Scheduling multiprocessor tasks - an overview, *European Journal on Operations Research*, 94 (1996), 215-230.

[11] J. Du and J. Leung, Complexity of scheduling parallel task systems, *SIAM Journal on Discrete Mathematics*, 2 (1989), 473-487.

[12] A. Feldmann, M.-Y. Kao, J. Sgall and S.H. Teng, Optimal online scheduling of parallel jobs with dependencies, *ACM Symposium on the Theory of Computing* (1993), 642-651.

[13] A. Feldmann, J. Sgall and S.H. Teng, Dynamic scheduling on parallel machines, *IEEE Symposium on Foundations of Computer Science* FOCS (1991), 111-120.

[14] M. Garey and D.S. Johnson, Complexity results for multiprocessor scheduling under resource constraints, *SIAM Journal on Computing* 4 (1975), 397-411.

[15] M. Garey and D.S. Johnson, Computers and Intractability, A Guide to the Theory of NP-completeness, Freeman, San Francisco, 1979.

[16] M. Grötschel, L. Lovász and A. Schrijver, Geometric Algorithms and Combinatorial Optimization, Springer, Berlin, 1988.

[17] L.G. Khachian, A polynomial time algorithm in linear programming, *Soviet Mathematics Doklady* 20 (1979), 191-194.

[18] C. Kenyon and E. Remila, Approximate strip packing, *IEEE Symposium on Foundations of Computer Science* FOCS (1996), 31-36.

[19] Y. Kopidakis and V. Zissimopoulos, An approximation scheme for scheduling independent jobs into subcubes of a hypercube of fixed dimension, *Theoretical Computer Science* (1997), 265-273.

[20] Y. Kopidakis, Approximabilite des Problemes d 'Ordonnancement dans les Systemes Paralleles, These de Doctorat, Universite de Paris-Sud (1997).

[21] K. Li and K.H. Cheng, On 3-dimensional packing, *SIAM Journal on Computing* 19 (1990), 847-867.

[22] W. Ludwig and P. Tiwari, Scheduling malleable and nonmalleable parallel tasks, *ACM-SIAM Symposium on Discrete Algorithms* SODA (1994), 167-176.

[23] A. Schrijver, Theory of Linear and Integer Programming, Wiley, Chichester, 1986.

[24] A. Steinberg, A strip packing algorithm with absolute performance bound 2, *SIAM Journal on Computing* 26 (1997), 401-409.

[25] J. Turek, J. Wolf and P. Yu, Approximate algorithms for scheduling parallelizable tasks, *ACM Symposium on Parallel Algorithms and Architectures* SPAA (1992), 323-332.

[26] Y. Zhu and M. Ahuja, On job scheduling on a hypercube, *IEEE Transactions on Parallel Distributed Systems* 4 (1993), 62-69.

MAX-PLANCK-INSTITUTE FOR COMPUTER SCIENCE, IM STADTWALD, 66 123 SAARBRÜCKEN, GERMANY

E-mail address: jansen@mpi-sb.mpg.de

DIMACS Series in Discrete Mathematics
and Theoretical Computer Science
Volume **43**, 1999

Blocking Behaviors of Broadcast Switching Networks in Random Traffics

Dongsoo S. Kim

ABSTRACT. This paper presents the probabilistic behavior of three-stage Clos networks for broadcast communications in random traffic and random routing strategy. The switch modules in middle and output stage are assumed to be able to duplicate messages, but the input stage switches are to lack of the fanout capability. Although the simple routing strategy theoretically requires an excessively large number of middle switches to provide nonblocking paths to a compatible request, the probabilistic model gives an observation that the possibility to block the compatible request dramatically decreases and approaches zero as the number of middle switches is much less than the one the nonblocking networks require. The analytic model is also corroborated by performing simulations based on a random traffic generator and a random routing.

1. Introduction

There have been many efforts to extend the Clos network [**1**], which was initially developed for point-to-point communications, to accommodate multicasting and/or broadcasting connection requests [**2, 3, 4, 6, 8**]. The nonblocking broadcast Clos networks, however, required an excessively large number of middle stage switches, which made the switching system so expensive and complicate.

A three-stage Clos network denoted by $\mathcal{C}(n_1, r_1, n_2, r_2, m)$ has r_1 input stage switches, m middle stage switches and r_2 output stage switches. Each input stage switch has n_1 external input ports (inlets) and has connections to every middle stage switches such that there are $r_1 \times m$ ***I**-interstage* links between the input stage and the middle stage. Each output stage switch has n_2 external output ports (outlets) and is connected to every middle stage switches such that there are $m \times r_2$ ***O**-interstage* links between the middle stage and the output stage. A Clos network is called symmetric if n_1 and r_1 are same as n_2 and r_2, respectively, which is denoted as $\mathcal{C}(n, r, m)$. Figure 1 shows a symmetric Clos network $\mathcal{C}(3, 4, 5)$.

We assume a switch module is nonblocking and has a full fanout capability so that a connection request can be represented as $C = (x, Y)$, where x is an input switch and Y is the set of destination output switches. The cardinality of Y is called a *degree of fanout* of the request. A set of connection requests are said to be *compatible* if sums of occurrences of each input switch and each output switch

1991 *Mathematics Subject Classification.* Primary 94A05, 05C70; Secondary 90B12, 05C50.
Key words and phrases. Clos networks, broadcast, blocking probability.

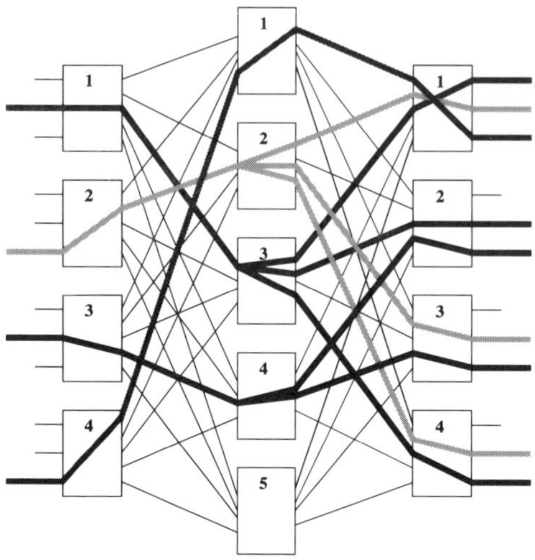

$$C_1 = (1, \{1, 2, 4\})$$
$$C_2 = (2, \{1, 3, 4\})$$
$$C_3 = (3, \{2, 3\})$$
$$C_4 = (4, \{1\})$$

FIGURE 1. Clos switching network $\mathcal{C}(3, 4, 5)$.

are not greater than n_1 and n_2, respectively. A connection request is compatible to the existing network if all requests including the new request are compatible. For a new connection request, a *route* is a set of path connecting the input port to the set of the output ports. A connection request is *satisfiable* or *routable* if one can find a route on which all interstage links can be exclusively taken over. In Figure 1, for example, $C_{\text{new}} = (4, \{2, 3, 4\})$ is compatible to the existing configuration but $C'_{\text{new}} = (4, \{1, 2, 3, 4\})$ is not because output switch 1 has no free port to be assigned.

Masson and Jordon [**6**] first studied the nonblocking broadcast on the 3-stage Clos networks. Hwang [**2, 3**] introduced rearrangeability and nonblocking property in the networks under various conditions and fanout restrictions. The first study which allow any stage switches to fanout was carried out by Yang and Masson [**8**]. They proved that if the number of middle switches is $O(n \log r / \log \log r)$, the network is nonblocking provided that all requests use at most k middle switches for some constant k. On the other hand, there have been several efforts to find out the blocking behavior of the switching network for point-to-point communication [**5, 7**]. Blocking for the broadcast communication was considered inevitable because construction of nonblocking network is too expensive.

In this paper, we will introduce an analytical model to estimate the blocking probability of broadcast connection requests in Clos switching networks. The routing algorithm tries to find a single middle switch whose links to all destination switches do not carry any requests (lazy fanout algorithm). If it can not find such a middle switch, it manifest the request to be blocked.

2. Blocking Probability of Broadcast Networks

In this section, we will prove the blocking probability of 3-stage Clos networks for broadcast communications under the lazy fanout algorithm.

THEOREM 1 (Hwang '79). *A Clos network $\mathcal{C}(n, r, m)$ is nonblocking for a connection requests of fanout d under the lazy fanout strategy if and only if $m > (d+1)(n-1)$*

PROOF. Let us consider a compatible request of fanout d. The maximum number of blocking **I**-interstage is $n-1$ due to the compatibility. For each output switch, the maximum number of blocking **O**-interstage is $n-1$. At the worst case, the blocking **O**-interstage are disjoint so that the number of blocking links is at most $d(n-1)$. If we are given one more middle switch in addition to the maximum blocking switches from input stage and output stage, the network is nonblocking. □

Let us assume a broadcast switching network $\mathcal{C}(n_1, r_1, n_2, r_2, m)$ has input and output network utilization of a and b, respectively. Let $\bar{d}$ be the mean fanout of all requests in the switching systems. an_1r_1 input ports are busy and hence it makes $an_1r_1\bar{d} = bn_2r_2$ output ports busy. The network utilization can be denoted as $\max(a, b)$. In this case, each link in **I**-interstage has a probability of $p_1 = an_1/m$ to be busy and each link in **O**-interstage has $p_2 = bn_2/m$ to be busy. For a symmetric Clos network, a is always smaller than b because $\bar{d}$ is greater than 1.

2.1. Probability of k Empty Rows. Let us first focus on the **O**-interstage links. For a compatible request, destination output switches have at most $n-1$ busy outlets. We will model the blocking property with occupancy problem in rectangular matrix cells. In an $m \times d$ matrix whose each cell can be occupied by a ball with probability p_2 and each column could have at most $n-1$ balls, a row is said to be *empty* if the row contains no ball. $p_d(m, p_2, n, k)$ denotes the probability that there exist k empty rows in the $m \times d$ matrix.

DEFINITION 1. *A_h^d denotes the event that there are h balls in the d-th column which means the output switch d has h busy **O**-interstage links. E_k^d denotes the event that there are k empty rows in d columns which means there are k middle switches to route a request of fanout d. B_j denotes the event that an input switch for a broadcast request has already j busy **I**-interstage links.*

LEMMA 1.

$$p_1(m, p_2, n, k) = \frac{\binom{m}{m-k} {p_2}^{m-k} {q_2}^k}{\sum\limits_{i=0}^{n-1} \binom{m}{i} {p_2}^i {q_2}^{m-i}},$$

where $q_2 = 1 - p_2$.

PROOF. Assume that the number of balls$(m-k)$ in the cells follows a binomial distribution (m, p_2). We can obtain the following conditional probability under a condition that the column contains at most $n-1$ balls,

$$\begin{aligned} p_1(m, p_2, n, k) &= \Pr(A_{m-k}^1 \mid m-k \le n-1) \\ &= \frac{\Pr(A_{m-k}^1 \cap m-k \le n-1)}{\Pr(m-k \le n-1)} \qquad (1) \\ &= \frac{\binom{m}{m-k} {p_2}^{m-k} {q_2}^k}{\sum_{i=0}^{n-1} \binom{m}{i} {p_2}^i {q_2}^{m-i}} \end{aligned}$$

□

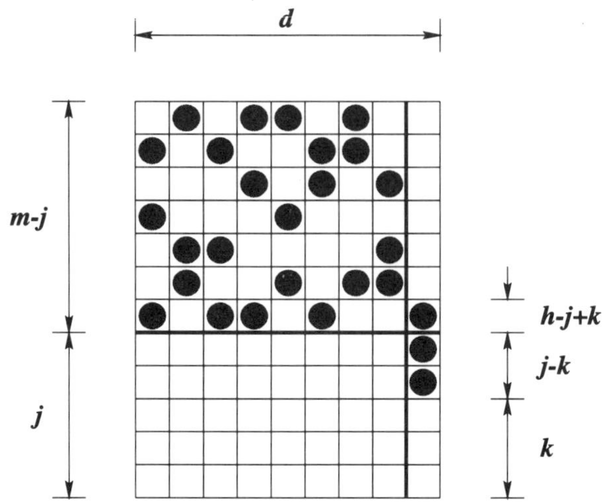

FIGURE 2. k empty rows in $m \times d$ matrix

LEMMA 2. *Given two independent events E_j^{d-1} and A_h^d which are the events that the first $d-1$ columns has j empty rows and the d-th column has h balls in $m \times d$ matrix as shown in Figure 2, the conditional probability of k empty rows, $\Pr(E_k^d \mid A_h^d \cap E_j^{d-1})$, is*

$$\Pr(E_k^d \mid A_h^d \cap E_j^{d-1}) = \frac{\binom{j}{j-k}\binom{m-j}{h-j+k}}{\binom{m}{h}}$$

PROOF. There are a total $\binom{m}{h}$ ways to put h balls into m cells of the d-th column. The number of ways to put $j-k$ balls among h into j cells which are the empty rows in the $(d-1)$ columns is $\binom{j}{j-k}$. The number of ways to put the remaining $h-j+k$ balls into $m-j$ cells is $\binom{m-j}{h-j+k}$. □

Because of the compatibility of the new request, each column (each output switch) has at most $n-1$ balls (busy links) so that the conditional probability $\Pr(A_h^d \mid h \le n-1)$, for short $\Pr(A_h^d)$, is obtained as

$$\Pr(A_h^d) = \Pr(A_h^d \mid h \le n-1) = \frac{\Pr(A_h^d \cap h \le n-1)}{\Pr(h \le n-1)} = \frac{\binom{m}{h}{p_2}^h {q_2}^{m-h}}{\sum_{i=0}^{n-1}\binom{m}{i}{p_2}^i {q_2}^{m-i}} \tag{2}$$

LEMMA 3. *The probability of k empty rows in $m \times d$ matrix, $P_d(m, a, n, k)$, is*

$$P_d(m, p_2, n, k)$$

$$= \frac{\displaystyle\sum_{h=0}^{\min(n-1,m-k)} \sum_{j=k}^{\min(m,k+h)} \binom{m}{h}{p_2}^h {q_2}^{m-h} P_{d-1}(m, p_2, n, j)\frac{\binom{j}{j-k}\binom{m-j}{h-j+k}}{\binom{m}{h}}}{\displaystyle\sum_{j=0}^{n-1}\binom{m}{j}{p_2}^j {q_2}^{m-j}}$$

PROOF. The probability of j empty rows in $m \times (d-1)$ matrix is $\Pr(E_j^{d-1}) = P_{d-1}(m, p_2, n, j)$. Then,

$$\begin{aligned} P_d(m, p_2, n, k) &= \sum_{\text{all } h, j} \Pr(E_k^d \cap A_h^d \cap E_j^{d-1}) \\ &= \sum_{\text{all } h, j} \Pr(E_k^d \mid A_h^d \cap E_j^{d-1}) \Pr(A_h^d \cap E_j^{d-1}) \\ &= \sum_{\text{all } h, j} \Pr(E_k^d \mid A_h^d \cap E_j^{d-1}) \Pr(A_h^d) \Pr(E_j^{d-1}) \end{aligned}$$

As shown before, the d-th column should have at most $n-1$ balls. Moreover, if the column has more than $m-k$ balls, the matrix can not have k empty rows. Thus, the value of h is

$$0 \le h \le \min(n-1, m-k)$$

In order to have k empty rows in $m \times d$ matrix, the first $(d-1)$ column must have at least k empty rows as you can see in the Figure 2 so that we can obtain the value of j,

$$j \ge 0, \quad m-j \ge 0, \quad h-j+k \ge 0, \quad j-k \ge 0$$

and finally

$$k \le j \le \min(m, k+h)$$

□

2.2. Blocking Probability of Requests with Fanout d. Let us consider an input switch which has the new request with fanout d, whose blocking probability is what we want to find. We know the input switch could have at most $n-1$ requests due to the compatibility, and hence at most $(n-1)$ **I**-interstage links can carry requests. Then, the probability of j busy **I**-interstage links, $\Pr(B_j \mid j \le n-1)$ for short $\Pr(B_j)$, can be obtain as follows:

$$\Pr(B_j) = \Pr(B_j \mid j \le n-1) = \frac{\Pr(B_j \cap j \le n-1)}{\Pr(j \le n-1)} = \frac{\binom{m}{j} {p_1}^j {q_1}^{m-j}}{\sum_{i=0}^{n-1} \binom{m}{i} {p_1}^i {q_1}^{m-i}} \tag{3}$$

For the request of fanout d, assume that there are j busy **I**-interstage links so that j middle switches are not accessible from the input switch, whereas k middle switches are empty by carrying no request to the set of destinations. The blocking probability of the request can be obtained by considering that the set of k middle switches is the subset of j unavailable middle switches.

$$\Pr(k \text{ are the subset of } j \mid B_j \cap E_k^d) = \frac{\binom{j}{j-k}\binom{m-j}{m-j}}{\binom{m}{m-k}} = \frac{\binom{j}{j-k}}{\binom{m}{m-k}} \tag{4}$$

THEOREM 2. *The blocking probability of a request with fanout d is*

$$P_B(d) = \frac{\sum_{j=0}^{n-1} \sum_{k=0}^{\min(j,m)} \binom{m}{j} {p_1}^j {q_1}^{m-j} P_d(m, p_2, n, k) \frac{\binom{j}{j-k}}{\binom{m}{m-k}}}{\sum_{i=0}^{n-1} \binom{m}{i} {p_1}^i {q_1}^{m-i}}$$

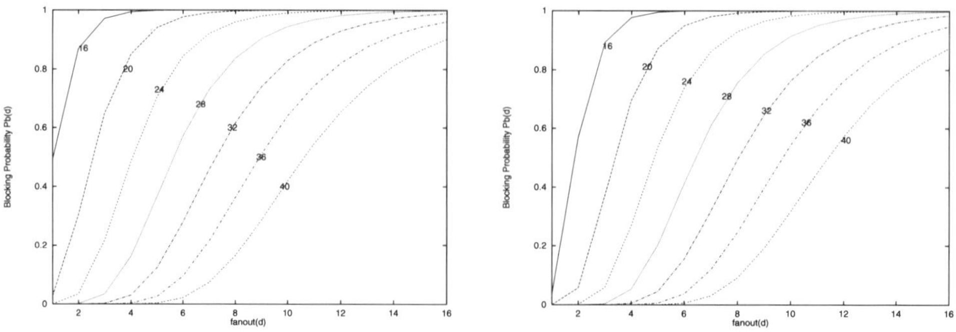

(a) Almost all requests are point-to-point

(b) Fanout of requests are distributed and their mean fanout $\bar{d} = 4$

FIGURE 3. Blocking probabilities for each fanout of broadcast Clos networks $\mathcal{C}(16, 16, m)$ for $16 \le m \le 40$ at 80% network utilization. The digits on the lines are indicating the number of middle stage switches.

PROOF. Assume that j middle switches are not accessible and k middle switches are empty. The input switch can carry at most $n-1$ requests due to the compatibility so that j is not greater than $n-1$. In addition to, we know k should not be greater than j, otherwise k middles can not be the subset of j middles. Thus, we can get that

$$0 \le j \le n-1 \qquad \text{and} \qquad 0 \le k \le \min(j, m) \tag{5}$$

The events of individual links to be busy are assumed to be independent. From Lemma 3, the blocking probability of the request with fanout d, $P_B(d)$, is obtained as

$$\begin{aligned} P_B(d) &= \sum_{j=0}^{n-1} \sum_{k=0}^{\min(j,m)} \Pr(k \text{ are the subset of } j \mid B_j \cap E_k^d) \Pr(B_j \cap E_k^d) \\ &= \sum_{j=0}^{n-1} \sum_{k=0}^{\min(j,m)} \Pr(k \text{ are the subset of } j \mid B_j \cap E_k^d) \Pr(B_j) \Pr(E_k^d) \\ &= \frac{\sum_{j=0}^{n-1} \sum_{k=0}^{\min(j,m)} \binom{m}{j} {p_1}^j {q_1}^{m-j} P_d(m, p_2, n, k) \frac{\binom{j}{j-k}}{\binom{m}{m-k}}}{\sum_{i=0}^{n-1} \binom{m}{i} {p_1}^i {q_1}^{m-i}} \end{aligned}$$

□

Figure 3 shows the blocking probabilities of requests with different fanout depending on the number of middle stage switches and the mean fanout of requests in the switching systems. The probability rapidly increases as the degree of fanout becomes larger and forms S-shapes. The steepness of probability becomes obtuse as the number of middle switches increases. The probability for $\bar{d} = 1$ is larger than

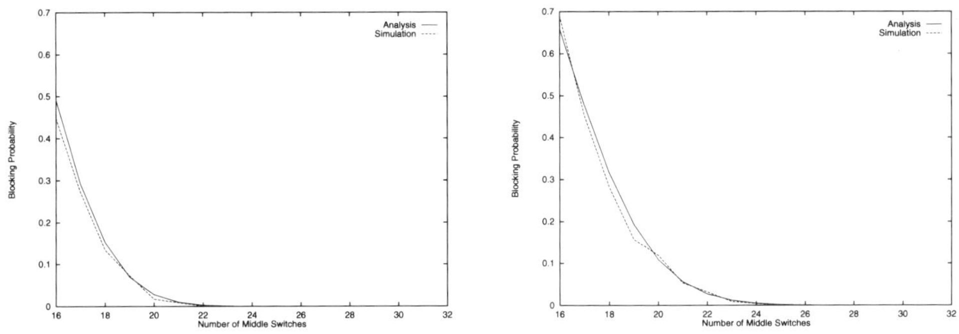

(a) all requests are point-to-point connections

(b) all requests have fanout 2

FIGURE 4. Blocking probability of broadcast Clos networks $\mathcal{C}(16, 16, m)$ for $16 \leq m \leq 32$. The analytic results were calculated for 80% network utilization .

the one for $\bar{d} = 4$ at the same fanout and the same number of middles. The reason is that the number of busy **I**-interstage links at the higher mean fanout is smaller than at the lower mean fanout even though the number of busy **O**-interstage links are same for both cases because network utilization was assumed to be identical. Therefore, p_2 are same, but p_1 for the higher mean fanout is smaller than for the lower mean fanout.

3. Experimental

In this section, the analytic model we presented will be verified by performing network simulations. The simulator consists of a random request generator and a random router. The random request generator creates compatible requests whose fanout are determined by a given distribution. However, the source and the set of destinations of the new request are randomly determined. The random router tries to find a middle stage switch to route the new request. If no middle switch satisfy the request, it is said to be blocked and discarded. If more than one middle switches can be used for the request, the router choose one arbitrarily. Because our purpose of simulation is to verify the analytic model, we do not employ any strategy to enhance the routability. The blocking probability is expected to be enhanced if we apply some routing strategy such as packing.

Figure 4 shows the blocking probabilities for fixed fanout when the network utilization is 80%. In Figure 4(a), the blocking probability for point-to-point communications was shown. Even though the Clos network is deterministically nonblocking at $m = 31$, we can find that the blocking probability practically become close to zero at $m \geq 24$. Figure 4(b) shows the blocking probability when all requests have fanout 2. By theorem 1, the Clos network is nonblocking at $m = 46$. However, the blocking probability approaches zero when $m \geq 28$.

Let $f(d)$ be the frequency of incoming requests with fanout d, which the router tried to route. Note that $\sum df(d)$ is not equal to $\bar{d}$ which was seen in the previous section because $\bar{d}$ was the mean fanout of the routed requests. Since the blocking

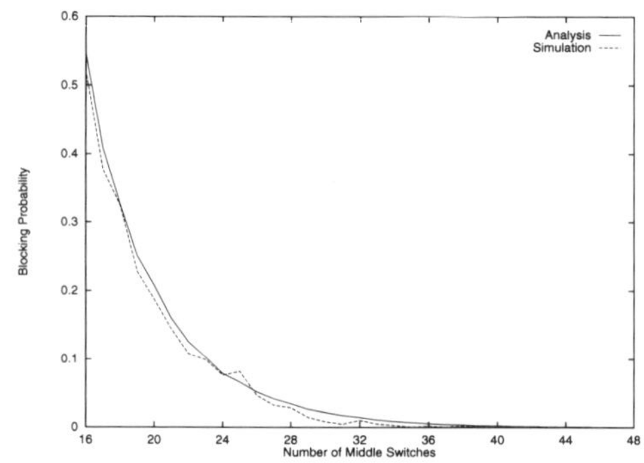

FIGURE 5. Blocking probability of broadcast Clos networks $\mathcal{C}(16,16,m)$ for $16 \leq m \leq 48$ when requests are exponentially distributed.

probabilities become spread widely depending on the degree of fanout, the fanout distribution of routed requests do not possibly follow $f(d)$. By the way, the blocking probability of all possible requests is obtained that

(6)

$$\Pr(\text{Block}) = \sum_{d=1}^{r} f(d) P_B(d) = \sum_{d=1}^{r} f(d) \frac{\sum_{j=0}^{n-1} \sum_{k=0}^{\min(j,m)} \binom{m}{j} {p_1}^j {q_1}^{m-j} P_d(m, p_2, n, k) \frac{\binom{j}{j-k}}{\binom{m}{m-k}}}{\sum_{i=0}^{n-1} \binom{m}{i} {p_1}^i {q_1}^{m-i}}$$

Figure 5 shows the total blocking probability when the mean fanout is small in the range of 1 and 2. When the mean fanout is relatively large, A_h^d is not independent to E_j^{d-1} in the previous section. For example, if almost all requests are large fanout, the blocking probability will more depend on other factor — the number of requests in the switching system. Assume all requests are of fanout 8 in the Clos network $\mathcal{C}(16,16,m)$, the maximum number of requests the switch can handle is $16 \cdot 16/8 = 32$ so that all requests are nonblocking if the switch has 32 middle stage switches. Our model did not consider this case but the probability is conservative so that real blocking probability would not be larger than our model in case of high mean fanout.

4. Conclusion

In this paper, the blocking probability of broadcast Clos switching networks were studied. The lazy fanout routing algorithm find a single middle switch to satisfy an incoming broadcast connection requests. Even though this algorithm theoretically needs $O(nr)$ middle stage switches to be nonblocking, we found that the blocking probability of the Clos network with small mean fanout approaches zero at the much smaller number of middles in random traffic patterns. Simulation results were also presented to consolidate the analytic model.

References

[1] C. Clos. A study of non-blocking switching networks. *Bell Syst. Tech. J.*, 32:406–424, Mar. 1953.

[2] F. K. Hwang. Rearrangeability of multiconnection three-stage networks. *Networks*, 2:301–306, 1972.
[3] F. K. Hwang. Three-stage multi-connection networks which are nonblocking in the wide sense. *Bell Syst. Tech. J.*, 58(10):1283–1287, Dec. 1979.
[4] D. M. Koppelman and A. Y. Oruç. Nonblocking broadcast switching networks. *IEEE Trans. on Info.*, 40(1):278–284, Jan. 1992.
[5] C. Y. Lee. Analysis of switching networks. *Bell Syst. Tech. J.*, 34(6):1287–1315, Nov. 1955.
[6] G. M. Masson and B. W. Jordan, Jr. Generalized multistage connection networks. *Networks*, 2:192–209, 1972.
[7] Y. Yang and N. H. Kessler. Modeling the blocking behavior of clos networks. In *DIMACS Workshop on Network Switching*, 1997.
[8] Y. Yang and G. M. Masson. Nonblocking broadcast switching networks. *IEEE Trans. on Comp.*, 40(9):1005–1015, Sept. 1991.

Department of Computer Science and Engineering, University of Minnesota, Minneapolis, MN 55455

E-mail address: `dkim@cs.umn.edu`

DIMACS Series in Discrete Mathematics
and Theoretical Computer Science
Volume **43**, 1999

Greedy randomized adaptive search procedures for the Steiner problem in graphs

S. L. Martins, P. M. Pardalos, M. G. C. Resende, and C. C. Ribeiro

ABSTRACT. We describe four versions of a Greedy Randomized Adaptive Search Procedure (GRASP) for finding approximate solutions of general instances of the Steiner Problem in Graphs. Different construction and local search algorithms are presented. Preliminary computational results with one of the versions on a variety of test problems are reported. On the majority of instances from the OR-Library, a set of standard test problems, the GRASP produced optimal solutions. On those that optimal solutions were not found, the GRASP found good quality approximate solutions.

1. Introduction

Posed independently by Hakimi [**21**] and Levin [**30**], the Steiner problem in graphs (SPG) consists in connecting a subset of given nodes on a graph with the minimum cost tree. The SPG has also many equivalent formulations as an integer program [**22**] or as a continuous nonconvex global optimization problem [**26**]. Karp [**25**] showed earlier that the SPG decision problem is NP-complete in general. Even for some restrictions like grid graphs [**19**], bipartite graphs [**20**], chordal and split graphs [**46**], the problem remains NP-complete. Moreover, on directed graphs, Provan and Ball [**38**] showed that the variant of the rooted Steiner arborescence of a directed graph is NP-complete. NP-completeness still sticks to the problem with planar acyclic digraphs with indegree and outdegree at most two [**37**]. However, there are some classes of instances for which polynomial time algorithms exist. In what follows, we give a short survey of existing algorithms and heuristics for the Steiner problem in graphs.

All known exact SPG algorithms for general graphs are in some way enumerative algorithms. However, they differ in how the enumeration is done and how clever their strategies are to avoid total enumeration. In order to avoid total enumeration, topological and other properties of general graphs can be used. Another alternative is to develop good lower and upper bounds in order to reduce the possible search space. Two simple approaches are proposed by Hakimi [**21**]. While one

---

1991 *Mathematics Subject Classification.* Primary 90C27; Secondary 90B12, 68R05, 90C35, 94C15.

Key words and phrases. Combinatorial optimization, Steiner problem, local search, GRASP, network design.

of them is based on spanning tree enumeration, the other approach is a topology enumeration recursive algorithm. Another algorithm, which generates spanning trees for a derived problem in order of increasing total length until a solution to the original problem can be inferred, is proposed by Balakrishnan and Patel [**2**]. Also, implicit enumeration approaches are given by Shore, Foulds, and Gibbons [**41**], and Yang and Wing [**50**], where the set of all trees spanning the set of given points is, in a systematic way, separated into smaller subsets to be analyzed, using upper and lower bounds, in order to determine whether or not they contain the optimal feasible solution. Along another line, straightforward dynamic programming approaches are proposed by Dreyfus and Wagner [**10**], and Levin [**30**]. A branch-and-bound approach that uses heuristics to provide good lower bounds and to choose the next edge for consideration in the backtracking process is proposed by Foulds and Gibbons [**17**], and Shore, Foulds, and Gibbons [**41**]. Other approaches, including branch-and-bound, are proposed by Beasley [**3, 4**], Aneja [**1**], Maculan [**32**], Lucena [**31**], Wong [**48**], and Chopra, Gorres, and Rao [**7**]. Exact algorithms for the SPG can benefit from preprocessing the graph so that the size of the problem to solve is reduced. Reduction techniques are discussed in Beasley [**3**], Balakrishnan and Patel [**2**], Duin and Volgenant [**13, 14**], Voss [**44, 45**], Iwainsky et al. [**24**], Chopra, Gorres, and Rao [**7**], Khoury and Pardalos [**27**], and Duin [**11, 12**].

Many heuristics have been proposed to find approximate solutions for the SPG. For example, in Aneja [**1**], a greedy variation of a set covering algorithm is presented, and an MST-based procedure is used as a shortcut in the dual ascent algorithm given by Wong [**48**]. Another MST-based heuristic was proposed by Choukmane [**8**] and independently by Plesník [**36**], Kou, Markowsky, and Berman [**28**], Bharath-Kumar and Jaffe [**6**], and Iwainsky et al. [**24**]. A heuristic based on Kruskal's MST algorithm is given by Wu, Widmayer, and Wong [**49**]. Another heuristic approach based on Prim's MST algorithm is given by Takahashi and Matsuyama [**42**]. While all of the above MTS-based approaches implicitly choose the Steiner nodes (optional nodes in the Steiner tree), an average distance heuristic that selects those nodes is given by Rayward-Smith [**39**]; refer also to Rayward-Smith and Clare [**40**], and Foulds and Rayward-Smith [**18**]. A different heuristic based on neighborhood contraction techniques is provided by Plesník [**36**]. The reader is referred to Winter [**47**], Hwang and Richards [**22**], Hwang, Richards, and Winter [**23**], and Cieslik [**9**] for surveys.

2. GRASP for the Steiner problem in graphs

Approximate solutions for the Steiner minimal tree problem can be obtained by many techniques, including node, spanning tree, and path-based approaches. In this section, we apply the concepts of GRASP to the approximate solution of the Steiner problem in graphs, using a spanning tree based construction phase and a node-based local search phase. We also propose a path-based construction phase and a path-based local search. Combining these two construction phases with the two local search phases, yields four versions of GRASP.

2.1. GRASP. A greedy randomized adaptive search procedure (GRASP) [**15, 16**] can be seen as a metaheuristic which captures good features of pure greedy algorithms (e.g. fast local search convergence and good quality solutions) and also of random construction procedures (e.g. diversification). Each iteration consists of the construction phase, the local search phase and, if necessary, the incumbent solution update. In the construction phase, a feasible solution is built, one element at a time. At each construction iteration, the next element to be added is determined by ordering all elements in a candidate list with respect to a greedy function that estimates the benefit of selecting each element. The probabilistic component of a GRASP is characterized by randomly choosing one of the best candidates in the list, but usually not the top one.

The solutions generated by a GRASP construction are not guaranteed to be locally optimal. Hence, it is almost always beneficial to apply local search to attempt to improve each constructed solution. A local search algorithm works in an iterative fashion by successively replacing the current solution by a better one from its neighborhood. It terminates when there are no better solutions in the neighborhood. Success for a local search algorithm depends on the suitable choice of a neighborhood structure, efficient neighborhood search techniques, and the starting solution. The GRASP construction phase plays an important role with respect to this last point, since it produces good starting solutions for local search. The customization of these generic principles into an approximate algorithm for the Steiner problem in graphs is described in the following.

Before we describe the algorithms, we define some notation. Let $G = (V, E)$ be a connected undirected graph, where V is the set of nodes and E denotes the set of edges. Given a non-negative weight function $w : E \rightarrow \mathbb{R}$ associated with its edges and a subset $X \subseteq V$ of terminal nodes, the Steiner problem SPG(V,E,w,X) consists in finding a minimum weighted connected subgraph of G spanning all terminal nodes in X. The solution of SPG(V,E,w,X) is a Steiner minimum tree (SMT). The non-terminal nodes that end up in the SMT are called *Steiner nodes.*

A graph $H = (V', E')$ is said to be a subgraph of $G = (V, E)$ if $V' \subseteq V$ and E' is a subset of the edges in E having both extremities in V'. This graph H is said to span a subset $U \subseteq V$ of the nodes in G if $U \subseteq V'$. For any subset of nodes $W \subseteq V$, the edge subset $E(W) = \{(i, j) \in E \mid i \in W, j \in W\}$ defines the induced subgraph $G(W) = (W, E(W))$ in G by the nodes in W.

We denote by $T_{V,E,w}(X)$ the Steiner minimum tree solving the Steiner problem SPG(V, E, w, X) formulated above. Given the graph $G = (V, E)$ with non-negative weights w associated with its edges, the minimum spanning tree problem MSTP(V, E, w) consists in finding a minimum weighted subtree of G spanning all nodes in V. This problem can be seen as particular case of the Steiner problem SPG(V, E, w, X), in which $X = V$. Accordingly, we denote by $T_{V,E,w}(V)$ the minimum spanning tree solving MSTP(V, E, w). We can associate a feasible solution of the Steiner problem SPG(V, E, w, X) with each subset $S \subseteq V \setminus X$ of Steiner nodes such that the graph $G(S \cup X) = (S \cup X, E(S \cup X))$ is connected, given by a minimum spanning tree solving problem MSTP$(S \cup X, E(S \cup X), w)$. Let S^* be the set of Steiner nodes in the optimal solution of SPG(V, E, w, X). The optimal solution $T_{V,E,w}(X)$ is a minimum spanning tree of the graph induced in G by the node set $S^* \cup X$, i.e., the solution to the minimum spanning tree problem MSTP$(S^* \cup X, E(S^* \cup X), w)$.

In the following, solutions of the Steiner problem SPG(V, E, w, X) will be characterized by the associated set of Steiner nodes and one of the corresponding minimum spanning trees. Accordingly, the search for the Steiner minimum tree $T_{V,E,w}(X)$ will be reduced to the search for the optimal set S^* of Steiner nodes.

2.1.1. *Spanning tree based construction phase.* The spanning tree based construction phase of GRASP makes use of the distance network heuristic, suggested by Choukmane [8], Iwainsky, Canuto, Taraszow and Villa [**24**], Kou, Markowsky and Berman [**28**], and Plesník [**36**], with time complexity $O(|X||V|^2)$. Mehlhorn [**34**] proposed a modification of the original version, leading to a procedure using simple data structures and presenting an improved time complexity. For every terminal node $i \in X$, let $N(i)$ be the subset of non-terminal nodes of V that are closer to i than to any other terminal node. A non-terminal node may belong to the neighborhood of more than one terminal node. The first step of this procedure consists in the computation of a graph $G' = (X, E')$, where $E' = \{(i,j), i,j \in X \mid \exists (k,\ell) \in E, k \in N(i), \ell \in N(j)\}$. Then, a weight $w'_{ij} = \min\{d(i,k) + w_{k\ell} + d(\ell, j) \mid (k,\ell) \in E, k \in N(i), \ell \in N(j)\}$ is associated with each edge $(i,j) \in E'$, where $d(a,b)$ denotes the shortest path from a to b in the original graph $G = (V, E)$ in terms of the weights w. Next, Kruskal's greedy algorithm [**29**] is used to solve the minimum spanning tree problem MSTP(X, E', w'). This algorithm works as follows. A minimum spanning tree is constructed, one edge at a time, until all nodes are connected. At each iteration, the edge with the least weight is selected for insertion in the tree, from among those whose insertion does not create a cycle in the current solution. The edges in the minimum spanning tree $T_{X,E',w'}(X)$ so obtained are replaced by the edges in the corresponding shortest paths in the original graph G. Neighborhoods $N(i), \forall i = 1, \ldots, |X|$ can be computed in time $O(|E| + |V| \log |V|)$ [**34**], which is the same complexity of the minimum spanning tree computation. Then, the overall complexity of the distance heuristic network with Mehlhorn's improvements is only $O(|E| + |V| \log |V|)$, which is much better than the original bound.

As discussed in the previous section, the construction phase of GRASP relies on randomization to build different solutions at different iterations. Graph $G' = (X, E')$ is created only once and does not change throughout all computations. In order to add randomization to Mehlhorn's version of the distance network heuristic, we make the following modification in Kruskal's algorithm. Instead of selecting the feasible edge with the least weight, we build a restricted candidate list with all edges $(i,j) \in E'$ such that $w'_{ij} \leq w'_{\min} + \alpha(w'_{\max} - w'_{\min})$, where $0 \leq \alpha \leq 1$ and $w'_{\min}$ and $w'_{\max}$ denote, respectively, the least and the largest weights among all edges still unselected to form the minimum spanning tree. Then, an edge is selected at random from the restricted candidate list. The operations associated with the construction phase of our GRASP are implemented in lines 2 and 4 of the pseudo-code of algorithm `GRASP_for_SPG` outlined in Figure 1.

2.1.2. *Node-based local search.* Since the solution produced by the construction phase is not necessarily a local optimum, local search can be applied as an attempt to improve it. The first step towards the implementation of a local search procedure consists in identifying an appropriate neighborhood definition.

Let S be the set of Steiner nodes in the current Steiner tree. We have noticed in Section 2.1 that each subset S of Steiner nodes can be associated with a feasible solution of the Steiner problem SPG(V, E, w, X), given by a minimum spanning tree $T_{S\cup X, E(S\cup X), w}(S \cup X)$ solving problem MSTP$(S \cup X, E(S \cup X), w)$. Moreover, let $W(T)$ denote the weight of a tree T. The neighbors of a solution characterized by its set S of Steiner nodes are defined by all sets of Steiner nodes which can be obtained either by adding to S a new non-terminal node, or by eliminating from S one of its Steiner nodes.

Given the current tree $T_{S\cup X,E(S\cup X),w}(S\cup X)$ and a non-terminal node $s\in\{V\setminus X\}\setminus S$, the computation of neighbor $T_{S\cup\{s\}\cup X,E(S\cup\{s\}\cup X),w}(S\cup\{s\}\cup X)$ obtained by the insertion of s into the current set S of Steiner nodes can be done in $O(|V|)$ average time, using the algorithm proposed by Minoux [**35**]. For each non-terminal node $t\in S$, neighbor $T_{S\setminus\{t\}\cup X,E(S\setminus\{t\}\cup X),w}(S\setminus\{t\}\cup X)$ obtained by the elimination of t from the current set S of Steiner nodes is computed by Kruskal's algorithm as the solution of the minimum spanning tree problem MSTP$(S\setminus\{t\}\cup X,E(S\setminus\{t\}\cup X),w)$.

In order to speedup the local search, since the computational time associated with the evaluation of all insertion moves is likely to be much smaller than that of the elimination moves, only the insertion moves are evaluated in a first pass. The evaluation of elimination moves is performed only if there are no improving insertion moves.

2.1.3. *Algorithm description.* We now describe the GRASP consisting of a a spanning tree based construction phase and a node-based local search phase. The pseudo-code with the complete description of procedure `GRASP_for_SPG` for the Steiner problem in graphs is given in Figure 1. The procedure takes as input the original graph $G=(V,E)$, the set X of terminal nodes, the edge weights w, the restricted candidate list parameter α $(0\le\alpha\le 1)$, a seed for the pseudo random number generator, and the number of iterations (`max_iterations`). Graph G' is computed in line 2. The GRASP procedure is repeated `max_iterations` times. In each iteration, a greedy randomized solution T is constructed in line 4 using the randomized version of Kruskal's algorithm. Let S be the set of Steiner nodes in T.

Next, the local search attempts to produce a better solution. In line 5 we initialize the best set of Steiner nodes as those in the current solution, and the weight of the best neighbor as that of the current solution. The loop from line 6 to 11 searches for the best insertion move. In line 7 we compute the minimum spanning tree T^{+s} associated with problem MSTP$(S\cup\{s\}\cup X,E(S\cup\{s\}\cup X),w)$ defined by the insertion of node s into the current set of Steiner nodes. Let $W(T^{+s})$ be its weight. In line 8 we check whether the new solution T^{+s} improves the weight of the current best neighbor. The best set of Steiner nodes and the weight of the best neighbor are updated in line 9. Once all insertion moves have been evaluated, we check in line 12 whether an improving neighbor has been found. If this is the case, the set of Steiner nodes, the current Steiner tree and its weight are updated in line 13, and the local search resumes from this new current solution.

If no improving insertion move is found, then the elimination moves are evaluated. In line 15 we reinitialize the best set of Steiner nodes as those in the current solution, and the weight of the best neighbor as that of the current solution. We check in line 17 whether the graph $G^{-t}=((S\setminus\{t\})\cup X,E((S\setminus\{t\})\cup X))$ obtained by the elimination of node t is connected or not. If it is connected, we compute in line 18 the minimum spanning tree T^{-t} associated with problem MSTP$((S\setminus\{t\})\cup X,E((S\setminus\{t\})\cup X),w)$ defined by the elimination of node t from the current set of Steiner nodes. Again, let $W(T^{-t})$ be its weight. In line 19 we check whether the new solution T^{-t} improves the weight of the current best neighbor. Once again, the best set of Steiner nodes and the weight of the best neighbor are updated in line 20. Once all elimination moves have been evaluated, we check in line 24 whether an improving neighbor has been found. If this is the case, the set of Steiner nodes, the current Steiner tree and its weight are updated in line 25, and the local search resumes from this new current solution.

procedure GRASP_for_SPG(V, E, w, X, α,seed,max_iterations)
1 *best_value* $\leftarrow \infty$;
2 Compute graph $G' = (X, E')$ and weights $w'_{ij}, \forall (i,j) \in E'$;
3 **for** $k = 1, \ldots,$ max_iterations **do**
/* Construction phase */
4 Apply a randomized version of Kruskal's algorithm to obtain a spanning tree T of $G' = (X, E')$ with S as its set of Steiner nodes;
/* Insertion moves */
5 *best_set* $\leftarrow S$; *best_weight* $\leftarrow W(T)$;
6 **for all** $s \in (V \setminus X) \setminus S$ **do**
7 Compute the minimum spanning tree T^{+s};
8 **if** $W(T^{+s}) <$ *best_weight* **then do**
9 *best_set* $\leftarrow S \cup \{s\}$; *best_weight* $\leftarrow W(T^{+s})$;
10 **end then**;
11 **end for**;
12 **if** *best_weight* $< W(T)$ **then do**
13 $S \leftarrow S \cup \{s\}$; $T \leftarrow T^{+s}$; $W(T) \leftarrow W(T^{+s})$; go to line 5;
14 **end then**;
/* Elimination moves */
15 *best_set* $\leftarrow S$; *best_weight* $\leftarrow W(T)$;
16 **for all** $t \in S$ **do**
17 **if** $G^{-t} = ((S \setminus \{t\}) \cup X, E((S \setminus \{t\}) \cup X))$ is connected **then do**
18 Compute the minimum spanning tree T^{-t};
19 **if** $W(T^{-t}) <$ *best_weight* **then do**
20 *best_set* $\leftarrow S \setminus \{t\}$; *best_weight* $\leftarrow W(T^{-t})$;
21 **end then**;
22 **end then**;
23 **end for**;
24 **if** *best_weight* $< W(T)$ **then do**
25 $S \leftarrow S \setminus \{t\}$; $T \leftarrow T^{-t}$; $W(T) \leftarrow W(T^{-t})$; go to line 5;
26 **end then**;
/* Best solution update */
27 **if** $W(T) <$ *best_value* **then do**
28 $S^* \leftarrow S$; $T^* \leftarrow T$; *best_value* $\leftarrow W(T)$;
29 **end then**;
30 **end for**;
31 **return** S^*, T^*;
end GRASP_for_SPG;

FIGURE 1. Pseudo-code of the sequential GRASP procedure for the Steiner problem in graphs

If the solution found at the end of the local search phase is better than the best solution found so far, we update in line 28 the best set of Steiner nodes, the current Steiner tree and its weight. The best set S^* of Steiner nodes and the best Steiner tree T^* are returned in line 31.

2.1.4. *Acceleration scheme.* In order to accelerate the local search phase, we implemented a faster evaluation scheme for the insertion moves (lines 5-14 in procedure `GRASP_for_SPG` in Figure 1). The basic idea consists in keeping a candidate list with promising insertion moves, which is periodically updated.

In the first GRASP iteration, we build a list containing the `k_best` improving insertion moves, which is kept in nondecreasing order of the associated move values. At each following iteration, let S be the set of Steiner nodes in the current solution. Instead of reevaluating all insertion moves, we just take the node s corresponding to the first move in this candidate list and reevaluate the weight of a minimum spanning tree associated with the insertion of this node into the current set of Steiner nodes. If this move reveals itself to be an improving one, then the current solution is updated, the move is eliminated from the candidate list, and the local search resumes from the new set $S \cup \{s\}$ of Steiner nodes. Otherwise, if the move is not an improving one, it is eliminated from the candidate list and the next candidate move is evaluated. Once the candidate list becomes empty, a new full iteration is performed, all insertion moves are evaluated, and the candidate list is rebuilt.

TABLE 1. Effect of the acceleration scheme on OR-Library problems of series C

	`GRASP_for_SPG`		With acceleration	
Problem	$W(T^*)$	secs	$W(T^*)$	secs
C.01	85	4.79	85	4.51
C.02	144	5.05	144	4.58
C.03	754	82.16	754	32.77
C.04	1079	116.03	1080	46.07
C.05	1579	148.73	1579	57.93
C.06	55	4.55	55	4.39
C.07	102	5.87	102	5.47
C.08	509	91.80	509	38.29
C.09	707	164.61	707	61.41
C.10	1093	270.07	1093	114.21
C.11	32	3.95	32	3.99
C.12	46	6.47	46	6.14
C.13	261	104.20	260	45.89
C.14	324	171.57	324	81.48
C.15	556	335.22	556	158.02
C.16	11	5.50	11	5.49
C.17	18	7.65	18	6.34
C.18	115	104.66	116	56.97
C.19	147	188.99	147	131.20
C.20	267	485.02	267	337.78

We present in Table 1 some computational results illustrating the efficiency of the above acceleration scheme on an SGI Challenge (20 196 MHZ MIPS R10000 processors) with 6144 Mbytes of RAM. For each of the 20 series C test problems from the OR-Library [**5**], we present the weight $W(T^*)$ of the best solution found and the computation time in seconds (secs) obtained by (i) a straightforward version of algorithm `GRASP_for_SPG` with a random values for α, and (ii) the same algorithm using the above acceleration scheme with `k_best` $= 100$. Each algorithm was run for 10 iterations. These results show that this technique significantly reduced the computational times, even attaining a speedup factor of 2.68 in the case of problem C.09, with almost no loss in terms of solution quality.

```
procedure PATH_TM(V,E,w,X)
1     Compute shortest paths from terminal nodes to nodes in V;
2     Select a terminal node v, at random, and set T ← {v};
3     while there are terminal nodes not in T do
4         Γ ← {x | x ∈ X and x ∉ T};
5         u ← argmin{dist(x,T) | x ∈ Γ};
6         T ← T ∪ {shortest path from u to T};
7     end while;
end PATH_TM;
```

FIGURE 2. A path-based construction procedure

2.2. Other versions of GRASP. We now propose alternative construction and local search phases for a GRASP for the Steiner problem in graphs. Both algorithms are path-based.

2.2.1. *Path-based construction phase.* This construction procedure uses a randomized version of the algorithm described in the pseudo-code in Figure 2. This algorithm was proposed by Takahashi and Matsuyama [**42**].

The randomization is applied at line 5, where instead of selecting the closest terminal node, a restricted candidate list with close nodes is constructed and a node is selected at random from this list.

2.2.2. *Path-based local search.* We begin with some definitions. A *key-node* is a Steiner node with degree at least three. A *key-path* is a path in a Steiner tree T of which all intermediate nodes are Steiner nodes with degree two in T, and whose end nodes are either terminal or key-nodes. A Steiner tree has a most $|S|-2$ key-nodes and $2|S|-3$ key-paths. A minimum Steiner tree consists of key-paths that are shortest paths between key-nodes or terminals. We use the key-path based local search, proposed by Verhoeven, Severens, and Aarts [**43**].

Let $T = \{l_1, l_2, \ldots, l_K\}$ be a Steiner tree, where each l_i, $i = 1, \ldots, K$, denotes a key-path. Also, let C_i and C_i' denote the two components that result from the removal of the key-path l_i from T. Any local search requires the definition of a neighborhood. In this case, we define the neighborhood of the current tree T as

$$N(T) = \{C_i \cup C_i' \cup \mathrm{sp}(C_i, C_i') \mid i = 1, \ldots, K\},$$

where $\mathrm{sp}(C_i, C_i')$ is the shortest path between C_i and C_i'. Observe that $C_i \cup C_i' \cup \{l_i\} = T$ and $N(T)$ contains at most $2|S|-3$ neighbors.

The pseudo-code of the path-based local search is given in Figure 3. Given a set of K key-paths, the procedure considers the removal of each key-path in the loop in lines 4–12. The removal of key-path l_i results in the decomposition of the tree into components C_i and C_i'. In line 5 the length of the shortest path from C_i to C_i' is compared to the length of key-path l_i. If the shortest path is smaller than the length of the key-path, a new tree is set up by removing key-path l_i and replacing it with the shortest path from C_i to C_i' in line 6. If the new tree is not a set of key-paths, a new set of key-paths that make up the tree is computed in lines 7–9. In case there was an improvement, an indicator is set in line 10, to restart the local search.

```
procedure PATH_LS(V,E,w,X,T = {l_1, l_2, ... , l_K})
    improve ← .true.;
    while improve do
        improve = .false.;
        for i = 1, 2, ... , K do
            if sp(C_i, C'_i) < length(l_i) then do
                T ← T \ {l_i} ∪ sp(l_i);
                if necessary then do
                    update T to be a set of key-paths;
                end then;
                improve ← .true.;
            end then;
        end for;
    end while;
end PATH_LS;
```

FIGURE 3. A path-based local search

Note that solutions only have neighbors with lower or equal cost. A replacement of a key-path in T can lead to the same Steiner tree if no shorter path exists. This implies that local minima have no neighbors.

3. Preliminary experimental results

In this section, we report on preliminary computational results with an implementation of the GRASP described in Subsection 2.1, i.e. using the spanning tree based construction procedure and the vertex-based local search.

We considered the 60 instances from classes C, D, and E of Beasley's OR-Library [5]. The experiments were done on a Silicon Graphics Challenge computer with 20 196 MHz MIPS R10000 processors and 6.144 Gbytes of memory. The code was executed each time on a single processor.

The programs were written in the C programming language and were compiled with the SGI C compiler `cc` using compiler flags `-O3 -64`. User running times were measured with the system routine `getrusage`.

All instances were solved with identical parameter settings. The restricted candidate list parameter α was selected at random using the uniform distribution in the interval $[0, 1]$. Each GRASP iteration used the same parameter during the entire iteration. The acceleration scheme described in Subsubsection 2.1.4 was activated with the parameter `k_best` $= 100$.

Table 2 summarizes the computational results. For each instance, the table lists the number of nodes, arcs, and terminals in the graph, the optimal or best known solution for that instance, the weight $W(T^*)$ of the best solution T^* found, the iteration in which the best solution was found, and the average running time per iteration in CPU seconds. We make the following observations regarding the computational results:

- In 47 out of 60 test problems, the GRASP produced an optimal solution. This broke down to 20 out of 20 for series C, 16 out of 20 for series D, and 11 out of 20 for series E.

TABLE 2. Statistics for OR-Library problems of series C, D, and E

Problem	Nodes	Arcs	Terminals	Opt	$W(T^*)$	itr. T^* found	avg. secs/itr
C.01	500	625	5	85	85	1	0.35
C.02	500	625	10	144	144	1	0.35
C.03	500	625	83	574	574	19	4.19
C.04	500	625	125	1079	1079	2	4.37
C.05	500	625	250	1579	1579	1	6.33
C.06	500	1000	5	55	55	1	0.29
C.07	500	1000	10	102	102	1	0.38
C.08	500	1000	83	509	509	1	3.58
C.09	500	1000	125	707	707	6	6.46
C.10	500	1000	250	1093	1093	1	12.04
C.11	500	2500	5	32	32	1	0.29
C.12	500	2500	10	46	46	1	0.46
C.13	500	2500	83	258	258	60	4.92
C.14	500	2500	125	323	323	60	8.82
C.15	500	2500	250	556	556	1	18.47
C.16	500	12500	5	11	11	1	0.27
C.17	500	12500	10	18	18	1	0.34
C.18	500	12500	83	113	113	772	6.43
C.19	500	12500	125	146	146	386	14.76
C.20	500	12500	250	267	267	1	17.50
D.01	1000	1250	5	106	106	1	0.99
D.02	1000	1250	10	220	220	1	1.47
D.03	1000	1250	167	1565	1565	70	22.10
D.04	1000	1250	250	1935	1935	1	26.37
D.05	1000	1250	500	3250	3250	2	49.12
D.06	1000	2000	5	67	68	1	1.04
D.07	1000	2000	10	103	103	1	1.27
D.08	1000	2000	167	1072	1072	263	30.28
D.09	1000	2000	250	1448	1448	33	51.69
D.10	1000	2000	500	2110	2110	2	129.45
D.11	1000	5000	5	29	29	1	0.87
D.12	1000	5000	10	42	42	1	1.00
D.13	1000	5000	167	500	500	1	35.77
D.14	1000	5000	250	667	669	18	57.31
D.15	1000	5000	500	1116	1116	1	23.85
D.16	1000	25000	5	13	13	1	1.37
D.17	1000	25000	10	23	23	1	1.56
D.18	1000	25000	167	223	228	20	58.79
D.19	1000	25000	250	310	315	1	122.48
D.20	1000	25000	500	537	537	1	209.53
E.01	2500	3125	5	111	111	1	5.03
E.02	2500	3125	10	214	214	3	6.66
E.03	2500	3125	417	4013	4014	46	353.35
E.04	2500	3125	625	5101	5102	34	689.68
E.05	2500	3125	1250	8128	8128	1	2113.88
E.06	2500	5000	5	73	73	1	4.85
E.07	2500	5000	10	145	145	1	7.65
E.08	2500	5000	417	2640	2645	2	525.25
E.09	2500	5000	625	3604	3605	20	1476.68
E.10	2500	5000	1250	5600	5600	1	5582.31
E.11	2500	12500	5	34	34	1	5.00
E.12	2500	12500	10	67	67	1	6.99
E.13	2500	12500	417	1280	1291	1	813.23
E.14	2500	12500	625	1732	1735	7	1526.08
E.15	2500	12500	1250	2784	2788	14	8915.61
E.16	2500	62500	5	15	15	1	9.16
E.17	2500	62500	10	25	25	1	9.54
E.18	2500	62500	417	564	573	68	1326.42
E.19	2500	62500	625	758	768	18	3228.76
E.20	2500	62500	1250	1342	1342	1	12565.35

- Of the 13 instances for which sub-optimal solutions were produced, in four instances GRASP was off by 1 unit, in six instances GRASP was off by 2 to 5 units, and in three it was off by 6 to 11 units. The largest percentage relative error (difference between solution found and optimal solution divided by the optimal) was 2.24% for D.18. On only five instances was the percentage relative error greater than 1%.
- In 37 out of the 60 instances, the GRASP produced the best solution in the first iteration. In only three of those 37 instances did GRASP not find an optimal solution.
- We compare the results obtained by GRASP in classes D and E with those found with the key-path based local search approach in Verhoeven, Severens, and Aarts [**43**]. In all cases, the GRASP solutions were as good or better than the key-path based local search solutions. In 26 out of the 40 instances, the GRASP solution is better than the key-path based solution. In the few cases in which the local search solution matches the solution produced by GRASP, the number of terminal nodes is very small (five or ten). However, we should point out that the average running time per iteration of the key-path based local search approach is substantially smaller than that of the average GRASP iteration.

4. Concluding remarks

We described greedy randomized adaptive search procedures for the Steiner problem in graphs. One implementation of our procedures was tested on 60 standard test problems and was shown to find good quality solutions. It found optimal solutions for 47 instances and was never off by more than 2.24% of the optimal. Running times increase with the number of nodes, arcs, and terminals.

We compare the results obtained by GRASP in classes D and E with those found with the key-path based local search approach in Verhoeven, Severens, and Aarts [**43**]. In all cases, the GRASP solutions were as good as or better than the key-path based local search solutions. We noted that the average running time per iteration of the key-path based local search approach is substantially smaller than that of the average GRASP iteration. This is an indication that the GRASP based on the key-path local search approach could be more efficient than the current version tested using a node-based local search. We have already implemented the key-path based local search procedure and are in the process of integrating it to the GRASP code. Computational results on the new version will be reported soon.

References

[1] Y.P. Aneja. An integer linear programming approach to the Steiner problem in graphs. *Networks*, 10:167–178, 1980.

[2] A. Balakrishnan and N.R. Patel. Problem reduction methods and a tree generation algorithm for the Steiner network problem. *Networks*, 17:65–85, 1987.

[3] J.E. Beasley. An algorithm for the Steiner problem in graphs. *Networks*, 14:147–159, 1984.

[4] J.E. Beasley. An SST-based algorithm for the Steiner problem on graphs. *Networks*, 19:1–16, 1989.

[5] J.E. Beasley. OR-Library: Distributing test problems by electronic mail. *J. of Oper. Res. Soc.*, 41:1069–1072, 1990.

[6] K. Bharath-Kumar and J.M. Jaffe. Routing to multiple destinations in computer networks. *IEEE Trans. Commun.*, COM-31:343-351, 1983.

[7] S. Chopra, E.R. Gorres, and M.R. Rao. Solving the Steiner tree problem on graphs using branch and cut. *ORSA Journal on Computing*, 4:320–335, 1992.

[8] E.-A. Choukhmane. Une heuristique pour le problème de l'arbre de Steiner. *RAIRO Recherche Opérationnelle*, 12:207–212, 1978.

[9] D. Cieslik. *Steiner minimal trees*, Kluwer Academic Publishers, Dordrecht, 1998.
[10] S.E. Dreyfus and R.A. Wagner. The Steiner problem in graphs. *Networks*, 1:195–207, 1972.
[11] C.W. Duin. Steiner's problem in graphs. *Ph.D. Thesis*, University of Amsterdam, 1993.
[12] C.W. Duin. Reducing the graphical Steiner problem with a sensitivity test. In *Network Design: Connectivity and Facilities Location*, P.M. Pardalos and D.-Z. Du (Eds.), *DIMACS Series on Discrete Mathematics and Theoretical Computer Science*, 40:79–107, 1998.
[13] C.W. Duin and A. Volgenant. Some generalizations of the Steiner problem in graphs. *Networks*, 17:353–364, 1987.
[14] C.W. Duin and A. Volgenant. Reduction tests for the Steiner problem in graphs. *Networks*, 19:549–567, 1989.
[15] T.A. Feo and M.G.C. Resende. A probabilistic heuristic for a computationally difficult set covering problem. *Operations Research Letters*, 8:67–71, 1989.
[16] T.A. Feo and M.G.C. Resende. Greedy randomized adaptive search procedures. *Journal of Global Optimization*, 6:109–133, 1995.
[17] L.R. Foulds and P.B. Gibbons. A branch and bound approach to the Steiner problem in graphs. *Proceedings of the 14th Annual Conference of the Operational Research Society of New Zealand*, 1:61–70, 1978.
[18] L.R. Foulds and J.V. Rayward-Smith. Steiner problems in graphs: Algorithms and applications. *Eng. Optimization*, 7:7–16, 1983.
[19] M.R. Garey and D.S. Johnson. The rectilinear Steiner tree problem is NP-complete. *SIAM J. Appl. Math.*, 32:826–834, 1977.
[20] M.R. Garey and D.S. Johnson. *Computers and Intractability*, W. H. Freeman, San Francisco, 1979.
[21] S.L. Hakimi. Steiner's problem in graphs and its implications. *Networks*, 1:113–133, 1971.
[22] F.K. Hwang and D.S. Richards. Steiner tree Problems. *Networks*, 2:55–89, 1992.
[23] F.K. Hwang, D.S. Richards, and P. Winter. *The Steiner Tree Problem,* Elsevier, Amsterdam, 1992.
[24] A. Iwainsky, E. Canuto, O. Taraszow, and A. Villa. Network decomposition for the optimization of connection structures. *Networks*, 16:205–235, 1986.
[25] R.M. Karp. Reducibility among combinatorial problems. In *Complexity of Computer Communications*, R.E. Miller and J.W. Thatcher (Eds.), Plenum Press, New York, 85–103, 1972.
[26] B.N. Khoury, P.M. Pardalos, and D.W. Hearn. Equivalent formulations for the Steiner problem in graphs. In *Network Optimization Problems,* P.M. Pardalos and D.-Z. Du (Eds.), World Scientific, 111–124, 1993.
[27] B.N. Khoury and P.M. Pardalos. A heuristic for the Steiner problem on graphs. *Comp. Opt. & Appl.*, 6:5–14, 1996.
[28] L. Kou, G. Markowsky, and L. Berman. A fast algorithm for Steiner trees. *Acta Info.*, 15:141–145, 1981.
[29] J.B. Kruskal. On the shortest apanning subtree of a graph and the traveling salesman problem. *Proceedings of the American Mathematical Society*, 7:48–50, 1956.
[30] A. Levin. Algorithms for the shortest connection of a group of graph vertices. *Soviet Math. Doklady*, 12:1477–1481, 1971.
[31] A. Lucena. Steiner problem in graphs: Lagrangean relaxation and cutting planes. *COAL Bulletin*, 21:2–7, 1992.
[32] N. Maculan. The Steiner problem in graphs. *Ann. Discrete Math.*, 31:185–222, 1987.
[33] T.L. Magnanti and R.T. Wong. Network design and transportation planning: Models and algorithms. *Trans. Science*, 18:1–55, 1984.
[34] K. Mehlhorn. A faster approximation for the Steiner problem in graphs. *Information Processing Letters*, 27:125–128, 1988.
[35] M. Minoux. Efficient greedy heuristics for Steiner tree problems using reoptimization and supermodularity. *INFOR*, 28:221–233, 1990.
[36] J. Plesník. A bound for the Steiner tree problem in graphs. *Math. Slovoca*, 31:155-163, 1981.
[37] J.S. Provan. A polynomial algorithm for the Steiner tree problem on terminal-planar graphs. Technical Report UNC/ORSA/Tech. Rep.-83/10, Dep. Operations Research, University of North Carolina, Chapel Hill, 1983.
[38] J.S. Provan and M.O. Ball. Computing network reliability. *Operations Res.*, 32:516–526, 1984.
[39] V.J. Rayward-Smith. The computation of nearly minimal Steiner trees in graphs. *Int. J. Math. Ed. Sci. Technol.*, 14:15–23, 1983.
[40] V.J. Rayward-Smith and A. Clare. On finding Steiner vertices. *Networks*, 16:283–294, 1986.
[41] M.L. Shore, L.R. Foulds, and P.B. Gibbons. An algorithm for the Steiner problem in graphs. *Networks*, 12:323–333, 1982.
[42] H. Takahashi and A. Matsuyama. An approximate solution for the Steiner problem in graphs. *Math. Jpn.*, 24:573–577, 1980.
[43] M.G.A. Verhoeven, M.E.M. Severens, and E.H.L. Aarts. Local search for Steiner trees in graphs. In *Modern Heuristics Search Methods*, V. J. Rayward-Smith et al. (Eds.), John Wiley, 117–129, 1996.

[44] S. Voss. Steiner's problem in directed graphs: Logical tests in a branch and bound algorithm. *NATO Advanced Research Workshop on Topological Network Design*, Copenhagen, 1989.
[45] S. Voss. A reduction-based algorithm for the Steiner problem in graphs. *Methods of Operations Res.* 58:239–252, 1989.
[46] K. White, M. Farber, and W. Pulleyblank. Steiner trees, connected domination and strongly chordal graphs. *Networks*, 15:109–124, 1985.
[47] P. Winter. Steiner problem in networks: A survey. *Networks*, 17:129–167, 1987.
[48] R.T. Wong. A dual ascent approach to Steiner tree problems on a directed graph. *Mathematical Programming*, 28:271–287, 1984.
[49] Y.F. Wu, P. Widmayer, and C. K. Wong. A faster approximation algorithm for the Steiner problem in graphs. *Acta Info.*, 23:223–229, 1986.
[50] Y.Y. Yang and O. Wing. An algorithm for the wiring problem. *Digest IEEE Int. Symp. Electrical Networks*, 14–15, 1971.

(S. L. Martins) Department of Computer Science, Catholic University of Rio de Janeiro, Rio de Janeiro, RJ 22453-900 Brazil
E-mail address, S. L. Martins: simone@inf.puc-rio.br

(P. M. Pardalos) Center for Applied Optimization, Department of Industrial and Systems Engineering, University of Florida, Gainesville, FL 32611 USA
E-mail address, P. M. Pardalos: pardalos@ufl.edu

(M. G. C. Resende) Information Sciences Research Center, AT&T Labs – Research, Florham Park, NJ 07932 USA
E-mail address, M. G. C. Resende: mgcr@research.att.com

(C. C. Ribeiro) Department of Computer Science, Catholic University of Rio de Janeiro, Rio de Janeiro, RJ 22453-900 Brazil
E-mail address, C. C. Ribeiro: celso@inf.puc-rio.br

DIMACS Series in Discrete Mathematics
and Theoretical Computer Science
Volume **43**, 1999

On the Mixing Time of the Triangulation Walk and other Catalan Structures

Lisa McShine and Prasad Tetali

1. Introduction

Consider a graph on the set of triangulations of a convex polygon with n sides, wherein two triangulations are considered adjacent if (and only if) one can be obtained from the other by "flipping" an internal diagonal – note that each internal diagonal in a triangulation is in a unique quadrilateral, and by "flipping" we mean replacing one diagonal by the other diagonal of the same quadrilateral. A random walk can be easily defined on this graph with the property that, eventually, the probability of being at any given triangulation is independent of the choice of triangulation. This gives a Monte Carlo method of generating a triangulation of a convex n-gon *uniformly at random*. However, the efficiency of such a scheme depends crucially on the rate of approach to stationarity of the random walk. Although there are other ways of obtaining randomly such a triangulation in an efficient way, the analysis of this particular scheme has remained open.

The main objective in this paper is to show that $O(n^5 \log(n/\epsilon))$ steps are sufficient to get close (i.e. within ϵ in variation distance) to the stationary distribution, which happens to be uniform over the set of triangulations. Independently, Molloy et al. [**11**] have recently shown that at least $\Omega(n^{3/2})$ steps are necessary and also that $O(n^{23} \log(n/\epsilon))$ steps are sufficient. While our upper bound is much more reasonable and our proof much simpler, we believe the truth to be closer to their lower bound.

The idea in [**11**] in getting an upper bound on the mixing time was to bound the so-called *conductance* of the Markov chain, a technique first introduced by Jerrum and Sinclair. Here we use the comparison technique due to Diaconis and Saloff-Coste [**3**]. Recently this technique has also been used to analyze the mixing time of certain Markov chains on tilings in [**12**]. Before describing our idea further, we need to define some other *Catalan*

1991 *Mathematics Subject Classification.* 68R05, 68Q25, 60J10, 05A05.

structures – a term we use to denote any of the possible combinatorial structures whose counting sequence is the sequence of Catalan numbers, given by $c_n = \binom{2n}{n}/(n+1)$, for $n \geq 1$, and $c_0 = 1$. Some other examples of Catalan structures include the set of binary trees on n internal nodes and the set of *Dyck paths* (also called *mountain-valley* diagrams) of length $2n$. (See Fig. 1 for illustrations and Section 2 for precise definitions.) Each such Catalan structure offers an *interchange graph* in a natural way, with an appropriate definition of a local move or a local interchange similar to the move described above for the graph on triangulations of convex n-gon. (See Fig. 2 for illustration.)

Our strategy can now be summarized as follows. We use the fact that the interchange graph on triangulations of a convex n-gon is isomorphic to the interchange graph on binary trees with $n-2$ internal nodes; this fact was established in [**14**]. We then compare, *à la* Diaconis–Saloff-Coste, the chain on binary trees with the chain on Dyck paths. That is, we use the upper bound on the mixing time of the chain on Dyck paths as established by [**16**] (see also [**10**]) to get an upper bound on the mixing time of the chain on binary trees (equivalently, the chain on triangulations).

Dobrow and Fill ([**4**], [**5**]) analyzed a Markov chain on the state space of all binary trees on n nodes, wherein the transitions between states were defined in terms of a certain *move-to-front* rule. However, the Markov chain described in [**4**] does not yield the uniform distribution as the stationary distribution. Although in principle one can generate a binary tree uniformly at random using a more direct approach (as done in [**8**], [**1**],...), it is conceivable that there exists a faster method via the generation of an equivalent Catalan structure. Note that each Catalan structure offers its own Markov chain – the interchange graphs corresponding to different Catalan structures need not be (and typically are not) isomorphic; hence the corresponding random walks on the interchange graphs can behave in significantly different ways. Finally, uniform random generation of triangulations of *non-convex* polygons is apparently of sufficient interest to the community of computer graphics and computational geometry, because so far for these problems, there are no known rigorous ways of efficient random generation. It remains to be seen whether Monte Carlo Markov chain techniques will help in the non-convex case, and in the more general case of triangulations using n points (in general position) in the plane. A particularly appealing feature of the Markov chain approach to the problem of random generation is that it uses much less space, and typically much smaller (random/pseudo-random) numbers in the simulation.

In a different spirit, the nature of the present study is akin to the research which arose out of the analysis of Markov chains based on various card-shuffling schemes. Several researchers (most notably P. Diaconis) have succeeded in obtaining sharp estimates on the rates of mixing of chains based on such schemes. In fact, the comparison technique (see [**2**], [**3**]) was a result of such investigations. There does not seem to be analogous work on

Markov chains on Catalan structures, and we view the present contribution as the beginning of a systematic study.

The following presentation is organized as follows. In Section 2 we describe the background material on Catalan structures and Markov chains relevant to our work. In Section 3 we present the proof of the bound on the mixing time of the Markov chain(s) in question. In the final section, we conclude with some remarks on further work in progress.

2. Preliminaries

In this section we first give a brief description of some Catalan structures, and associated interchange graphs on them. We then describe relevant results from the literature on rapidly mixing Markov chains which we need in the next section to analyze the chain on binary trees on n internal nodes.

2.1. Catalan structures. The nth Catalan number, c_n, satisfies the following recurrence relation (see [**7**]).

$$c_n = c_0 c_{n-1} + c_1 c_{n-2} + \cdots + c_{n-1} c_0,$$

where $c_0 = 1$, and $c_1 = 1$. Also recall that for $n \geq 1$, we have $c_n = \binom{2n}{n}/(n+1)$. There are several interesting survey papers (see e.g. [**6**], [**7**]) on Catalan sequences which describe various combinatorial interpretations of these sequences. Perhaps the best source is a list of 60 or so interpretations in [**15**]. We shall henceforth use the term *Catalan structures* to mean any of the possible combinatorial structures whose counting sequence satisfies the recurrence relation describing the Catalan sequence. Some examples include the following (see Fig. 1 for $n = 3$).

Consider a convex polygon K, with $n+2$ vertices, labeled $1, 2, \ldots, n+2$, clockwise. A triangulation of K is a dissection of K into n triangles, using nonintersecting diagonals of K. The number of such triangulations is c_n.

For the purpose of this article, a binary tree of size n is a rooted tree with n *internal nodes* (those with two descendants) and $n+1$ *external nodes* or *leaves* (those with no descendants). It is well known that there are c_n such binary trees with n internal nodes.

A Dyck path from $(0,0)$ to $(2n,0)$ is a lattice path with steps $(1,1)$ and $(1,-1)$ never falling below the x-axis.

Finally, label $2n$ equally spaced points around the circumference of a circle; join the points in pairs by n nonintersecting chords. The number of such Dyck paths and such chord diagrams is also c_n. We recommend that the reader refers to a lovely exposition of some of the Catalan structures by Martin Gardner [**6**], who also describes interesting bijections between these structures.

Each Catalan structure offers its own *interchange graph* in a natural way. The Markov chains which we will study are all random walks on these interchange graphs. The principle behind the definition of each interchange graph is the same:

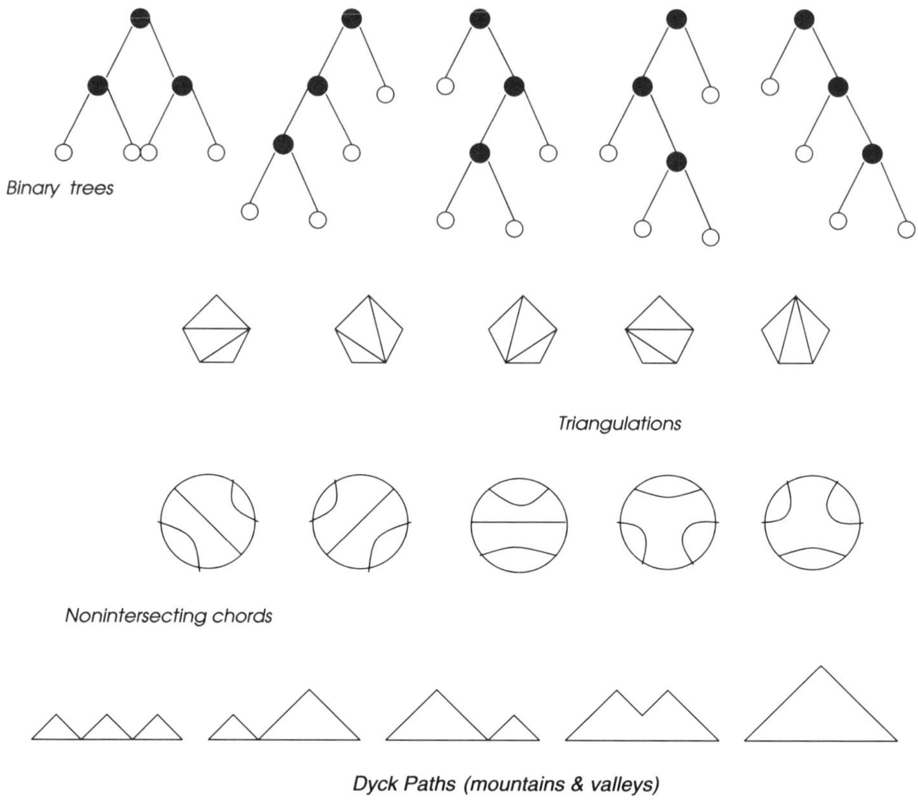

FIGURE 1. Examples of Catalan structures for $c_3 = 5$

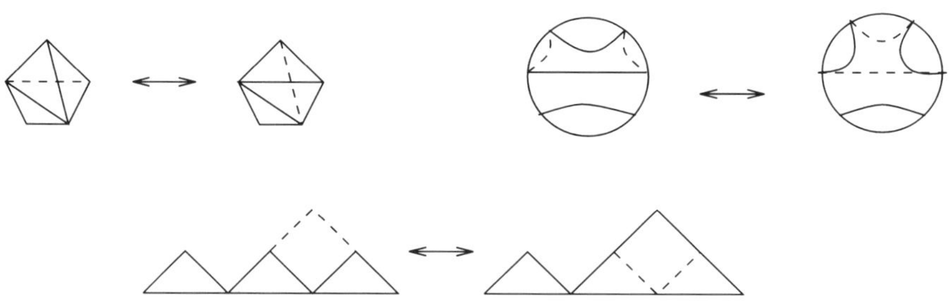

FIGURE 2. Local moves defining interchange graphs

> The vertices of the graph are labeled with the elements of a particular Catalan structure (of size n), and two vertices are adjacent in the graph if a natural (local) operation transforms the element corresponding to one of the vertices into that of the other.

Examples of such local operations/moves are illustrated in Figs. 2 and 4. Suppose the structure is the set of triangulations of a convex $(n+2)$-gon. The set of triangulations forms the vertex set of the interchange graph, and two triangulations are adjacent if one can be obtained from the other by a diagonal flip, as described in [**14**]. Every diagonal in a triangulation of a convex polygon defines a quadrilateral. A diagonal flip replaces that diagonal with the other diagonal of the same quadrilateral. Sleator et al. [**14**] obtained, *inter alia*, tight upper and lower bounds (of $2n-6$) on the diameter of this interchange graph and other results on triangulations of the sphere (see [**9**] for a simpler proof).

Two binary trees with n internal nodes are adjacent if one can be transformed into the other by applying the rotation operation. A rotation at a node is defined as shown in Fig. 4. Sleator et al. also showed that this graph is isomorphic to the previous one on triangulations of a convex $(n+2)$-gon.

Similarly, in the collection of Dyck paths of length $2n$, two elements are adjacent if one may be changed into the other by flipping a peak into a valley (that is, changing $(1,1),(1,-1)$ to $(1,-1),(1,1)$) or a valley into a peak (that is, changing $(1,-1),(1,1)$ to $(1,1),(1,-1)$). It is easy to see that the diameter of this graph is precisely $n(n-1)/2$.

For the set of nonintersecting chords in a circle, a valid way to define interchanges is to pick a pair of chords and to replace them with two new chords obtained by matching the endpoints of the original chords, if and only if such an exchange results in a valid nonintersecting chord diagram. It is easy to show that this yields a connected undirected graph, and in fact (although less obvious) a straightforward proof by induction on n can be given to show that the diameter of such an interchange graph is $n-1$, where n is the number of chords. (Also see the remarks in Conclusions.)

2.2. Comparison of mixing times of Markov chains. Let (Ω, P, π) denote an ergodic (that is, irreducible and aperiodic) Markov chain, with finite state space Ω, transition probability matrix P, and stationary distribution π. We assume that P is reversible, that is, for all $x, y \in \Omega$,

$$\pi(x)P(x,y) = \pi(y)P(y,x)$$

We will also assume that we are considering discrete-time Markov chains. Then, for $x, y \in \Omega$, $t \in \mathbb{Z}^+$, $P^t(x,y)$ denotes the t-step probability of going from x to y. The time the Markov chain takes to be close to the stationary distribution, starting from state x, can be measured by the *variation distance*,

$$\Delta_x(t) = \frac{1}{2}\sum_{y\in\Omega} |P^t(x,y) - \pi(y)|.$$

The variation distance from the worst state is denoted by

$$\Delta(t) = \max_{x\in\Omega} \Delta_x(t).$$

For $\epsilon > 0$ (usually $0 < \epsilon < 1$), the *mixing time* from state x is defined by

$$\tau_x(\epsilon) = \min\{t : \Delta_x(t') \leq \epsilon, \forall t' \geq t\}.$$

The mixing time of the Markov chain is the mixing time from the worst state,

$$\tau(\epsilon) = \max_{x \in \Omega} \tau_x(\epsilon).$$

In the following, the mixing time of a Markov chain will always refer to the mixing time from a worst state.

Let $1 = \lambda_0 > \lambda_1 \geq \lambda_2 \geq \cdots \geq \lambda_{|\Omega|-1} > -1$ denote the eigenvalues of P. The following result (see [**13**]) shows the relationship between mixing time and maximum eigenvalue. Strictly speaking λ_1 should be replaced by $\lambda_{\max} = \max(\lambda_1, \lambda_{|\Omega|-1})$, but in the Markov chains we will be describing we ensure that $\lambda_1 > \lambda_{|\Omega|-1} > 0$ by adding self-loops of probability at least $1/2$.

THEOREM 1 (Sinclair). *For* $\epsilon > 0$,

(i)

$$\forall x \in \Omega, \tau_x(\epsilon) \leq \frac{1}{1-\lambda_1} \log\left(\frac{1}{\pi(x)\epsilon}\right),$$

(ii)

$$\tau(\epsilon) = \max_{x \in \Omega} \tau_x(\epsilon) \geq \frac{\lambda_1}{2(1-\lambda_1)}.$$

Let P, $\tilde{P}$ denote two reversible Markov chains on the same state space Ω, with the same stationary distribution π. Diaconis and Saloff-Coste [**3**] provide the following geometric bound between $\lambda_1(P)$, $\lambda_1(\tilde{P})$. In fact the result compares Dirichlet forms associated with P and $\tilde{P}$, so the result applies to all non-trivial eigenvalues, not just λ_1, and also to the log-Sobolev constants of P and $\tilde{P}$. Also, as stated in [**3**], the two chains need not share the same stationary distribution, it suffices if they have comparable distributions.

Let $\tilde{P}$ denote the Markov chain with known eigenvalues (or known mixing time), and let P denote the chain whose mixing time we would like to bound, by comparison with $\tilde{P}$.

Let $E(P) = \{(x,y) : P(x,y) > 0\}$ and $E(\tilde{P}) = \{(x,y) : \tilde{P}(x,y) > 0\}$ denote the sets of edges of the two chains, viewed as directed graphs.

For each (x,y) with $\tilde{P}(x,y) > 0$, define a path γ_{xy} using a fixed sequence of states, $x = x_0, x_1, \ldots, x_{k-1}, x_k = y$, with $P(x_i, x_{i+1}) > 0$. The length of such a path is denoted by $|\gamma_{xy}|$ and $|\gamma_{xy}| = k$.

Further let $\Gamma(z,w) = \{(x,y) \in E(\tilde{P}) : (z,w) \in \gamma_{xy}\}$ denote the set of paths (in P) which use the edge (z,w).

PROPOSITION 2 (Diaconis–Saloff-Coste). *With the above notation we have*

$$1 - \lambda_1(P) \geq \frac{1}{A(\Gamma)}(1 - \lambda_1(\tilde{P})),$$

where

$$A(\Gamma) = \max_{(z,w)\in E(P)} \left\{ \frac{1}{\pi(z)P(z,w)} \sum_{\Gamma(z,w)} |\gamma_{xy}|\pi(x)\tilde{P}(x,y) \right\}.$$

Crucially, note that $A(\Gamma)$ depends on our choice of paths $\Gamma = \{\gamma_{xy}\}$, and that we need define these paths only between pairs of states which are *adjacent in the known chain.*

The strategy used in comparing mixing times based on the above proposition is described in [**12**]. First begin with a bound on the known mixing time of a chain. Part (ii) of Theorem 1 allows us to lower bound the spectral gap of such a chain. Then Proposition 2 is used to lower bound the spectral gap of an unknown chain by bounding the parameter $A(\Gamma)$, for a carefully chosen Γ. This then gives a bound on the mixing time of the unknown chain via part (i) of Theorem 1. The technique is summarized in the following proposition.

PROPOSITION 3 (Randall–Tetali). *Let* $\tilde{\tau}(\epsilon)$, $\tau(\epsilon)$ *denote the mixing times of* $\tilde{P}$ *and* P *respectively, and let* $\pi_* = \min_{x\in\Omega}\pi(x)$. *Then with* $A(\Gamma)$ *defined as in Proposition 2: for* $0 < \epsilon < 1$,

$$\tau(\epsilon) \le \frac{4\log\left(\frac{1}{\epsilon\pi_*}\right)}{\log\left(\frac{1}{2\epsilon}\right)} A(\Gamma)\tilde{\tau}(\epsilon).$$

3. An upper bound on the mixing time of binary trees

Sleator et al. [**14**] gave a bijection between the set of triangulations of a convex $(n+2)$-gon and binary trees with n internal nodes so that if two triangulations differ by a diagonal flip, the corresponding binary trees differ by a rotation operation. This correspondence shows that the two interchange graphs are isomorphic. In turn it allows us to study the random walk on the interchange graph on binary trees on n internal nodes, which we will denote (following [**14**]) $RG(n)$ and from this deduce the result for triangulations.

The transition probabilities of the Markov chain on $RG(n)$ are defined as follows. For two distinct binary trees z and w,

$$\begin{aligned} P(z,w) &= 1/[2(n-1)], \text{ if } (z,w)\in E(RG(n)) \\ P(z,z) &= 1/2. \end{aligned}$$

The transition probabilities of the Markov chain on $DG(n)$, the interchange graph for Dyck paths of length $2n$, are defined as follows. For two distinct Dyck paths x and y,

$$\begin{aligned} \tilde{P}(x,y) &= 1/[2(2n-3)] \text{ if } (x,y)\in E(DG(n)) \\ \tilde{P}(x,x) &= 1-\textstyle\sum_{y\sim x} P(x,y) \ge 1/2 \end{aligned}$$

It is easy to verify that the two Markov chains as defined above do indeed satisfy the reversibility (i.e. detailed-balance) condition and also that they share the uniform distribution as the stationary distribution π. So for $x \in \tilde{\Omega}$ and $z \in \Omega$, we have $\tilde{\pi}(x) = \pi(z) = 1/c_n$.

Let $\tau(\epsilon)$ and $\tilde{\tau}(\epsilon)$ denote the mixing times of the Markov chains on $RG(n)$ and $DG(n)$ respectively. First we may deduce from [**16**] that the Markov chain on $DG(n)$ has mixing time

$$\tilde{\tau}(\epsilon) = O(n^3(\log n + \log 1/\epsilon)) \ , \ 0 < \epsilon < 1$$

Let our known chain, $\tilde{P}$, is the chain with the set of Dyck paths of length $2n$ as the state space, denoted $\tilde{\Omega}$; our unknown chain, P, is the chain with the set of binary trees with n internal nodes as the state space, denoted Ω. We would like to use Proposition 3 to get a bound on the mixing time $\tau(\epsilon)$. First note that Proposition 3 (and 2) require the state spaces $\tilde{\Omega}$ and Ω of $\tilde{P}$ and P, respectively, to be the same. In the present case, although they are not the same, the propositions are still applicable in view of the fact that we are able to define a bijection $f : \tilde{\Omega} \to \Omega$ between them.

• **Bijection via binary strings.** The bijection is easiest to describe through a bijection from each to another Catalan structure – the set of binary strings of length $2n$ with equal number of 1's and 0's, wherein the number of 1's in each string is always greater than or equal to the number of 0's as we count from left to right in the string. The bijection is illustrated in Fig. 3.

Given a Dyck path of length $2n$, a segment of slope $+1$ corresponds to a 1 and a segment of slope -1 corresponds to a 0. Given a binary string on n internal nodes, label the *left edges* (edges leading to left descendants) with a 1 and the *right edges* with a 0. Now the corresponding binary string is the one obtained by reading the labels as the tree is traversed, recursively, starting from the root, first visiting the left subtree and then the right subtree. Note that this provides a binary string with the above mentioned property, since every node (in the binary tree) which has a right descendant also has a left descendant. It is also easy to see that to each binary string with the above property, there is a unique Dyck path and a unique binary tree which corresponds to it.

The canonical path γ_{xy} of the comparison technique, is now a path in P, which can be described as a sequence of states, $f(x) = z_0, z_1, \dots, z_{k-1}, z_k = f(y)$, for $(x, y) \in E(DG(n))$. The description of the paths will be simplified by using the above binary string representation of Dyck paths from now on.

Following [**14**], we will state some definitions. A *subtree* of a binary tree is either a single node or a binary tree with at least one internal node. Subtrees will be denoted by T_i, and this will stand both for the subtree and the binary string representation of T_i. The *depth* of a node in a binary tree is the length of the shortest path from the root to that node. It is convenient to view these binary trees as binary search trees with labels on the nodes, namely, with the property that the label of a node is bigger than the labels of all the nodes in its left subtree, and smaller than the labels of

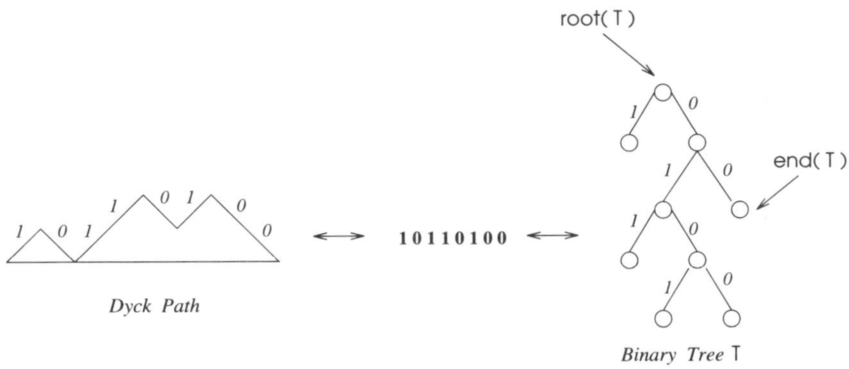

FIGURE 3. Bijection via binary strings

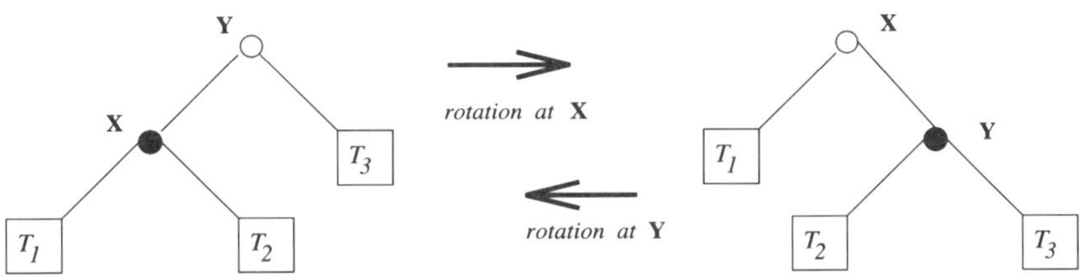

FIGURE 4. Local moves on binary trees

the nodes in its right subtree. This gives a natural ordering on the nodes of a tree. The rotation operation mentioned earlier preserves this node ordering, not surprisingly, since the rotations were invented as a way of restructuring binary search trees. Specifically, the rotation operation is defined as shown in Fig. 4.

Rotations do not change the number of internal nodes, and for a tree with n internal nodes, there are $n-1$ rotations possible at any time, one for each internal node, except for the root. The rotation at a node brings the node one step closer to the root, thus decreasing the depth of that node by exactly one. Although a rotation changes the local structure of the tree, it leaves the rest of the tree intact.

Referring to Fig. 4 above, a rotation at X decreases the depth of nodes in subtree T_1 by 1, increases the depth of nodes in subtree T_3 by 1, and leaves the depth of all other nodes the same. In the same way, a rotation at Y decreases the depth of nodes in subtree T_3 by 1, increases the depth of nodes in subtree T_1 by 1, and leaves the depth of all other nodes the same. We call each rotation either an X-rotation or a Y-rotation depending on whether it is of the form of a rotation at X in Fig. 4 above or of the form of a rotation at Y respectively.

• **Canonical paths** Γ. We need to define $\Gamma = \{\gamma_{xy}\}$, for $(x, y) \in E(\tilde{P})$. This can be done in a natural way, once we analyze a transition from x to

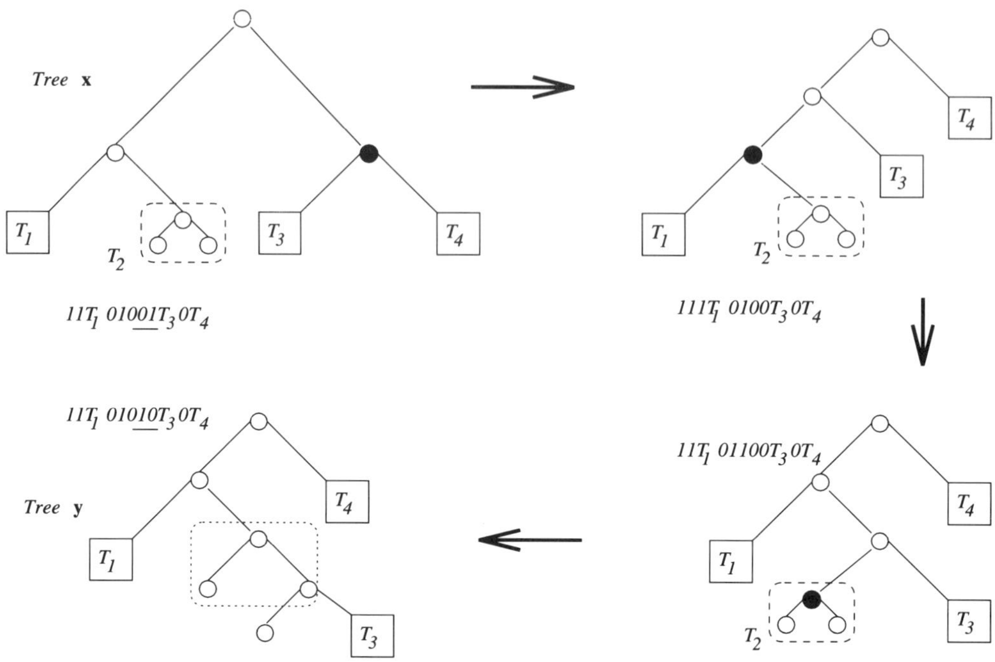

FIGURE 5. Canonical path from $\mathbf{x}$ to $\mathbf{y}$, differing in a 01–10 flip

y in $DG(n)$, according to whether it is a $01 \to 10$ flip or a $10 \to 01$ flip, and interpret the flip in terms of the corresponding binary trees, $f(x)$ and $f(y)$.

The easy case is if $f(x)$ and $f(y)$ differ by a single rotation – if $f(x) \sim f(y)$ in P, then γ_{xy} is simply the edge $(f(x), f(y)) \in E(P)$. But in general, $(x,y) \in E(\tilde{P})$ does not imply $(f(x), f(y)) \in E(P)$. See for example, Fig. 5, our final illustration which characterizes the differences in two binary trees, which have adjacent representations as Dyck paths. In such a case, we will define a unique sequence of rotations which transforms $f(x)$ into $f(y)$, and the corresponding sequence of edges in P which forms the corresponding canonical path γ_{xy}. We shall do this for the case when $x \to y$ is a $01 \to 10$ flip, and in the other case the path is just the reverse of the path in this case; we are justified in doing this since the interchange graphs can be viewed as undirected graphs.

To simplify the discussion, we will now introduce some new terms which are illustrated in Fig. 3. The *root* of a subtree T_i, $root(T_i)$, is the top vertex of that subtree. The *end* of a subtree T_i, $end(T_i)$, is the rightmost vertex of T_i, or equivalently, the node with the largest label when the tree T_i is viewed as a binary search tree. (A sure way to find the $end(T)$, starting from the $root(T)$ for any tree T, is to keep taking the right branch for as long as possible!)

The following observation is key to understanding Γ and to bounding $A(\Gamma)$. A $01 \to 10$ flip moves a particular left subtree (T_3 in Fig. 5) hanging from the right child of some node N to being the rightmost subtree of the left child of N. This is easy to verify by considering the binary strings corresponding to x and y, which differ in a $01 \to 10$ flip, and then by constructing $f(x)$ and $f(y)$. We call such a subtree the *characteristic subtree* of that particular $01 \to 10$ flip. In Fig. 5, T_3 is the characteristic subtree of the flip which takes x to y. Note that such a subtree can also be a single node. (For the $10 \to 01$ flip, just the opposite happens, defining a characteristic subtree in a similar way.) Thus an edge $(x, y) \in E(\tilde{P})$ is characterized by a unique subtree of the binary trees represented by $f(x)$ and $f(y)$.

The canonical path γ_{xy} is the unique sequence of rotations which transforms the binary tree $f(x)$ into $f(y)$ – the first rotation is performed at the parent of the root of the characteristic subtree, and then every subsequent rotation is at the sibling of the root of the characteristic subtree. (In Fig. 5 the black nodes denote the nodes at which rotations are performed; also for convenience, the trees are called **x** and **y**, rather than $f(x)$ and $f(y)$.)

• **Bounding** $A(\Gamma)$. We will show that $A(\Gamma)$ is at most $O(n)$, by arguing that the length of a canonical path is at most n, and that each rotation in a binary tree is used by at most one canonical path γ_{xy}, where $x \to y$ is a $01 \to 10$ flip, and by at most one canonical path corresponding to a $10 \to 01$ flip.

First notice that in a path corresponding to a $01 \to 10$ flip, the depth of the root of the characteristic subtree remains the same after the first rotation, but increases by exactly one with every subsequent rotation. At most $n - 2$ internal nodes can participate (by being siblings of the root of the characteristic subtree) in increasing the depth – the grandparent of the characteristic subtree and any nodes in the right subtree of the right child of the grandparent are the nodes where a rotation is not performed in such a canonical path. This shows that the length of a canonical path can be at most $n - 1$. (The argument for a $10 \to 01$ flip is analogous.)

Secondly, consider an arbitrary rotation $(z, w) \in E(P)$. Whether it is an X-rotation or a Y-rotation, there are always at most two choices for a subtree to play the role of a characteristic subtree. Referring to Fig. 4, if (z, w) is an X-rotation, then either T_3 is the characteristic subtree of a $01 \to 10$ flip or T_2 is the characteristic subtree of a $10 \to 01$ flip. The rotation (z, w) and the choice of either T_2 or T_3 as the characteristic subtree, uniquely identifies the pair (x, y) such that $(z, w) \in \gamma_{xy}$. Referring once again to Fig. 4, if (z, w) is an X-rotation, then T_3 will eventually end up as the right subtree of *end*(T_2), giving us y. Knowing y and the fact that T_3 is the characteristic subtree uniquely determines x. If on the other hand, T_2 were to be the characteristic subtree, then w is in fact y, and x can be uniquely determined given that $x \to y$ is now a $10 \to 01$ flip.

Thus, for a fixed $\Gamma(z, w)$, $|\Gamma(z, w)| \leq 2$, and when $(z, w) \in \gamma_{xy}$, $|\gamma_{xy}| \leq n$.

We have $\pi(x) = \pi(z) = (n+1)/\binom{2n}{n}$. Also, $P(z,w) = 1/(2n-2)$, for all $(z,w) \in E(P)$, and $\tilde{P}(x,y) = 1/(4n-6)$, for all $(x,y) \in E(\tilde{P})$. Thus,

$$\begin{aligned} A(\Gamma) &= \max_{(z,w)\in E(P)} \left\{ \frac{1}{\pi(z)P(z,w)} \sum_{\Gamma(z,w)} |\gamma_{xy}|\pi(x)\tilde{P}(x,y) \right\} \\ &\le \frac{2n-3}{n-1} 2n = O(n). \end{aligned}$$

Applying Proposition 3, with the known bound on the mixing time of $\tilde{P}$, we can now bound the mixing time of P, the Markov chain on binary trees:

$$\begin{aligned} \tau(\epsilon) &\le \frac{4\log\left(\frac{1}{\epsilon\pi_*}\right)}{\log\left(\frac{1}{2\epsilon}\right)} n(n^3(\log n + \log 1/\epsilon)) \\ &= \frac{4\left(\log(1/\epsilon) + \log(c_n)\right) n^4 (\log n + \log 1/\epsilon)}{\log(1/2) + \log(1/\epsilon)} \\ &= O\left(\frac{n^5 \log n}{\log(1/\epsilon)} + n^4 \log n + n^5 + n^5 \log n\right) \\ &= O(n^5((\log n + \log 1/\epsilon)), \end{aligned}$$

thus establishing the following theorem.

THEOREM 4. *The mixing time of the Markov chain on triangulations of a convex $(n+2)$-gon (equivalently, on binary trees with n internal nodes) satisfies, for $0 < \epsilon < 1$,*

$$\tau(\epsilon) = O\left(n^5 \log(n/\epsilon)\right).$$

4. Conclusions

It is to be noted that, while Wilson showed his (upper) bound is tight up to a multiplicative constant, we do not believe our bound to be tight for the chain on triangulations. To get a better estimate we believe that one either needs a direct argument (say, via coupling) or a better comparison – comparison with a faster Catalan structure. Our present candidate for such a potentially faster chain is the one on the set of nonintersecting chord diagrams with $2n$ equally spaced points on a circle. It can be shown, with a proof by induction on n, that the diameter of this graph is $n-1$, whereas the diameter of $DG(n)$ is $\Theta(n^2)$. It can also be shown, with a suitable bijection, that $DG(n)$ is a proper subgraph of this graph, for $n \ge 3$. This is part of the rationale for our belief that the graph on nonintersecting chord diagrams is a better comparison candidate. Work is in progress on this problem. Of course, by comparison with $DG(n)$, we can always get *some* bound on the mixing time of this candidate problem: for example, by choosing our paths in the obvious way — adjacency in $DG(n)$ gives adjacency in the chord interchange graph, making each path a path of *length one* — we get $A(\Gamma)$ is $O(1)$, and so, using Proposition 3, an upper bound for the mixing time of the Markov chain on the chord interchange graph is $O(n^4(\log n + \log 1/\epsilon))$. But surely, a much better bound is the truth!

We also intend to make a systematic study of several other Catalan structures, with the hope of obtaining tight bounds on the diameters, the eigenvalues, and the rates of mixing of Markov chains on the associated interchange graphs.

Acknowledgments. The second author is grateful to Bob Dobrow for bringing the triangulation walk to his (the author's) attention, and for useful discussions. The authors also thank Dana Randall and Bill Steiger for valuable comments.

References

[1] M. Atkinson and J. Sack, Generating binary trees at random, *Info. Proc. Letters*, **41** (1992), 21-23.

[2] P. Diaconis and L. Saloff-Coste, Comparison technique for random walks on finite groups, *Ann. Probability*, **21** (1993), 2131-2156.

[3] P. Diaconis and L. Saloff-Coste. Comparison theorems for reversible Markov chains. *Ann. Appl. Probability* **3**, 1993, pp. 696-730.

[4] R.P. Dobrow and J.A. Fill, On the Markov chain for the move-to-root rule for binary search trees, *Ann. Appl. Prob.* **5** (1995), 1-19.

[5] R.P. Dobrow and J.A. Fill, Rates of convergence for the move-to-root Markov chain for binary search trees, *Ann. Appl. Prob.* **5** (1995), 20-36.

[6] M. Gardner, Time travel and other mathematical bewilderments, Freeman, New York, NY (1987).

[7] R. Graham, D. Knuth, and O. Patashnik, *Concrete Mathematics: a foundation for computer science*, Addison-Wesley (1979), Massachusetts.

[8] B. Johnsen, Generating binary trees with uniform probability, *BIT* **31** (1991), 15-31.

[9] C. Kenyon and W. Thurston, Rotation distance using flows, preprint.

[10] M. Luby, D. Randall, and A. Sinclair, Markov chain algorithms for planar lattice structures, *Proc. of the 36th IEEE Symp. on Foundations of Computer Science*, (1995).

[11] M. Molloy, B. Reed, and W. Steiger, On the mixing rate of the triangulation walk, preprint (November, 1997).

[12] D. Randall and P. Tetali. Analyzing Glauber dynamics by comparison of Markov chains. Proceedings of *LATIN'98*; also to appear in *Lecture Notes In Computer Science*.

[13] A.J. Sinclair. *Algorithms for random generation and & counting: a Markov chain approach*. Birkhäuser, Boston, 1993, p.61.

[14] D. Sleator, R. Tarjan and W. Thurston. Rotation distance, triangulations, and hyperbolic geometry. *J.AMS.* **1**, 1988.

[15] R. Stanley. "Problems on Catalan and related numbers" in *Enumerative Combinatorics*, Volume II. Version of 12 November 1997, to be published by Cambridge University Press.

[16] D. Wilson. Mixing times of lozenge tiling and card shuffling Markov chains, draft of manuscript.

School of Mathematics, Georgia Tech, Atlanta, GA 30332-0160
E-mail address: `mcshine@math.gatech.edu`

School of Mathematics, Georgia Tech, Atlanta, GA 30332-0160
E-mail address: `tetali@math.gatech.edu`

DIMACS Series in Discrete Mathematics
and Theoretical Computer Science
Volume **43**, 1999

Bayesian Approach for Randomization of Heuristic Algorithms of Discrete Programming

Jonas Mockus, Audris Mockus, and Linas Mockus

ABSTRACT. Discrete optimization problems are often solved using "heuristics" (expert opinions defining how to solve a family of problems). The paper is about ways to speed up the search by combining several heuristics involving randomization. Using expert knowledge a prior distribution of optimization results as functions of heuristic decision rules is defined and is continuously updated while solving a particular problem. This approach (BHA or Bayesian Heuristic Approach) is different from the traditional Bayesian Approach (BA) where the prior distribution is defined on a set of functions to be minimized.

The paper focuses on the main objective of BHA that is improving any given heuristic by "mixing" it with other decision rules. In addition to providing almost sure convergence such mixed decision rules often outperform (in terms of speed) even the best heuristics as judged by the considered examples. However, the final results of BHA depend on the quality of the specific heuristic. That means the BHA should be regarded as a tool for enhancing the best heuristics but not for replacing them.

The paper is concluded by a short discussion of Dynamic Visualization Approach (DVA). The goal of DVA is to exploit heuristics directly, bypassing any formal mathematical framework.

The purpose of the paper is to inform the authors inventing and applying various heuristics and about the possibilities and limitations of BHA hoping that they will improve their heuristics using this powerful tool.

1. General Ideas of Bayesian Heuristic Approach

The traditional numerical analysis considers optimization algorithms which guarantee some accuracy for all functions to be optimized. This includes the exact algorithms (that is the worst case analysis). Limiting the maximal error requires a computational effort that in many cases increases exponentially with the size of the problem. The alternative is average case analysis where the average error is made as small as possible. The average is taken over a set of functions to be optimized. The average case analysis is called the Bayesian Approach (BA) [**Dia88, Moc89**].

There are several ways of applying the BA in optimization. The Direct Bayesian Approach (DBA) is defined by fixing a prior distribution P on a set of functions $f(x)$ and by minimizing the Bayesian risk function $R(x)$ [**DeG70, Moc89**]. The risk

1991 *Mathematics Subject Classification.* Primary 90C10,90C26; Secondary 62C10,65C05.

function describes the average deviation from the global minimum. The distribution P is regarded as a stochastic model of $f(x),\ x \in R^m$ where $f(x)$ might be a deterministic or a stochastic function. In the Gaussian case assuming (see [**Moc89**]) that the $(n+1)$th observation is the last one

$$R(x) = \frac{1}{\sqrt{2\pi} s_n(x)} \times$$

$$\int_{-\infty}^{+\infty} \min(c_n, z) e^{-\frac{1}{2}\left(\frac{y - m_n(x)}{s_n(x)}\right)^2} dz. \tag{1.1}$$

Here $c_n = \min_i z_i - \epsilon,\ z_i = f(x_i),\ m_n(x)$ is the conditional expectation given the values of $z_i,\ i = 1, ..., n,\ d_n(x)$ is the conditional variance, $\epsilon > 0$ is a correction parameter and

$$x_{n+1} = \arg\min_x R(x)$$

is a point minimizing the risk (called the Bayesian decision).

The objective of DBA (used mainly in continuous cases) is to provide as small average error as possible while keeping the convergence conditions.

The Bayesian Heuristic Approach (BHA) means fixing a prior distribution P on a set of functions $f_K(x)$ defining the best values obtained using K times some heuristic $h(x)$ to optimize a function $v(y)$ of variables $y \in R^n$ [**MEM$^+$97**]. Usually the components of y are discrete variables. The heuristic $h(x)$ defines an expert opinion about the decision priorities. It is assumed that the heuristics or their "mixture" depend on some continuous parameters $x \in R^m$, where $m < n$. We start the detailed discussion with the Direct Bayesian Approach (DBA) because that is the main instrument of the BHA.

2. Direct Bayesian Approach (DBA)

The Wiener process is common [**Kus64, Sal71, TZ89**] as a stochastic model applying the DBA in the one-dimensional case $m = 1$.

The Wiener model implies that almost all the sample functions $f(x)$ are continuous, that increments $f(x_4) - f(x_3)$ and $f(x_2) - f(x_1)$, $x_1 < x_2 < x_3 < x_4$ are stochastically independent, and that $f(x)$ is Gaussian $(0, x)$ at any fixed $x > 0$. Note that the Wiener process originally provided a mathematical model of a particle in Brownian motion.

The Wiener model is extended to multi-dimensional case, too [**Moc89**]. However, simple approximate stochastic models are preferable if $m > 1$. These models are designed by replacing the traditional Kolmogorov consistency conditions because they require the inversion of matrices of nth order for computing the conditional expectation $m_n(x)$ and variance $d_n(x)$. The favorable exception are the Markov processes including the Wiener one. Extending the Wiener process to $m > 1$ the Markovian property disappears.

Replacing the regular consistency conditions by:
- continuity of the risk function $R(x)$
- convergence of x_n to the global minimum
- simplicity of expressions of $m_n(x)$ and $s_n(x)$

the following simple expression of $R(x)$ is obtained using the results of [**Moc89**].

$$R(x) = \min_{1 \le i \le n} z_i - \min_{1 \le i \le n} \frac{\|x - x_i\|^2}{z_i - c_n}.$$

The aim of the DBA is to minimize the expected deviation. In addition, DBA has some good asymptotic properties, too. It is shown in [**Moc89**] that

$$d^*/d_a = \left(\frac{f_a - f^* + \epsilon}{\epsilon}\right)^{1/2}, \ n \rightarrow \infty$$

where d^* is density of x_i around the global optimum f^*, d_a and f_a are average density of x_i and average value of $f(x)$, and ϵ is the correction parameter in expression (1.1). That means that DBA provides convergence to the global minimum for any continuous $f(x)$ and greater density of observations x_i around the global optimum if n is large. Note that the correction parameter ϵ has a similar influence as the temperature in simulated annealing. However, that is a superficial similarity since, using DBA the good asymptotic behavior should be regarded just as an interesting "by-product". The reason is that Bayesian decisions are applied for the small size samples where asymptotic properties are not noticeable.

Choosing the optimal point x_{n+1} for the next iteration using DBA one solves a complicated auxiliary optimization problem minimizing the expected deviation $R(x)$ from the global optimum (see Figure 1). That makes the DBA useful mainly for the computationally expensive functions of a few ($m < 20$) continuous variables. This happens in wide variety of problems such as maximization of yield of differential amplifier, optimization of mechanical system of shock absorber, optimization of composite laminates, estimation of parameters of immunological model and nonlinear time series, planning of extremal experiments on thermostable polymeric composition [**Moc89**].

Using DBA the expert knowledge is included by defining the prior distribution. In BHA the expert knowledge is involved by defining the heuristics and optimizing their parameters using DBA.

3. Bayesian Heuristic Approach (BHA)

Using DBA the expert knowledge is included by defining the prior distribution. In BHA the expert knowledge is involved by defining the heuristics and optimizing their parameters using DBA.

If the number of variables is large and the objective function is not expensive the Bayesian Heuristic Approach (BHA) is preferable. That is the case in many discrete optimization problems. As usual these problems are solved using heuristics based on an expert opinion. Heuristics often involve randomization procedures depending on some empirically defined parameters. The examples of such parameters are the initial temperature if the simulated annealing is applied or the probabilities of different randomization algorithms if their mixture is used. In these problems the DBA is a convenient tool for optimization of the continuous parameters of various heuristic techniques. That is called the Bayesian Heuristic Approach (BHA) [**MEM⁺97**].

The example of knapsack problem illustrates the basic principles of BHA in discrete optimization. Given a set of objects $j = 1, ..., n$ with values c_j and weights g_j, find the most valuable collection of limited weight.

$$\max_y v(y), \ v(y) = \sum_{j=1}^{n} c_j y_j, \ \sum_{j=1}^{n} g_j y_j \leq g.$$

Here the objective function $v(y)$ depends on n Boolean variables $y = (y_1, ..., y_n)$, where $y_j = 1$ if object j is in the collection, and $y_j = 0$ otherwise. The well known greedy heuristics $h_j = c_j/g_j$ is the specific value of object j. The greedy heuristic algorithm: "take the greatest feasible h_j", is very fast but it may get stuck in some non-optimal decision.

A way to force the heuristic algorithm out of such non-optimal decisions is by taking decision j with probability $r_j = \rho_x(h_j)$, where $\rho_x(h_j)$ is an increasing function of h_j and $x = (x_1, ..., x_m)$ is a parameter vector. The DBA is used to optimize the parameters x by minimizing the best result $f_K(x)$ obtained applying K times the randomized heuristic algorithm $\rho_x(h_j)$. That is the most expensive operation of BHA therefore the parallel computation of $f_K(x)$ should be used when possible reducing the computing time in proportion to a number of parallel processors.

Optimization of x adapts the heuristic algorithm $\rho_x(h_j)$ to a given problem. Let us illustrate the parameterization of $\rho_x(h_j)$ using three randomization functions: $r_i^l = h_i^l / \sum_j h_j^l,\ l = 0, 1, \infty$ Here the upper index $l = 0$ denotes the uniformly distributed component and $l = 1$ defines the linear component of randomization. The index ∞ denotes the pure heuristics with no randomization where $r_i^\infty = 1$ if $h_i = \max_j h_j$ and $r_i^\infty = 0$, otherwise. In this case parameter $x = (x_0, x_1, x_\infty)$ defines the probabilities of using randomizations $l = 0, 1, \infty$ correspondingly. The optimal x may be applied solving different but related problems, too [**MEM$^+$97**]. That is very important in the "on-line" optimization adapting the BHA algorithms to some unpredicted changes.

Another simple example of BHA application is by trying different permutations of some feasible solution y^0. In this case heuristics are defined as the difference $h_i = v(y^i) - v(y^0)$ between the permuted solution y^i and the original one y^0. The well known simulated annealing algorithm illustrates the parameterization of $\rho_x(h_j)$ depending on a single parameter x. Here the probability of accepting a worse solution is equal to $e^{-h_i/x}$, where x is the "annealing temperature".

The comparison of BHA with exact $B\&B$ algorithms solving a set of the flow-show problems shows the table 1 from [**MEM$^+$97**]. where S is the number of

TABLE 1. Comparing the BHA and the truncated B&B

$R = 100,\ K = 1,\ J = 10,\ S = 10$, and $O = 10$					
Technique	f_B	d_B	x_0	x_1	x_∞
BHA	6.18	0.13	0.28	0.45	0.26
CPLEX	12.23	0.00	—	—	—

tools, J is the number of jobs, O is the number of operations, f_B, x_0, x_1, x_∞ are the mean results, d_B is the variance, and "CPLEX" denotes the standard MILP technique truncated after 5000 iterations. The table shows that in the randomly generated flow-shop problems the average make-span obtained by BHA was almost twice less that obtained by the exact $B\&B$ procedure truncated at the same time as BHA . The important conclusion is that stopping the exact methods before they reach the exact solution is not a good way to obtain the approximate solution.

The BHA has been used to solve the batch scheduling [**MEM$^+$97**] and the clustering (parameter grouping) problems. In the clustering problem the only parameter x was the initial annealing temperature [**DS90**].

The main objective of BHA is improving any given heuristic by defining the best parameters and/or the best "mixtures" of different heuristics. The heuristic decision rules mixed and adapted by BHA often outperform (in terms of speed) even the best individual heuristics as judged by the considered examples. In addition, BHA provides almost sure convergence. However, the final results of BHA depend on the quality of the specific heuristics including the expert knowledge. That means the BHA should be regarded as a tool for enhancing the heuristics but not for replacing them.

Many well known optimization algorithms such as Genetic Algorithms (GA) [**Gol89**], GRASP [**MPPR98**], and Tabu Search (TS) [**Glo94**], may be regarded as generalized heuristics that can be improved using BHA. There are a number of heuristics tailored to fit specific problems. For example, the Gupta heuristic was the best one while applying BHA to the flow-shop problem [**MEM$^+$97**].

Genetic Algorithms [**Gol89**] is an important "source" of interesting and useful stochastic search heuristics. It is well known [**AV91**] that the results of the genetic algorithms depend on the mutation and cross-over parameters. The Bayesian Heuristic Approach could be used in optimizing those parameters.

In the GRASP system [**MPPR98**] the heuristic is repeated many times. During each iteration a greedy randomized solution is constructed and the neighborhood around that solution is searched for a local optimum. The "greedy" component constructs a solution, one element at a time until a solution is constructed. A possible application of the BHA in GRASP is in optimizing a random selection of a candidate to be in the solution because different random selection rules could be used and their best parameters should be defined. BHA might be useful as a local component, too, by randomizing the local decisions and optimizing the corresponding parameters.

In tabu search the issues of identifying best combinations of short and long term memory and best balances of intensification and diversification strategies may be obtained using BHA.

Hence it seems that the Bayesian Heuristics Approach may be considered when applying almost any stochastic or heuristic algorithm of discrete optimization. The proven convergence of a discrete search method (see, for example, [**And96**]) is an asset. Otherwise, the convergence conditions are provided tuning the BHA [**MEM$^+$97**] if needed.

4. Process Scheduling

4.1. Introduction. In [**MR96a**] a general approach to the short-term scheduling problem of multipurpose batch and continuous operations has been described. A mathematical programming formulation that takes into account a number of the features of realistic industrial problems has been presented. The formulation is based on a continuous time representation, in which the planning horizon is divided into a number of intervals of unequal and unknown duration resulting in a large mixed integer nonlinear program. In principle, the optimal solution of this MINLP can be obtained by standard exact Generalized Benders Decomposition, Outer Approximation, or Branch and Bound techniques. However, in practice, it is known that such problem structures often result in exponential growth of solution times with problem size, thus rendering most industrially-relevant problems intractable.

This work is an attempt to address the above problem. It was suggested in [**MR94**] that the Bayesian Heuristic approach could be employed for the solution of such problems. The BH framework allows us to adapt any specific heuristic for a given class of problems. Furthermore, by clever choice of a heuristic we can make the numerical algorithm more efficient. In [**MR96b**] a general simulated annealing heuristic is used. In this work we employ a specialized Material Requirements Planning heuristic tuned for a class of batch and continuous scheduling problems. Computational experiments suggest that the BH approach combined with nonuniform time discretization shows promise for the solution of batch and continuous scheduling problems. Section 4.2 summarizes the key aspects of the BH approach. Section 4.4 describes how the MRP heuristic [**Orl75**] is tuned to incorporate batch and continuous operations. Finally, section 4.8 illustrates the effectiveness of this technique vis-a-vis the standard branch and bound technique applied to the Uniform Time Discretization approach (UDM).

4.2. Bayesian Heuristic Approach in Process Scheduling. Algorithms of exponential complexity are usually required to obtain the exact solution of global and discrete optimization problems. Even in cases when an approximate solution which lies within some tight error bounds is acceptable, the exponential complexity often remains. The desire to guarantee satisfactory results for the worst case is an important factor in forcing exponential complexity. Therefore in practice the solution of many applied global and discrete optimization problems is often approached using heuristics.

Most decision processes consist of a number of steps. For example, in batch scheduling problems these steps are: selecting a task, selecting a suitable equipment unit to process the task, and determining the amount of material to be processed by this task. In each step, an object is selected from some decision set (for example, select a task from a given set of tasks which produce the necessary product). A heuristic is a set of rules used to perform a step. For the classical knapsack problem, for example, this heuristic might be to select an object with maximal specific price (price over weight ratio). It may be helpful to think of this process as descending the decision tree by a path predefined by heuristic rules. In Figure 1 (upper part) the descent in a decision tree using the rule of always choosing the leftmost node is illustrated.

The key idea of the BH approach is to randomize these heuristic rules. Instead of descending the decision tree only once, the solution is repeated many times by selecting different paths. Each path is selected by applying the heuristic rule with some parameterized probability. Since the set of parameters for each descent in the decision tree is different, each set represents a different solution replicate and thus the best of them can be chosen for retention. In Figure 14.2 for each replicate the left most node is chosen with some probability which is the same within a specific replicate but is different for each replicate.

The parameterization is another key feature of the BH approach. If we know or expect that some heuristic "works" well, then we may increase the efficiency of the search by randomizing the parameters of the heuristic. Instead of solving a multidimensional discrete optimization problem directly we tune the parameters of the randomized heuristic. This tuning process is a low dimensional continuous optimization problem. We propose to solve this tuning problem using the Bayesian method of global optimization [**Moc89**].

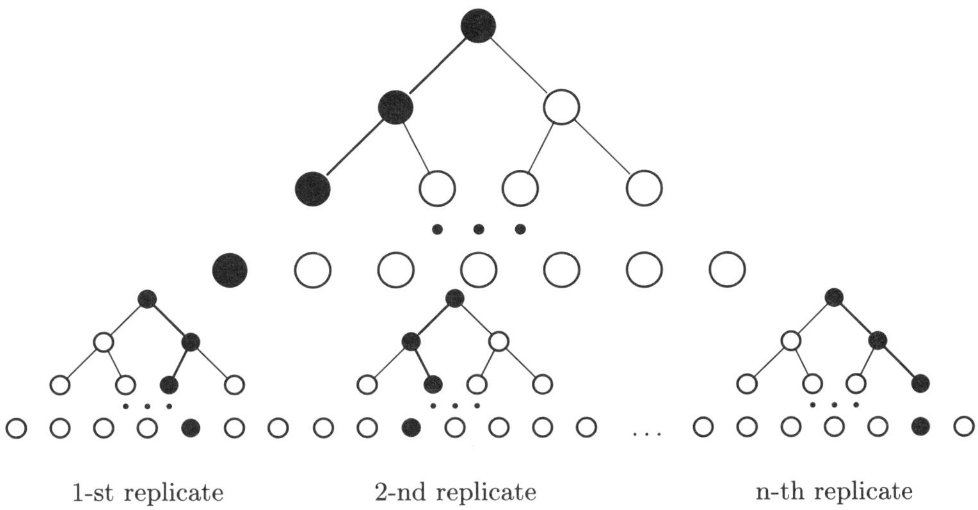

FIGURE 1. Decision tree
Deterministic descent (upper part). Randomized descents (lower part).

The main advantage of the worst case rigorous approach is that it yields error bounds on the solution. The main disadvantage is the focus on the worst possible case for a given class of problems. If this class is large, then in order to obtain sufficiently tight bounds many iterations may be required. This is the natural "cost" of such a guarantee.

The main advantage of the Bayesian approach is its focus on average case performance. An additional advantage of the BH approach is the possibility of including expert knowledge in a natural and convenient way. The potential ability to "learn" is also a positive feature of the BH approach. By learning we mean that the decision parameters which are optimal for some problems of the given class may be "good enough" for the rest of the class. The main disadvantage is that it is in general not possible to obtain and maintain guaranteed bounds on the quality of solution.

4.3. Penalty Function. The feasible region of the MINLP model defined in the first part of this series [**MR96a**] is quite complex. However, Bayesian global optimization methods are designed for a feasible region described by a hyper rectangle. Thus we need some device to transform the feasible region to a hyper rectangle. The penalty function is a convenient way to do this. Since both binary and continuous variables are present the penalty function must have two components: one for the violation of the binary constraints and one for violation of the continuous constraints. The right choice of penalty parameters is also important. If the values of the parameters are chosen to be too large, then the optimization problem may degenerate into the search for the "nearest" feasible decision. If the penalty parameters are chosen to be too small, then the constraints may be violated. We may reach some compromise by increasing the penalty parameters after each iteration. In general, the binary penalty parameter must be much greater than the continuous one, because while some violation of the continuous constraints may be permissible, the binary constraints must be satisfied. Of course, one may ignore the strict nature of the binary constraints at the initial stages of global optimization, when the optimum is still far away but then one must satisfy these constraints exactly when approaching the global optimum.

4.4. Material Requirement Planning Heuristic. MRP is an inventory management and production planning technique [**Orl75**] which, given a delivery schedule for a final products, determines the initiating times for all raw materials orders, for the production runs needed to prepare all required intermediate products, as well as the starting times for the production of the final product itself. Given a delivery time for a product shipment, each branch of the product processing tree is traced from the product in question and each component requirement is calculated. If the inventory is inadequate, then a production order is issued for that component. The tracing of each branch of the processing tree continues until all raw materials requirements have either been met or ordered.

For scheduling problems relevant to the chemical industry we have to extend this heuristic to handle unit and task assignment, batch size determination, and other features. For purposes of this paper we will employ simple and natural heuristic rules. Of course, these rules could be enhanced or augmented to give further improvements in the results. The following is the list of heuristic rules:

- *Product selection rule.* The production run of each product in a batch and continuous processing system has some due date. We begin with a product which has the earliest due date. The rationale for this is that knowledge of due dates close to start of the scheduling horizon is more concrete. When production of this product is scheduled we recursively also schedule the production of the intermediates required to produce it.
- *Equipment item selection rule.* To schedule production of a given product or intermediate it is necessary to select a task and an equipment unit to process this task. When there are few tasks producing the same product we randomly assign the quantity of a product that the given task has to produce. Then we select an equipment item which is available during the time interval closest to the due date (a unit may become unavailable for processing because it processes another task, it is shutdown for maintenance etc.).
- *Task start selection rule.* Usually the size of the unit availability interval is different than the task processing time. The rule will start a given task to end exactly at the end of this interval.

  For example, assume that task2 processing time on unit1 is 1h. Furthermore, the interval chosen in the previous step is from 3pm to 5pm. Then task2 is started at 4pm to end at 5pm.

4.5. Schedule Generation. The process of schedule generation using the randomized heuristic is essentially the same as that described in section 4.4. The only difference is that instead of using the heuristic rules deterministically we make our selection with some probability. Consider, for example, the product selection rule. Instead of selecting a product with smallest due date, we select the product with some probability r which is a function of its due date D_s. The probability r_s of selecting the product s is expressed as

$$r_s = xa_0 + (1-x)a_1h(d_s). \tag{4.1}$$

Here x is the parameter defined in the previous step of the BHA algorithm 1, M is the number of the products in the list , and the heuristic $h(d_s) = \exp\{-D_s\}$. We chosen an exponential function that the probability to select a product with late due date would be very small. As we see, the products with smallest due dates are chosen with highest probability, thus we preserve the character of the MRP heuristic while making it random. In a similar manner we deal with the other rules.

The generated schedule may be infeasible because of violations of the state capacity constraints. It is possible that the state capacity is exceeded or that a negative amount of material exists in a given state. These situations are handled by penalizing the variations from the minimal and the maximal state capacity values and adding this penalty to the objective function. The objective function is readily evaluated by calculating the storage and utility costs, the profit gained by satisfying demands less than raw material costs.

4.6. Bayesian Heuristic Algorithm. The key component of the BH framework is the randomized heuristic function. This function corresponds to the objective function in the global optimization. Once the randomized heuristic function is defined, we may optimize its parameters using the Bayesian global optimization method [**Moc89**] and as byproduct we obtain the optimum profit. For purposes of this paper we employ a parametric first order polynomial function of the Material Requirements Planning (MRP) heuristic as the randomized heuristic. The feasible region of this heuristic is quite complex, therefore a penalty function is used to transform the feasible region to a hyper rectangle. When using the MRP heuristic only the amount of material in the given state is allowed to exceed state capacity while all other constrains, binary and continuous, are not allowed to be violated. Thus we need only one penalty parameter for the continuous constraints. Theoretically, the value of the penalty parameter has to be much bigger than unit price of product to ensure a steep enough increase in penalty when we leave the feasible region. However, this value should not be too large so that the increase in penalty is quite smooth. Empirically we found the value of 500 to yield the best results for the test examples, although for other problems its value may be different.

In section 4.4 a version of the MRP heuristic for batch and continuous tasks was described . If d_i is some decision (select a product, a suitable equipment unit, or task start), then $h(d_i)$ is the heuristic function. The function $r(x, h(d_i))$ is a randomized heuristic function which gives the probability of the decision d_i. x is the randomization parameter. We used

$$r(x, h(d_i)) = xa_0 + (1-x)a_1 h(d_i)$$

where

$$a_0 = \frac{1}{M},$$
$$a_1 = \frac{1}{\sum_{i=1}^{M} h(d_i)},$$

and M is the number of possible decisions. The heuristic rules are as defined in section 4.4. Thus the Bayesian Heuristic Algorithm can be represented by the following simple steps:

Step 1. Fix parameter x using the global Bayesian method [**Moc89**]. [1]
Step 2. Generate a schedule by using the randomized MRP heuristic (see sections 4.4 and 4.5);
Step 3. Evaluate the schedule for this parameter;
Step 4. If the schedule is feasible (there is no penalty) and the value of the best schedule so far did not increase for 10 iterations then go to step 6;
Step 5. Go to step 1;
Step 6. Fix binary variables to the values given by the best schedule. [2] Substitute the values of these binary variables into the model to reduce the problem to a linear program. Then the solution of the linear program gives the exact starting times, batch sizes, and processing rates.

We see that the scheduler generates sequences and assignments while the LP model is producing the exact schedule. The LP model is derived from the Non-Uniform Discrete-Time Model (NUDM) model (see [**MR94**]) by substituting the sequencing and assignment variables which are fixed by the scheduler. This disaggregation of the scheduling problem allows us to solve the large scale MINLP problem by using a combination of heuristic algorithm and an efficient LP solver.

The key issue in using the BH approach is that together with the best schedule we also acquire the best randomization parameter x, or, in other words, we tailor the heuristic for a given class of problems. Initially we are parameterizing the heuristic, thus giving it a few extra degrees of freedom. By varying parameters in an intelligent way, so that the expected outcome is maximized, we find the parameter values which yields the best profit.

4.7. Relationship of the Bayesian Heuristic Approach with Simulated Annealing and Genetic Algorithms. The Bayesian Heuristic Approach has a direct and deep relationship with simulating annealing and genetic algorithms. For example, consider (4.1). It indicates that products with smaller due date are selected with higher probability. However, there is a non zero probability of selecting a product with a higher due date. This flexibility which is allowed by the heuristic randomization effectively increases the search space. In the simulated annealing case, the same effect is achieved by accepting a point with lower objective value.

Genetic algorithms through an evolution process proceed from some initial population A to final population B (see Figure 2).

This evolution process is basically managed by the mutation and crossover probabilities which are constant throughout the entire evolution process. In the Bayesian Heuristic Approach, in effect multiple evolutions occur. Initial population A evolves to some intermediate population I_1 via evolution which depends on some evolution parameter x_1, i. e. mutation and crossover probabilities are defined by this parameter. The intermediate population I_1 evolves to some other intermediate

[1]This provides the asymptotic convergence with probability one. That is the first feature of the Bayesian Heuristic Approach. For the convergence proof see [**MEM+97**].

We have to note, however, that parameter x is fixed based on the value of the profit of the previously generated schedules. Thus the second feature of a given global Bayesian method is that it fixes the next x to a value for which it expects to get maximal profit. Of course, it may be that this new x value does not give the best profit, thus we need to use additional iterations.

These two features were key reasons for using the global Bayesian method. The second feature makes the search more efficient than a pure random Monte-Carlo search while the first enables a more thorough exploration of the decision space.

[2]These variables correspond to the sequencing and assignment of tasks.

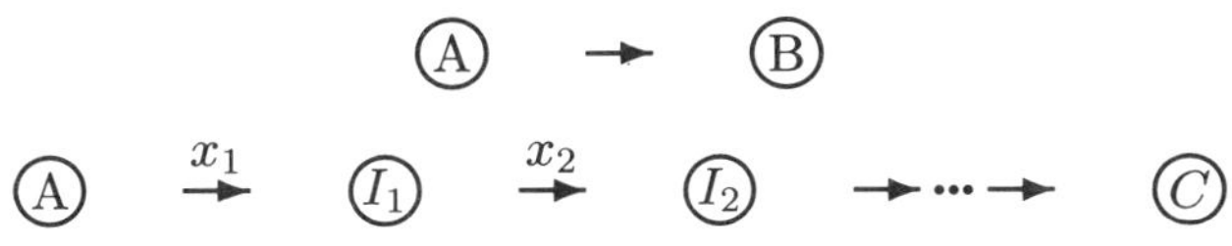

FIGURE 2. Bayesian approach versus genetic algorithms
Genetic algorithm. Single evolution (upper part). Bayesian approach. Multiple evolutions (lower part). Here x_i is chosen so that to maximize conditional expected fitness of i-th population.

population I_2 via evolution which depends on another evolution parameter x_2. This process is repeated many times. The best population is selected as the final population. Another important feature besides multiple evolutions is that each x_i is selected based on the fitness of the previous populations so as to maximize the conditional expected fitness of i-th population. In such a way, the process is not a purely random one but is aimed to maximize the fitness. By changing the mutation and crossover probabilities the tendency of genetic algorithms to produce local solutions is mitigated.

Another key difference of the BHA from simulated annealing and genetic algorithms is learning. In simulated annealing or genetic algorithms parameters, such as initial temperature or mutation probability etc., are found experimentally by studying a number of scheduling problems. However, in the Bayesian Heuristic Approach these parameters are tuned for each problem of a given class and do not need to be fixed for all classes. Thus, in contrast to simulating annealing and genetic algorithms, the Bayesian approach allows learning to occur for the problem at hand.

The further drawback of simulated annealing and genetic algorithms is that when applied to discrete or mixed integer problems they generate a large number of infeasible solutions. Thus a big percentage of the computation time is wasted in the generation of these infeasible solutions. The similar situation was observed when trying to employ simulation annealing heuristic within the Bayesian Heuristic Heuristic Approach [**MR96b**]. Because of this, only small size problems could be solved. Using the BHA with MRP heuristics one overcomes this problem and 70% percent of the generated schedules were feasible. This can be explained by the fact that MRP heuristic fits better to a given class of scheduling problems and thus generates mostly realizable schedules.

4.8. Scheduling Results. We compared the BHA algorithm with solution of the MILP uniform discretization model using branch and bound enumeration (B&B) [**ZPMR94**]. Results are reported for several test examples, both batch and continuous. The detailed numerical data for these examples is available electronically from the rcsplib account (rcsplib@ecn.purdue.edu) in the form of RCSPec language files.

We summarize the results in Table 2. Computational experiments were performed on a HP 9000/755 workstation using the commercially available CPLEX solver. The column "BH profit" shows the profit value obtained before performing the LP solution phase, i.e. the profit value given only by the statistical part of the algorithm. As we see, the LP solution does not significantly increase the value of

TABLE 2. B&B and BHA comparison for examples of batch and continuous processes

Problem	B&B Profit ($)	B&B Time (sec)	BHA Profit (%)	BHA Time (sec)	Number of replicates	BH profit (%)
batch1	3230	6	100	0.46	12	77.91
batch4	60534	1.7	99.99	2.90	13	93.79
batch3	105756	8.6	99.97	3.85	20	99.18
exIII	1400	66	100	3.65	15	100
cpctsp1	4724	0.6	100	0.06	11	100
cpctsp2	8122	188.5	100	0.20	11	100
cpctsp3	12015	*	99.60	3.01	16	99.6
cipac2	20879	*	100.54	31.8	29	100.39
cipac1	22800	824.6	107.77	27.3	29	107.77

the profit. However we can not draw definite conclusions about the need of the LP solution step from this case study alone. We have to note that while for batch processes (batch1, batch3, batch4, exIII, cpctsp) BHA gives solutions slightly worse than the branch & bound approach, for continuous processes BHA (cipac1, cipac2) gives better results than UDM. This can be explained by the inherent discrete nature of the UDM which is not suited for continuous tasks. Time variable can have only discrete values in the UDM case, while processing time of continuous tasks assumes continues values and thus can not be represented as discrete variables. NUDM works much better than UDM for problems with sequence dependent tasks (cipac2, cpctsp1, cpctsp2, cpctsp3) also. This is due to the fact that NUDM handles sequence dependent changeovers in a more efficient way, i.e. uses *resource-task* network representation as opposed to the *state-task* network representation used by the UDM. It is shown that constraints handling sequence dependent changeovers based on the *resource-task* network representation are tighter than those based on the *state-task* network representation . It is worth noting that with increasing number of sequence dependent tasks (cpctsp1, cpctsp2, cpctsp3) the solution time of B&B grows exponentially while the corresponding time of BHA grows only polynomial. * means that optimal solution for the examples cpctsp3 and cipac2 was not reached by B&B since the solution time was unreasonably large. The B&B tree was terminated after 20000 nodes. The computational time for the batch4 example is worse in BHA case due the fact that MRP heuristic is not very well suited for such processes (batch4 example contains zero wait states). It is possible to modify BHA account for the zero wait states by aggregating the two tasks connected by zero wait state together.

4.9. Scheduling Conclusions. In [**MR96a**] a general formulation of the short-term scheduling problem for complex multipurpose batch and continuous operations is presented. However, the size of the resulting MINLP raised serious concerns regarding the practical applicability of the NUDM. In this paper we provide an algorithm to overcome this limitations.

This algorithm is based on the BH approach to discrete optimization and a clever choice of heuristic is the key issue. If the heuristic is a very general one (as simulated annealing in [**MR96b**]) then the class of problems solved is large as is the computational time. If the heuristic is a special one (as MRP in our case) then the class of problems solved is smaller as is the computational time. Of course, problems which do not fit a given heuristic are solved also but the computational effort required can be unpredictably high (batch4 is a good example). Thus one of the directions for future work may well be the expert system which recognizes the structure of the problem and suggests a heuristic to solve this problem.

Usually processing data is uncertain (task processing times fluctuate around the mean value due to the quality of feed, due dates are not well known in advance etc.). The BH approach is a statistical framework and thus theoretically accommodates stochastic data. Thus the other promising future direction can be identified as an application of the BH approach to scheduling problems with uncertainty. NUDM allows one to model various uncertainties in time and size parameters without modifications since time and size are continuous variables (for the UDM uncertainty in processing times require major modifications).

5. Software for Global Optimization

5.1. Background. The global optimization software was initiated considering the results of international "competition" of different algorithms of global optimization (see [**DS78**]). The experience in real life optimization problems and some recent results were also used selecting the set of optimization algorithms. The set of algorithms of global optimization includes

- four versions of the Bayesian search,
- a version of clustering,
- a version of uniform deterministic grid,
- a version of pure Monte Carlo search.

Usually it is reasonable to start optimization by a global method and to finish it by some local method. An exception is two global algorithms: the Torn version of clustering [**TZ89**], and the Zilinskas version of the Bayesian technique [**TZ89**]. Both of these algorithms contain some simple local search algorithms. The local search is not necessary for those two methods, but it may be useful.

There are three local optimization methods:

- a method of variable metrics type with Lagrangian multipliers and penalty functions for constrained optimization of smooth functions (see [**Sch86**]),
- a method of simplex type of Nelder and Mead with penalty functions for constrained optimization of non-differentiable functions.
- a method of stochastic approximation type for "noisy" functions (see [**MOC89**]).

5.2. Application Areas. Each subroutine represents a global or a local method. The choice of method has to follow the idea that the computational complexity of the method should roughly correspond to that of the objective function:

- For computationally "expensive" functions the Bayesian methods could be recommended. Those methods need a large amount of auxiliary calculations to make each observation more efficient.
- For "cheap" functions the simple grid methods, like Monte Carlo or a uniform deterministic grid (see [**Sob67**]), can be better. Here observations are not so efficient, but auxiliary calculations are negligible. This explains a relative efficiency of simple methods when optimizing simple functions.

- The clustering techniques (see [**TZ89**]) may be the best choice, if we expect the number of local minima to be small.
- A relatively simple Bayesian technique is available [**TZ89**] for global optimization of one-dimensional functions;
- There are optimization problems where objective functions can be roughly represented as a sum of components depending on different variables. Here the Bayesian method of line search along the coordinates usually shows very good results. This method globally optimizes one variable at a time by one-dimensional Bayesian search. The difference of this method from other methods of global optimization is that it depends on the starting point. Thus a deviation from the global minimum can be made as small as desired by applying a multi-start procedure with different uniformly distributed starting points.

5.3. Constraints. All the global methods optimize in a rectangular region. Therefore we represent the linear and non-linear inequality constraints as some penalty functions. The same applies to the local method of stochastic approximation type. In local methods of simplex and variable metrics type the linear and the non-linear constraints can be defined directly. This may be done by constraint subroutines, supplied by the user in addition to the objective function.

5.4. Software Versions. The global optimization software is in four versions:

- portable Fortran Library,
- interactive software for Turbo C compiler and DOS operating system,
- interactive software for C++ compiler and UNIX operating system and X-Window system.
- interactive software for Java

One may notice a cycle of portability in this sequence of software versions. The sequence is started from by the portable Fortran library and is concluded by Java language. The two systems in between are more difficult to port. The Turbo C system is for DOS-compatible operating systems and C++ is for the UNIX environment. Fortran, Turbo C and C++ versions are described in [$\mathbf{MEM^+97}$]. Now we briefly consider the Java version.

5.5. Java Version.

5.5.1. *Global Minimizer for Java (GMJ).* The GMJ system [**Gry98**] is a class framework for implementing and testing global optimization algorithms ($METHODS$), functions to be optimized ($TASKS$), and and visual representations classes ($ANALYSIS$).

5.5.2. *Running GMJ.* The GMJ system can be run as a Java Applet or as Java Application. The advantage of running it as Java Applet is obvious-it can be used over the Web. The advantage of the Java Application is that the most recent Java features supported by the Java Development Kit (JDK) can be used.

5.5.3. *Configuring Applet.* The sample file *gmj.html* shows how to set configuration parameters of the *gmj* applet. The $<$ *applet* $>$ tag looks like this:

```
<applet code="lt.ktu.gmj.ui.GMJ.class"
                    codebase="Lib"
                    align="baseline"
                    width="450"
                    height="450"
```

```
                archive="gmj.jar">

<param name="TASKS"
value= "lt.ktu.gmj.tasks.Sin|Undefined">

<param  name="METHODS"
value="lt.ktu.gmj.methods.Mig1| lt.ktu.gmj.Bayes1">

<param name="ANALYSIS"
value= "lt.ktu.gmj.analysis.Convergence|
lt.ktu.gmj.analysis.Spectrum|lt.ktu.analysis.Projection">

</applet>
```

Here:

code specifies the applet class and should not be changed

codebase specifies the relative URL of the applet class archive file and user class files. The URL is relative to the location of the HTML file where the $< applet >$ tag resides or is absolute path.

archive lists the class archive files

width and *height* specifies the applet size, as it appears in the HTML page.

The parameter $TASKS$ lists tasks which are available when the applet loads. Complete class names (package and class name) are separated by | symbol.

The parameter $METHODS$ lists methods which are supported when the applet loads.

The parameter $ANALYSIS$ lists visual analysis classes which are supported when the applet loads.

5.5.4. *Configuring Stand-Alone Application.* The Java Runtime Environment (JRE) is required when running GMJ as a stand-alone application.

Before running GMJ, a $CLASSPATH$ should be configured, so that the GMJ classes can be found by the JRE. For Windows NT the command might look like this:

```
SET CLASSPATH=#CLASSPATH#;c:\gmj\gmj.jar
```

the application is started by loading the *lt.ktu.gmj.ui.GMJ* class:

```
java lt.ktu.gmj.ui.GMJ
```

Like the applet the application accepts three optional parameters: tasks, methods, and analysis objects.

5.5.5. *Display and Control.* GMJ displays a tab control that has three choices: method, task and operation. The appropriate pages can be selected by clicking on the page tabs. The detail discussion is in [**Gry98**].

5.6. Software Availability. The software .

The interactive UNIX C++ software (LINUX 1.2.8. version) and the library of portable Fortran subroutines (LINUX 1.2.8. and DOS versions) are included in [**MEM⁺97**], and available on *ftp : //optimum.mii.lt/pub*

6. Dynamic Visualization Approach (DVA)

We started the description of algorithms from the formal Direct Bayesian Approach (DBA). Then we considered semi-formal Bayesian Heuristic Approach

(BHA). An informal interactive optimization is needed if an optimization problem is not well defined. This may arise, for example, if the mathematical model, including the objective function, must be updated during the course of optimization process.

The informal interactive approach attempts to represent an optimization problem in a visual form that is domain specific and is intuitive to the domain expert. The visual representations can vary significantly across the domains. In the examples the dynamic visual representation of a smooth function in time and space turned out to be effective in several domains. The domain specific visual representation can efficiently convey information about a complex model and help make qualitative judgments about model's adequacy and optimality.

The efficiency of informal interactive optimization depends on dynamic visualization techniques. We regard dynamic visualization as an important tool using heuristics in an informal interactive way and thus will refer to it as a Dynamic Visualization Approach. Two basic techniques of dynamic visualization are considered in [**MEM$^+$97**]:

- space-time smoothing;
- image search.

Those techniques are explained in [**MEM$^+$97, EM96**] through real life examples. We mentioned Dynamic Visualization Approach (DVA) just to illustrate different ways of using heuristics, from pure formal Bayesian Approach (by priori distribution on a set of objective functions) to semi-formal Bayesian Heuristic Approach (by priori distribution on a set of parameters of randomized heuristics and their "mixtures") and pure interactive DVA (directly by the expert opinion using visual representation).

In the Java optimization system (see section 5.5) the possibility of the domain specific visualization is included by the $ANALYSIS$ classes. This way the Bayesian Heuristic Approach and the Dynamic Visualization are integrated. The simplest DV techniques such as convergence lines, projections and the objective function distributions are rather general. The more advanced visualization techniques are domain specific and should be developed while designing the $TASKS$ classes representing specific objective functions.

References

[And96] S. Andradottir. A global search method for discrete stochastic optimization. *SIAM Journal Optimization*, 6:513–530, 1996.

[AV91] I.P. Androulakis and V. Venkatasubramanian. A genetic algorithmic: Framework for process design and optimization. *Computers in Chemical Engineering*, 15:217–228, 1991.

[DeG70] M. DeGroot. *Optimal Statistical Decisions*. McGraw-Hill, New York, 1970.

[Dia88] P. Diaconis. Bayesian numerical analysis. In *Statistical Decision Theory and Related Topics*, pages 163–175. Springer Verlag, 1988.

[DS78] L.C.W. Dixon and G.P. Szego. *Towards global optimisation 2*. North Holland, Amsterdam, 1978.

[DS90] G. Dzemyda and E. Senkiene. Simulated annealing for parameter grouping. In *Transactions. Information Theory Statistical Decision Theory Random Processes*, pages 373–383, Praque, 1990.

[EM96] William Eddy and Audris Mockus. Dynamic visualization in modeling and optimization of ill defined problems, case studies and generalizations. In C. A. Floudas and Panos M. Pardalos, editors, *State of the Art in Global Optimization: Computational*

Methods and Applications, volume 7 of *Non-convex Optimization and its Applications*, pages 499–520. Kluwer Academic Publishers, 1996.

[Glo94] F. Glover. Tabu search: improved solution alternatives. In *Mathematical Programming. State of the Art 1994*, pages 64–92. University of Michigan, 1994.

[Gol89] D. E. Goldberg. *Genetic Algorithms in Search, Optimization, and Machine Learning*. Addison-Wesley, Reading, MA, 1989.

[Gry98] M. Grybauskas. Global minimizer for java (gmj) version 1.1. Technical report, Institute of Mathematics and Informatics, Working Paper, MII, Akademijos 5, Vilnius, Lithuania, 1998.

[Kus64] H.J. Kushner. A new method of locating the maximum point of an arbitrary multipeak curve in the presence of noise. *J. of Basic Engineering*, 86:97–100, 1964.

[MEM+97] J. Mockus, W. Eddy, A. Mockus, L. Mockus, and G. Reklaitis. *Bayesian Heuristic Approach to Discrete and Global Optimization*. Kluwer Academic Publishers, Dordrecht, The Netherlands, 1997.

[Moc89] J. Mockus. *Bayesian approach to global optimization*. Kluwer Academic Publishers, Dordrecht-London-Boston, 1989.

[MPPR98] T. Mavridou, P.M. Pardalos, L.S. Pitsoulis, and M.G.C. Resende. A GRASP for the biquadratic assignment problem. *European Journal of Operations Research*, 105:613–621, 1998.

[MR94] L. Mockus and G.V. Reklaitis. Mathematical programming formulation for scheduling of batch operations using nonuniform time discretization. In *AIChE annual meeting*, number 235d, San-Francisco, California, 1994.

[MR96a] L. Mockus and G. V. Reklaitis. Continuous time representation approach to batch and continuous process scheduling-ii. minlp formulation. submitted, 1996.

[MR96b] L. Mockus and G. V. Reklaitis. A new global optimization algorithm for batch process scheduling. In C. A. Floudas and Panos M. Pardalos, editors, *State of the Art in Global Optimization: Computational Methods and Applications*, volume 7 of *Non-convex Optimization and its Applications*, pages 521–538. Kluwer Academic Publishers, 1996.

[Orl75] J. Orlicky. *Material Requirement Planning*. McGraw-Hill, 1975.

[Sal71] V.R. Saltenis. On a method of multi-extremal optimization. *Automatics and Computers (Avtomatika i Vychislitelnayya Tekchnika)*, (3):33–38, 1971. (in Russian).

[Sch86] K. Schittkowski. Nlpql: A fortran subroutine solving constrained nonlinear programming problems. *Annals of Operations Research*, 5:485–500, 1985/86.

[Sob67] I.M. Sobolj. On a systematic search in a hypercube. *SIAM Journal on Numerical Analysis*, 16:790–793, 1967.

[TZ89] A. Torn and A. Zilinskas. *Global optimization*. Springer-Verlag, Berlin, 1989.

[ZPMR94] M. G. Zentner, J. F. Pekny, D. L. Miller, and G. V. Reklaitis. RCSP++: A scheduling system for the chemical process industry. In *Proc. PSE'94*, pages 491–495, Kyongju, Korea, 1994.

Department of Optimization, Institute of Mathematics and Informatics, Akademijos 4, Vilnius 2600, Lithuania
E-mail address: jonas@optimum.mii.lt

Software Production Research Department, Bell Laboratories Lucent Technologies Naperville, IL 60566-1444
E-mail address: audris@research.bell-labs.com

R&D, Monsanto, Skokie, IL 60077
E-mail address: LINAS.MOCKUS@monsanto.com

DIMACS Series in Discrete Mathematics
and Theoretical Computer Science
Volume **43**, 1999

On the Mixing Rate of the Triangulation Walk

Michael Molloy, Bruce Reed, and William Steiger

ABSTRACT. Let T_n denote the set of triangulations of a convex polygon K with n sides. We study the random walk on T_n whose transitions are "flips" of one of the $n-3$ internal diagonals of the current triangulation, the choice of diagonal being random. By bounding the conductance of this graph we show that the walk mixes rapidly, namely in time $O(n^\beta)$. A direct argument is given for the fact that the mixing rate is at least $\Omega(n^{3/2})$.

1. Introduction and Summary

Consider a convex polygon K on n points $p_0, \dots, p_{n-1}$, in clockwise order. A triangulation is a set of $n-3$ non-crossing diagonals $p_i p_j$ which partitions K into $n-2$ triangles. It is familiar (see e.g., [**7**]) that T_n, the set of all triangulations of K, satisfies

$$|T_n| = \gamma_{n-2} \equiv \frac{1}{n-1}\binom{2n-4}{n-2}, \tag{1}$$

the right hand side being the $(n-2)^{nd}$ Catalan number. We consider the graph G_n whose vertex set is $V = T_n$. Edges are triangulations which differ by one diagonal. Specifically, a diagonal of a triangulation $\tau \in V$ may be "flipped"; i.e., removed - creating a quadrilateral - and replaced in the opposite way to give a new triangulation τ' adjacent to τ in G_n. Thus G_n is $n-3$ regular.

It is natural to consider the random walk on G_n which moves to each neighbor of a vertex with equal chance. The key question is how quickly the walk converges to the steady-state distribution, in this case, the uniform. Often, there are practical reasons for studying such a random walk on a set S of combinatorial objects, for example to estimate the size of S or to generate an element of S at random. However we already know $|G_n| = \gamma_{n-2}$ and there already exist optimal algorithms to generate random triangulations [**3**], [**1**], [**2**]. But there are interesting sets which are related to T_n, e.g., $T_n(P)$, the set of triangulations of the non-convex, simple polygon P, and $T_n(K)$, the triangulations of the set K whose points are *not* in convex position; in both these cases the task of computing the size, and of generating a random element are open problems. By studying the walk on T_n we may learn how to study other sets of triangulations.

1991 *Mathematics Subject Classification.* Primary 60J15; Secondary 68R10.

Part of the work of M. Molloy was supported by an NSERC grant. M. Molloy and B. Reed were supported by NATO Collaborative Research Grant #CRG950235.

Part of the work of B. Reed was supported by a FAPESP grant.

Until recently, little was known about the triangulation walk. This is surprising because G_n is so familiar and well-studied [**7**], [**14**] (also [**4**], [**5**], [**6**], [**8**], [**12**], which are relevant because G_n is also the graph of rooted binary trees with $n-2$ internal nodes, adjacency defined by the rotation operation). In the present paper we will apply the method of conductance to show that the walk mixes rapidly. Write $\overline{S}$ for $T_n \backslash S$. We prove

THEOREM 1. *For any set $S \subset T_n$, there is a matching from S to $\overline{S}$ with*

$$\geq \frac{|S||\overline{S}|}{|T_n|n^\beta}$$

edges of G_n.

This would imply that the conductance is greater than $1/(2n^{\beta+1})$ and the mixing rate $O(n^{2\beta+3})$ (see e.g., [**13**]). We will prove this with $\beta = 11$ but think we can reduce it to 4 or 5. The importance of this result is that it established the fact that the walk mixes in polynomial time (see also [**11**]). More recently, McShine and Tetali [**9**] used a comparison technique, along with coupling bounds for a related chain, to obtain a mixing rate of about $O(n^5 \log n)$. While our mixing rate bound is much weaker, the conductance proof gives structural information about the triangulation graph that may be useful in other contexts.

In addition we show

THEOREM 2. *The mixing rate is at least $\Omega(n^{3/2})$.*

It is interesting here to compare this to the walk on the n-cube (similar to G_n, it is an n-regular graph with c^n vertices), which mixes in $O(n \log n)$ steps.

A key idea behind these results is a partitioning of T_n into sets in which all triangulations share a certain "central triangle" t. This allows us to apply induction on the 3 subpolygons defined by t. It also gives us the ability to move between sets of triangulations in a systematic way and to construct a matching with many edges from S to $\overline{S}$.

In the next section we collect some notations, definitions, and useful facts about T_n. The lower bound is proved in Section 3. We do this first because it is much more pleasant and easier than the proof of the conductance bound, which is sketched in Section 4.

2. Some Preliminaries

Again, G_n is the graph with vertex set $V = T_n$, the set of triangulations of the convex polygon K whose vertices are the points $p_0, \dots, p_{n-1}$, written in clockwise order; triangulations τ_1 and τ_2 in V are adjacent if τ_2 may be obtained by flipping one of the $n-3$ diagonals of τ_1. The triangulation walk moves from a vertex τ to one of its $n-3$ neighbors, each with equal probability. This gives an ergodic Markov chain $MC(n)$ which is reversible, and its stationary distribution is the uniform probability on V.

If $p_i p_j$ is a diagonal of a triangulation $\tau \in V$, $i > j$, define its length to be $\|p_i p_j\| = \min(i-j, n-i+j)$, the (fewest) number of successive edges of K between the endpoints. Every triangulation has a longest diagonal ℓ and it is clear that $n/3 \leq \|\ell\| \leq n/2$, the upper bound attainable only if n is even. The longest diagonal must be in a triangle $t = \Delta p_i p_j p_k$, which has the sum of the lengths of its three diagonals equal to n. We call t a *central triangle* and observe that it is

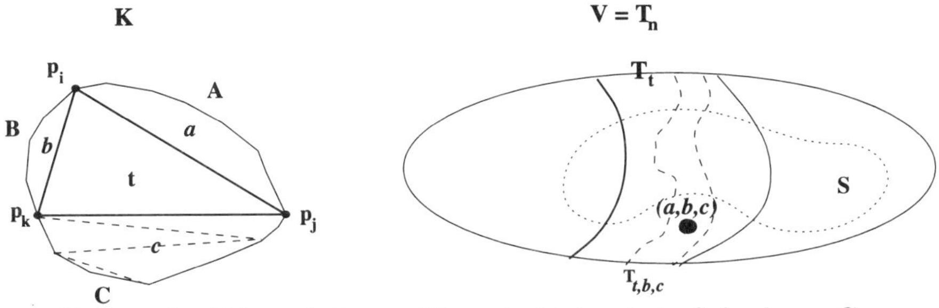

FIGURE 1. *Triangulations with central triangle t. Subpolygon C is triangulated by c. $\tau = (a, b, c)$ is in $\overline{S} \cap T_t$. $T_{t,b,c}$ denotes triangulations in T_t with b triangulating B and c triangulating C.*

unique unless n is even and τ has a diagonal ℓ with length $n/2$. In this case there are *two* such triangles, and to clarify the situation, we call $\ell = p_i p_{i+n/2}$ a *central edge* and regard it as a *degenerate central triangle.* The diagonals comprising a nondegenerate central triangle $t = \Delta p_i p_j p_k$ divide K into three subpolygons of sizes $\|p_i p_j\| + 1$, $\|p_j p_k\| + 1$, and $\|p_k p_i\| + 1$. Write

$$C_n = \{t = \Delta p_i p_j p_k : t \text{ a central triangle}\}$$

and observe that $|C_n| = \Theta(n^3)$. In fact $|C_n| \leq n^3/12$ for $n \geq 10$.

It will be useful to regard central triangles as vertices in a graph, $G(C_n)$. A non-degenerate central triangle $t = \Delta p_i p_j p_k$ is adjacent to another non-degenerate central triangle t' if t and t' have two points of K in common and the third points differ by exactly one. A degenerate central triangle $p_i p_{i+n/2}$ has four neighbors, $p_{i-1}p_{i-1+n/2}$; $p_{i+1}p_{i+1+n/2}$, $\Delta p_{i-1}p_{i+1}p_{i+n/2}$, and $\Delta p_i p_{i-1+n/2} p_{i+1+n/2}$. A central triangle has at most 6 neighbors in $G(C_n)$.

For each $t \in C_n$ we let T_t denote the triangulations τ which have t as their central triangle. The sets T_t partition V. Given a central triangle $t = \Delta p_i p_j p_k$, $i < j < k$, let A denote the subpolygon $p_i, p_{i+1}, \cdots, p_j, p_i$ defined by $p_i p_j$, B, the subpolygon defined by $p_i p_k$, and C, the subpolygon defined by $p_j p_i$ (see Figure 1). We write T_A, T_B, and T_C for the triangulations of A, B, and C, respectively. Observe that there are two degenerate cases here. The first is when n is even and t is degenerate. In this case we think of A as a polygon with one vertex and $|T_A| = 1$. In the other, n is odd and t contains an edge of K. In this case we think of A as a polygon with two vertices and again $|T_A| = 1$. We note that each triple $a \in T_A$, $b \in T_B$, $c \in T_C$ gives a distinct triangulation $\tau = (a, b, c) \in T_t$, and that

$$T_t = \bigcup_a \bigcup_b \bigcup_c (a, b, c),$$

so $|T_t| = |T_A||T_B||T_C|$. In fact suppose that $\|p_i p_j\| = q$, $\|p_j p_k\| = r$, and that $\|p_k p_i\| = s$, where $p + q + r = n$ and $n/3 \leq \max(q, r, s) \leq n/2$. This means that A has $q+1$ sides so (1) gives $|T_A| = \gamma_{q-1}$. Similarly $|T_B| = \gamma_{r-1}$ and $|T_C| = \gamma_{s-1}$, and we have

$$|T_t| = \gamma_{q-1}\gamma_{r-1}\gamma_{s-1}.$$

This expression is maximized when $q = 0$ and $r = s = n/2$ (a degenerate central triangle) and minimized when $q = r = s = n/3$. Since

$$\gamma_j \sim \frac{4^j}{\sqrt{\pi j^3}}$$

this implies

$$\frac{9}{n^3} \le \frac{|T_t|}{|V|} \le \frac{5}{n^{3/2}}. \tag{2}$$

One consequence is that for $n \ge 20$, $|T_t| \le |V|/16$ for degenerate central triangles.

A diagonal $p_i p_{i+2}$ of length 2 defines an *ear at* p_{i+1}, and every $\tau \in V$ has at least 2 (and at most $n/2$) ears. Define

$$E_i \equiv \{\tau \in V : \tau \text{ has an ear at } p_i\},\ i = 1, \dots, n,$$

the triangulations with ears at vertex p_i. Clearly $|E_i| = |T_{n-1}|$ so from (1),

$$|E_i| = \frac{1}{4}|T_n|\left(1 + \frac{3}{2n-5}\right). \tag{3}$$

The edge $p_0 p_1$ of K must be in some triangle of the triangulation, say $\Delta p_0 p_1 p_i$. According to (3), for about 1/4 of the triangulations in T_n, $p_i = 2$ (this puts an ear at p_1), and in about 1/4 of the triangulations $p_i = n-1$ (an ear at p_0). If a triangulation τ contains triangle $t = \Delta p_0 p_1 p_i$ we write $APEX_n(\tau) = \min(i, n-i)$, $2 \le i \le n-1$, the "distance" of the apex of t from its base, $p_0 p_1$. By (1), the number of triangulations with $APEX = i$ is

$$\gamma_{i-2}\gamma_{n-i-1}.$$

This implies that the average

$$A = \frac{1}{|T_n|}\sum_{\tau \in T_n} APEX_n(\tau) = O(\sqrt{n}); \tag{4}$$

also the variance $V = \sum_{\tau \in T_n}(A - APEX_n(\tau))^2 = O(n^{3/2})$. By symmetry, in a random triangulation, the apex of the triangle containing polygon edge $p_j p_{j+1}$ is expected to have distance $O(\sqrt{n})$ from that edge.

Finally, given central triangles t and t',

1. There is a path $P : t = u_0, u_1, \dots, u_k = t'$ in $G(C_n)$ with no interior u_i degenerate, and
2. $|T_{u_i}|$ is minimized on one of the endpoints unless both t and t' have a side of length ≤ 2. In this case we can guarantee that $\min(|T_{u_i}|, u_i \in P) \ge |T_t|/64$.

3. A Lower Bound on the Mixing Rate

We prove that the mixing rate is at least $\Omega(n^{3/2})$. Let $t = \Delta p_1 p_{n/2} p_{3n/4}$ be the initial central triangle and we start the chain at a random element $\tau \in T_t$. The walk continues while the central triangle avoids the interval I from $p_{2n/10}$ to $p_{3n/10}$, an interval of length $n/10$ (see Fig. 2). The important fact is that the subset $S \subset V$ whose central triangles have at least one vertex in I satisfies

$$P = \frac{|S|}{|V|} \ge c,$$

for some constant $c > 0$. Let N denote the number of steps of $MC(n)$ - starting at τ - before the walk hits a triangulation in S. We will show that

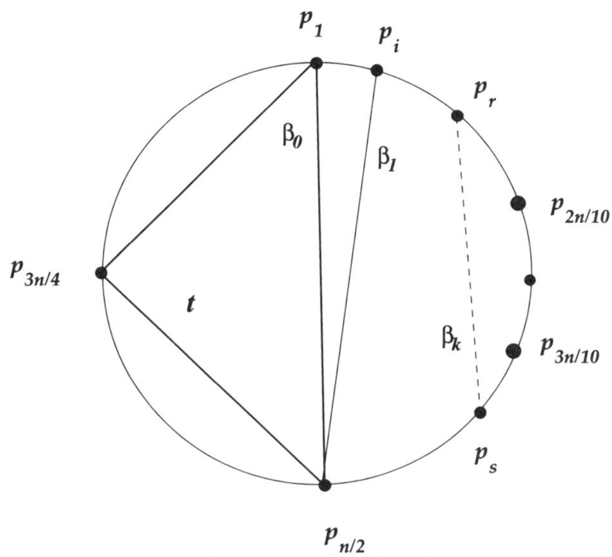

FIGURE 2. The Initial Central Triangle. $\beta_0 = p_1 p_{n/2}$ is the initial bounding edge; β_k is the edge after it has been flipped k times.

LEMMA 1. $E(N) \geq \Omega(n^{3/2})$.

From this it is easy to show that for any function $w(n) \uparrow \infty$,

$$\text{Prob}\left\{N < \frac{n^{3/2}}{w(n)}\right\} \to 0.$$

Since S has probability $P > 0$, the walk cannot be within ε of the uniform distribution in $n^{3/2}/w(n)$ steps. With a little more effort we can establish stronger concentration for N and eliminate $w(n)$.

PROOF. The initial central triangle t is on the left half of K. The edge $\beta = p_1 p_{n/2}$ is called the *bounding* edge: no vertex of t is to the right of it. Furthermore, the part of the triangulation to the right of the bounding edge is *random.* β is flipped with probability $1/(n-3)$ so $\geq n$ steps of the chain are expected before this flip. At the moment β is flipped, it is in a triangle that is part of a random triangulation of the right side of K, call it $\Delta p_1 p_i p_{n/2}$. If $i \in I$ the walk stops. Otherwise one of the subpolygons $p_1, \cdots, p_i, p_1$ or $p_i, \cdots, p_{n/2}, p_i$ includes I and the corresponding diagonal ($p_1 p_i$ or $p_i p_{n/2}$) becomes the new bounding edge β_1. According to (4), we expect one of the endpoints of the bounding edge to move towards I by at most $d\sqrt{n}$ vertices.

We apply the same argument inductively to the current bounding edge β_k. It will always be the case that (see Fig. 2)

1. The part of the current triangulation to the right of β_k is random.
2. The central triangle of the current triangulation is to the left of β_k.
3. At least n steps of the chain are expected before β_k is flipped.
4. If the walk continues, the new bounding edge β_{k+1} has one if its vertices closer to I than the corresponding vertex of β_k, and by an amount $\leq d\sqrt{n}$ vertices, expected.

Suppose the walk stops at the Mth step of the chain. The previous facts imply that $E(M) \geq d^{-1}n^{3/2}$. Since no central triangle has yet had a vertex in I, $N \geq M$ and we are finished. □

4. The Conductance Bound

The conductance $\Phi(G_n)$ is the minimum over all sets $S \subset V$, $|S| \leq |\overline{S}|$, of

$$\frac{1}{n-3}\frac{|\text{edges } S \to \overline{S}|}{|S|}.$$

Given $S \subset T_n$, write $r = |S|/|T_n|$ as its fraction of $V = T_n$. Theorem 1 asserts that for any $S \subset T_n$ there is a matching from S to $\overline{S}$ with

$$\geq \frac{|S||\overline{S}|}{|T_n|n^{\beta}} \tag{5}$$

edges of G_n and this implies that the conductance is greater than $1/(2n^{\beta+1})$.

For a central triangle $t \in C_n$ write $S_t = S \cap T_t$ and $r_t = |S_t|/|T_t|$ as its fraction of T_t. To prove (5) we need the following facts.

LEMMA 2. *For any central triangle t with $r_t \leq 12/23$, there is a matching with at least $\frac{2|S_t|}{n^{\beta}}$ edges from S_t to $\overline{S} \cap T_t$.*

LEMMA 3. *Consider central triangles t_1 with fraction $r_1 = |S_{t_1}|/|T_{t_1}|$ and t_2 with $r_2 = |S_{t_2}|/|T_{t_2}|$, and a path P (respectively cycle C) containing them in $G(C_n)$. Let MIN be the minimum of $|T_t|$ over all t on P (resp C). Then if no triangle internal to P (resp. on C) is degenerate, there is a matching between $S \cap (\cup_{t \in P(resp\ C)} T_t)$ and $\overline{S} \cap (\cup_{t\in P(resp\ C)} T_t)$ with*

$$\geq \frac{31|r_1 - r_2|MIN}{n^8} \left(resp\ \frac{40|r_1 - r_2|MIN}{n^8}\right)$$

edges. If there is a degenerate triangle interior to P (resp. on C) then we can find a matching with at least

$$\geq \frac{7|r_1 - r_2|MIN}{n^8} \left(resp\ \frac{8|r_1 - r_2|MIN}{n^8}\right)$$

edges.

LEMMA 4. *Consider adjacent triangles t_1 and t_2 in $G(C_n)$. Write $r_i = |S \cap T_{t_i}|/|T_{t_i}|$. If one of t_1 or t_2 is not degenerate then there is a matching between $S \cap (T_{t_1} \cup T_{t_2})$ and $\overline{S} \cap (T_{t_1} \cup T_{t_2})$ with*

$$\frac{63|r_1 - r_2| \min(|T_{t_1}|, |T_{t_2}|)}{n^8}$$

edges. If both t_1 and t_2 are degenerate then there is such a matching with at least

$$\frac{15|r_1 - r_2| \min(|T_{t_1}|, |T_{t_2}|)}{n^8}$$

edges.

Proof of Theorem 1: We prove (5) with $\beta = 11$. Because $r = |S|/|T_n| \leq 1/2$, the asserted size of the matching is less than 1 unless $\gamma_{n-2} > 2n^{11}$, so we proceed by induction on n, now assuming that $n \geq 35$.

If there were big matchings from $S \cap T_t$ to $\overline{S} \cap T_t$ for enough of the T_t we would be done. Let $Big = \{t \in C_n : r_t > 12/23\}$ and $T_{Big} = \bigcup_{t \in Big} T_t$. Applying Lemma 2 to central triangles *not* in Big and summing, we find there is a matching of size

$$2 \sum_{t \notin Big} \frac{|S \cap T_t|}{n^\beta} = 2 \frac{|S \cap \overline{T_{Big}}|}{n^\beta}.$$

The right hand side exceeds the asserted bound unless $|S \cap \overline{T_{Big}}| < |S|/2$, so we may assume $|S \cap T_{Big}| \geq |S|/2$.

This implies that there is a central triangle t_1 for which $r_{t_1} \geq 12/23$. Also, since $r \leq 1/2$, there is a $t_2 \in C_n$ with $r_{t_2} \leq 1/2$ and consequently we have a pair of central triangles with $|r_{t_1} - r_{t_2}| \geq 1/46$. We apply the second bound from Lemma 3. By (2), $MIN \geq 9|V|/(n^3)$, so there is a matching with at least

$$\frac{63|r_{t_1} - r_{t_2}||V|}{n^{11}}$$

edges. This says there is a matching with more than $63|S|/(23n^{11})$ edges. □

Proof of Lemma 3: Take a 2 colouring of the edges of the path (respectively three colouring of the cycle). Choose the colour class which maximizes $\sum |r_a - r_b|$, the sum over all $t_a t_b \in$ *colour class*. To get the desired result, sum over the edges of this colour class, applying Lemma 4 to each edge. □

Remark 1: Write $t_1 = \Delta p_i p_j p_k$ and $t_2 = p_i p_j p_{k'}$, $|k - k'| = 1$. If there were a big matching linking T_{t_1} to T_{t_2} then Lemma 4 would be true even without the $63/n^8$ factor. Unfortunately this never occurs; T_{t_1} and T_{t_2} are rarely even the same size. However there is a matching linking the subset of Y_1 of T_{t_1} in which $p_{k'}$ is an ear in the quotient polygon containing p_k with the subset Y_2 of T_{t_2} in which p_k is an ear in the quotient polygon containing $p_{k'}$. In fact the edges between these two sets form a matching. Furthermore, Y_1 contains about a quarter of T_{t_1} and Y_2 contains about a quarter of T_{t_2}. So, if $|S \cap Y_i|/|Y_i|$ were equal to r_i then Lemma 4 would be true with $63/n^8$, replaced by $1/4$. The only dificulty in the proof occurs when either r_1 is significantly different from $|S \cap Y_1|/|Y_1|$ or r_2 is significantly different from $|S \cap Y_2|/|Y_2|$.

To handle this situation, we apply the following lemma, which may have independent interest, to the two appropriate quotient polygons:

Fix m and write $S(m) = S \cap E_m$ for the elements of S with ears at p_m, To estimate the number of edges entering $S(m)$, define the dispersion

$$D_m(S) = \left| |S(m)| - |S| \frac{|E_m|}{|T_n|} \right|. \tag{6}$$

This measures the deviation from the expected number of elements of S with ears at p_m, assuming S were a randomly chosen set of size $|S|$.

LEMMA 5. **(The "Ear Lemma")** *For any $m = 1, \ldots, n$ there is a matching from S to $\overline{S}$ with at least $D_m(S)/n^8$ edges of G.*

PROOF. We recursively prove Lemmas 4 and 5 in the following way: We prove Lemma 4 for any n assuming Lemmas 4 and 5 for all $n' < n$. We prove Lemma 5 for any n assuming both lemmas for $n' < n$ and Lemma 4 for n. The proofs proceed by induction on central triangles. Here is a sketch.

We can take $|S(m)| \geq |S||E_m|/|T_n|$ in (6) since $D_m(S) = D_m(\overline{S})$ and the absolute value is redundant in one of these two cases. Given $t = \Delta p_i p_j p_k \in C_n$, the *ring* $R = R(t)$ is the set

$$\{t, \Delta p_{i+1}p_{j+1}p_{k+1}, \dots, \Delta p_{i-1}p_{j-1}p_{k-1}\}$$

of rotations of t (subscripts always mod$n$). Write

$$T_R = \bigcup_{t\in R} T_t$$

and $S_R = S \cap T_R$, and let $\Re$ be the set of all rings. A ring is *non-degenerate* if it consists of non-degenerate triangles.

For each $\ell = 1, \dots, \lfloor n/3+1 \rfloor$ the *level* X_ℓ is the set of $t \in C_n$ such that the minimum length quotient polygon has size ℓ, and we write

$$T_{X_\ell} = \bigcup_{t\in X_\ell} T_t$$

and $S_{X_\ell} = S \cap T_{X_\ell}$. We note that X_1 is empty if ℓ is odd and X_2 is empty if ℓ is even. We need to treat X_1 and X_2 differently from X_ℓ, $\ell \geq 3$, and we call then degenerate levels. Finally write for each $t \in C_n$, ring R, and ℓ

$$r = \frac{|S|}{|T_n|}, \quad r_t = \frac{|S_t|}{|T_t|}, \quad r_R = \frac{|S_R|}{|T_R|}, \quad r_\ell = \frac{|S_{X_\ell}|}{|T_{X_\ell}|},$$

and observe that

$$D_m(S) = \sum_{t\in C_n} \Big(|S_t \cap S(m)| - \frac{|S|}{|T_n|}|T_t \cap E_m|\Big). \tag{7}$$

If p_m is a vertex of $t = \Delta p_i p_j p_k$ then $|T_t \cap E_m| = 0$. Otherwise, letting λ_t denote the size of the quotient subpolygon containing p_m and using (3), it is $|T_t|(1+3/(2\lambda_t - 5))/4$, and we have

$$D_m(S) = \sum_{t\in C_n} \Big(|S_t \cap S(m)| - \frac{r|T_t|}{4}(1 + \frac{3}{2\lambda_t - 5})\Big),$$

where we understand that $\lambda_t = 1$ if t contains p_m. Clearly

$$D_m(S) = \sum_{t\in C_n} \Big(|S_t \cap S(m)| - \frac{r_t|T_t|}{4}(1 + \frac{3}{2\lambda_t - 5})\Big)$$

$$+\sum_{\ell=1}^{n/3} \sum_{R\in X_\ell} \sum_{t\in R} [(r_t - r_R) + (r_R - r_\ell) + (r_\ell - r)]\frac{1}{4}|T_t|(1 + \frac{3}{2\lambda_t - 5}),$$

and rewriting the second summations,

$$D_m(S) = \sum_{t\in C_n} \Big(|S_t \cap S(m)| - \frac{r_t|T_t|}{4}(1 + \frac{3}{2\lambda_t - 5})\Big)$$

$$+\frac{1}{4} \sum_{R\in\Re} \sum_{t\in R} (r_t - r_R)|T_t|(1 + \frac{3}{2\lambda_t - 5})$$

(8)

$$+\frac{1}{4}\sum_{\ell=1}^{n/3}\sum_{R\in X_\ell}\sum_{t\in R}(r_R - r_\ell)|T_t|(1+\frac{3}{2\lambda_t-5})$$

$$+\frac{1}{4}\sum_{\ell=1}^{n/3}\sum_{R\in X_\ell}\sum_{t\in R}(r_\ell - r)|T_t|(1+\frac{3}{2\lambda_t-5}).$$

We deal with these four summation terms one at a time. The idea is that if any are large, then the lemma is already true. On the other hand they cannot all be small.

Let σ_1 be the value of the first summation in (8) and we write

$$\sigma_1 = \sum_{t\in C_n} D^t(S),$$

where

$$D^t(S) = |S_t \cap S(m)| - \frac{r_t|T_t|}{4}(1+\frac{3}{2\lambda_t-5}).$$

We can interpret $D^t(S)$ as the dispersion in the quotient subpolygon containing p_m. Using Lemma 5 inductively and noting that the subpolygon has size $\lambda_t + 1 \le n/2$, $D^t(S)$ accounts for at least $D^t(S)/(n/2)^8$ edges from S_t to $\overline{S}\cap T_t$. Summing over C_n reveals that $D_m(S)$ accounts for at least $2^8\sigma_1/n^8$ edges from S to $\overline{S}$, so unless $\sigma_1 \le D_m(S)/2^8$, the ear lemma is already true.

Next consider

$$\sigma_2 = \frac{1}{4}\sum_{R\in\Re}\sum_{t\in R}(r_t - r_R)|T_t|(1+\frac{3}{2\lambda_t-5}),$$

the second summation term in (8), and consider a non-degenerate ring R. It consists of rotations of a triangle t with sides of length $a \le b \le c$, $a+b+c=n$, $n/3 \le c \le n/2$. For each t in R either (i) p_m is in t or p_m is in the quotient polygon of size (ii) $a+1$, (iii) $b+1$, or (iv) $c+1$. For each R, the term in σ_2 is

$$\sum_{t\in R}(r_t - r_R)\frac{|T_t|}{4}(1+\frac{3}{2\lambda_t-5}) = \sum_{t\in R} r_t\frac{|T_t|}{4} - r_R\sum_{t\in R}\frac{|T_t|}{4} + \sum_{t\in R}\frac{3}{2\lambda_t-5}\frac{|T_t|}{4}(r_t - r_R)$$

$$= \frac{|S_R|}{4} - \frac{|S_R|}{4} + \sum_{t\in R}\frac{3}{2\lambda_t-5}\frac{|T_t|}{4}(r_t - r_R).$$

Now, for each non-degenerate ring R, we let

$$D_R(S) = \frac{1}{4}\sum_{t\in R}\frac{3}{2\lambda_t-5}|T_t|(r_t - r_R)$$

be the term in the σ_2 sum which corresponds to R. For each non-degenerate R, we can find a cycle C_R in $G(C_n)$ containing no degenerate triangles, which contains all of R, and has exactly $3|R|$ vertices all of which are either in R or adjacent to an element of R. We call C_R, the *extended ring* for R. Each central triangle is in at most 7 extended rings since it has at most six neighbours in $G(C_n)$. Let $T_{C_R} = \bigcup_{t\in C_R} T_t$.

CLAIM 1. *(i) For each non-degenerate ring R, there is a matching between $S\cap T_{C_R}$ and $\overline{S}\cap T_{C_R}$ with at least $21D_R(S)/20$ edges. (ii) For the (at most one) degenerate ring R there is a matching from $S\cap T_R$ to $\overline{S}\cap T_R$ with at least $\sum_{t\in R}\frac{3}{2\lambda_t-5}|T_t|(r_t - r_R)$ edges.*

This claim is proved by choosing t_1 and t_2 which respectively maximize and minimize r_t over C_R and then applying Lemma 3 to t_1,t_2, and C_R. We omit the details.

Since each central triangle is in at most seven extended rings, the union of these matchings over the non-degenerate triangles has maximum degree seven, and hence has an eight colouring. Adding the edges of the disjoint matching from the degenerate ring to those of the largest color class, and summing the contributions of all the rings, we see that $D_m(S)$ accounts for at least $2\sigma_2/n^8$ edges from S to $\overline{S}$ so unless $\sigma_2 \leq D_m(S)/2$, Lemma 5 is already true.

Next we can apply Lemma 3 in a similar but more complicated fashion to show that the third term in (8), σ_3, must be less than $D_m(S)/400$, and finally, that σ_4, the fourth summation in (8), is less than $5D_m(S)/12$. Putting these statements together, the ear lemma is already true unless

$$D_m(S) \leq \frac{D_m(S)}{2^8} + \frac{D_m(S)}{2} + \frac{D_m(S)}{400} + \frac{5D_m(S)}{12} < D_m(S).$$

an impossibility that completes the proof. □

Proof of Lemma 4: We apply the Ear Lemma inductively, assuming it is true for polygons with at most $n-1$ sides. Suppose first the two triangles are non-degenerate. Let t_1, t_2, Y_1, Y_2 be as in Remark 1. Let $r_i' = |S \cap Y_i|/|Y_i|$. If $|r_1' - r_2'| \geq 64|r_1 - r_2|/n^8$, then the result holds by considering the matching linking Y_1 and Y_2. Otherwise, $|r_1 - r_1'| + |r_2 - r_2'| \geq (1 - (\frac{65}{n^8}))|r_1 - r_2|$. Since Y_i is just E_m for some ear in one of the quotient polygons defined by t_i, the result holds by applying the Ear Lemma to these two quotient polygon (we need to sum over all pairs of triangulations of the other polygons but this is straightforward). We get a 2^8 factor bonus because the quotient polygons are smaller. This explains the 63 in Lemma 4 which is not present in the Ear Lemma. If exactly one of the t_i is degenerate, we proceed as above, but now the matching is between a subset of T_{t_1} where we fix an ear in both quotient polygons. However the matching hits all of T_{t_2} so $MIN = |T_{t_2}|$ and the proof technique works. If t_1 and t_2 are both degenerate, then the fact that we consider two ears - not one - accounts for the loss of a factor of 4 in the result. We omit the details. □

Proof of Lemma 2: This is a straightforward, but detailed induction. We sketch a part of the argument to illustrate the idea.

If t is a central triangle with $r_t \leq 12/23$, we need to show that there is an S_t to $\overline{S_t}$ matching with at least $2|S_t|/n^\beta$ edges. Without loss of generality (see Fig. 1) we can take $|T_A| \geq |T_B| \geq |T_C|$. To simplify the calculations, we assume for now that t is not degenerate and that it does not have all three sides equal. Thus, A and B have fewer than $n/2$ vertices and C has fewer than $n/3$ vertices.

We need some notation. For $a \in T_A$, $b \in T_B$, and $c \in T_C$, let $T_{t,a,\cdot,\cdot}$ be the set of triangulations which agree with a on A. $T_{t,\cdot,b,\cdot}$ and $T_{t,\cdot,\cdot,c}$ are defined similarily. Further, $T_{t,a,b,\cdot}$ denotes those triangulations which agree with a on A and b on B. The restriction of these triangulations to C is the set T_C, so $|T_{t,a,b,\cdot}| = |T_C|$. We define $T_{t,a,\cdot,c}$ and $T_{t,\cdot,b,c}$ analogously. Finally note that $|T_{t,a,\cdot,\cdot}| = |T_B||T_C|$, $|T_{t,\cdot,b,\cdot}| = |T_A||T_C|$, and $|T_{t,\cdot,\cdot,c}| = |T_B||T_A|$. Also, $|T_{t,\cdot,b,c}| = |T_A|$ and $|T_{t,a,\cdot,c}| = |T_B|$. For each index set I defined above the intersection of S with T_I is denoted S_I, and $\overline{S_I} = T_I \backslash S_I$.

Now, for $a \in T_A$, we say that a is *dense* if $|S_{t,a,\cdot,\cdot}| \geq 3|T_{t,a,\cdot,\cdot}|/4$; otherwise a is *sparse*. First suppose (**Case 1**) that $\sum_{a\ sparse} |S_{t,a,\cdot,\cdot}| \geq |S_t|/4$. In this case, it is enough to show that for any sparse a there is a matching with at least

$$\frac{8|S_{t,a,\cdot,\cdot}||\overline{S_t}|}{n^\beta} \tag{9}$$

edges from $S_{t,a,\cdot,\cdot}$ to $\overline{S_{t,a,\cdot,\cdot}}$. Lemma 2 would then follow by summing over the sparse a.

To prove that (9) holds, fix some sparse a. For $b \in T_B$, we say that b is *a-dense* if $|S_{t,a,b,\cdot}| \geq 31|T_{t,a,b,\cdot}|/32$. Otherwise b is *a-sparse*. We must have either

1. $\sum_{b\ a-sparse} |S_{t,a,b,\cdot}| \geq |S_{t,a,\cdot,\cdot}|/32$ or
2. $\sum_{b\ a-sparse} |S_{t,a,b,\cdot}| \leq |S_{t,a,\cdot,\cdot}|/32$

In the first case we apply induction to T_C and see that for each a-sparse b there is a matching $M_{t,a,b}$ between $S_{t,a,b,\cdot}$ and $\overline{S_{t,a,b,\cdot}}$ with at least

$$\frac{|S_{t,a,b,\cdot}||\overline{S_{t,a,b,\cdot}}|}{|T_C||C|^\beta} \geq \frac{3^\beta|S_{t,a,\cdot,\cdot}|}{32n^\beta}$$

edges. Summing over the a-sparse b, we obtain (9).

In the second case, we let B_1 be the set of a-dense b and B_2, the set of a-sparse b. For each $c \in T_C$ define

$$EX_1(c) = \{b \in B_1 : (a,b,c) \notin S\}$$

and

$$EX_2(c) = \{b \in B_2 : (a,b,c) \in S\},$$

the *exceptional sets* for B_1 and B_2. We argue that these sets are likely to be small and then apply induction to B aargue that there is a big $S_{t,a,\cdot,c}$ to $\overline{S_{t,a,\cdot,c}}$ matching for a fixed fraction of $c \in T_C$. Summing over these c we again obtain (9). The case where $\sum_{a\ sparse} |S_{t,a,\cdot,\cdot}| \leq |S_t|/4$ is argued similarly. □

References

[1] D. Arnold and M. Sleep. Uniform Random Number Generation of n Balanced Parenthesis Strings. *ACM Trans. on Prog. Languages and Systems 2*, 122-128, 1980.

[2] M. Atkinson and J. Sack. Generating Binary Trees at Random. *Inf. Proc. Letters 41*, 21-23, 1992.

[3] L. Devroye, P. Flajolet, F. Hurtado, M. Noy, and W. Steiger. Properties of Random Triangulations and Trees. *Discrete and Comp. Geom.* (to appear) 1998

[4] L. Devroye. A Note on the Height of Binary Search Trees. *J.A.C.M. 33*, 489-498, 1986.

[5] L. Devroye. Application of the Theory of Records in the Study of Random Trees. *Acta Informatica 26*, 123-130, 1988.

[6] P. Flajolet and A. Odlyzko. The Average Height of Binary Trees and Other Simple Trees. *J. Comp. Syst. Sciences, 25*, 171-213, 1982.

[7] F. Hurtado and M. Noy. The Graph of Triangulations of a Convex Polygon. *ACM Symposium on Computational Geometry (to appear)*, 1996.

[8] H. Mahmoud. *Evolution of Random Search Trees.* John Wiley, 1992.

[9] L. McShine and P. Tetali. On the Mixing Time of the Triangulation Walk and Other Catalan Structures. *Proceedings of the DIMACS Workshop on Randomization Methods in Algorithm Design*, P. Pardalos, S. Rajasekaran, and J. Rolim, eds., Amer Math. Soc., Providence, RI, 1998.

[10] A. Meir and J. Moon. On the Altitude of Nodes in Random Trees. *Canad. J. Math. 30*, 997-1015, 1978.

[11] M. Molloy, B. Reed, and W. Steiger. On the Mixing Rate of the Triangulation Walk. Abstract 924-05-225 *Abstracts of Papers Presented to the AMS, Vol. 18, No. 3*, p. 432, June 1997.

[12] B. Pittel. On Growing Random Binary Trees. *J. Math. Anal. Appl. 103*, 461-480, 1984.
[13] A. Sinclair. Algorithms for Random Generation and Counting: A Markov Chain Approach. *Birkhauser, Boston, 1992*
[14] D. Sleator, R. Tarjan, and W. Thurston. Rotation Distance, Triangulations, and Hyperbolic Geometry. *J. Amer. Math. Soc. 1*, 647-681, 1988.

Department of Computer Science, University of Toronto, Toronto, Ontario
E-mail address: `molloy@cs.toronto.edu`

CNRS, Paris, France
Current address: IME, University of Sao Paulo, Sao Paulo, Brazil
E-mail address: `reed@lug.ibp.fr`

Department of Computer Science, Rutgers University, New Brunswick, NJ
E-mail address: `steiger@cs.rutgers,edu`

DIMACS Series in Discrete Mathematics
and Theoretical Computer Science
Volume **43**, 1999

WHEN AND HOW n CHOOSE k

IGOR PAK

ABSTRACT. We discuss different ways of generating k-subsets of an n-set, as well as of permutations, matrices and flags over the finite field $\mathbb{F}_q$. Connections with the Markov Chain approach and Bruhat decomposition are established. The black box groups approach is defined. No previous knowledge of the subject is assumed.

INTRODUCTION

The problem of choosing one or few objects among many has a long history and probably existed since the beginning of the human era (e. g. *"Choose twelve men from among the people,..."* Joshua 4:2). Historically this choice was mostly rational and random choice was considered to be a bad solution. Times have changed, however. Problems have become enormously complex and computers have intruded into virtually every part of human activity. In many cases random solution has become desirable, if not the only possibility. Which means that it's about time we understand the nature of a random choice.

The ultimate goal of this paper is to explain the meaning of the words *"pick at random a k-subset on an n-set"* and *"pick at random a k-dimensional subspace"*. Basically this paper is about details of the procedure of actually choosing this subset or subspace in real time. Somewhat unexpectedly these very details lead to some interesting and nontrivial questions.

For the past several years the author has discussed results in this paper with people working in different areas such as Combinatorics, Algebra, Probability Theory and Computer Science. It was quite striking to see that very few people actually understood the existing connections. Finally the author decided to write a self-contained review paper which would provide a basis for the mutual understanding. We restrict ourselves to study of the *classical cases* such as symmetric group S_n, k-subsets of $\{1, \dots, n\}$, full linear group $GL(n, \mathbb{F}_q)$ and k-dimensional subspaces of $\mathbb{F}_q^n$. In each of these cases we try to develop and compare various generating algorithms. While most of the results are well known, some are new and may be of interest.

Here is a general setup of a problem. We are interested in generating a uniformly chosen element from some finite set $\mathcal{S}$. If $\mathcal{S}$ has a global description, we can use it to get the desired random elements. One can think of it as reducing our randomization procedure to several coin flips. The problem is to optimize the number of steps in

---

1991 *Mathematics Subject Classification.* Primary 20-04, 05-04; Secondary 20B40, 20G40, 60G40, 60J15, 68R05.

Supported by NSF Postdoctoral Research Fellowship.

the algorithm as well as the number of coin flips required (which are also called *random bits*). The first half of the paper describes how to do that in the classical cases.

Suppose now that the global structure of our set $\mathcal{S}$ is too complicated to be used efficiently in order to generate random elements. The usual way to proceed then, is to define a set of local transformation, each of them easy to describe, and then run a random walk by choosing a local transformation at random, independently at each step. For example, consider a group G and its set of generators S. We can start at the identity and run a random walk using multiplications by a random generator.

Under mild conditions the random walk will converge, often extremely fast, to a uniform distribution. Thus after walking long enough we get a nearly uniform random elements in $\mathcal{S}$. Now the problem is to find how long is "long enough", which is to find a mixing time for our random walk. Unfortunately even in nice cases this becomes a hard problem.

There is another way to look at this problem. Consider a stopping rule which observes the walk and stops it based on the previous observations. If the stopping state of this rule is uniform, this will give us a new algorithm for generating random elements in $\mathcal{S}$. On the other hand, if the stopping state is uniform even conditioned on the stopping time, the expected stopping time is roughly the mixing time (see §6.1 the precise formulation). Such a stopping time is called a *strong uniform time*. Now, given such a strong uniform time we solve both problems at once (of generating a random element and of finding the mixing time of a random walk). In section 6 we present several constructions of strong uniform times.

Finally, suppose we are given only the local structure of the set $\mathcal{S}$ and little or no global description. For example, suppose our group G is defined by its set of generators S lying in a bigger group $G_0 \supset G$. This situation is known as a *black box group*. Now we have to run some kind of random walk to achieve near-uniformity. The theoretically important algorithm due to Babai shows that this can be done in $O(log^5|G|)$ steps. Another important question to ask is whether a black box group is isomorphic to a given group. While this question cannot be answered efficiently in general, it has a fast solution in nice cases.

The paper is constructed in the following way. We do not highlight the results into theorems or lemmas but rather divide the body of the paper into sections, subsections and subsubsections, each with a certain purpose. The sections and subsections inside the sections are made to be increasing in difficulty. The remarks and some references are provided at the end of the sections.

The paper is written in a manner in which there is less emphasis on the results and more on explanation of how one should approach the problem. Some of the technical details and generalizations were left out or just briefly mentioned. The idea was to make this paper "easy reading" rather than overwhelm it with results.

The first section of the paper is introductory; it contains several basic definitions, notations and some elementary computations. In the second section we discuss several elementary generating algorithms. In particular, we present a "smart" way of choosing a k-subset in an n-set without keeping too much data in the memory. Although virtually all the results in this section are pure classics these days, this section cannot be skipped without loss of understanding.

In section 3 we show how to choose a permutation or a k-subset from a particular non-uniform distribution. The results of this section will be used later for random sampling of complete flags and of k-dimensional subspaces. In Section 4 we introduce the concept of *Schubert cells* and prove several basic results about their structure. Later, in section 5 we use Schubert cell decomposition to generate random flags and subspaces over a finite field.

In section 6 we study some natural random walks on permutations, k-subsets and k-dimensional subspaces. We show that the mixing time for these walk is of the same order (up to a logarithmic factor) as the complexity of the algorithm to generate the corresponding combinatorial objects. This confirms the general rule that natural random walks converge extremely fast.

Finally, in section 7 we briefly describe Babai's algorithm and give a heuristic for solving recognition problem for black box groups isomorphic to symmetric group and to special linear group.

Acknowledgments

I would like to express my gratitude to Persi Diaconis for introducing me to the subject and giving me a feel for randomness. I am indebted to David Aldous, Jim Fill, Michael Rabin, Ronitt Rubinfeld, Gian–Carlo Rota, Richard Stanley, Herb Wilf and Peter Winkler for the interesting and helpful conversations. Special thanks to Sergey Bratus, Gene Cooperman, Jason Fulman and Nathan Lulov for remarks on the manuscript.

This research started when the author was a graduate student at Harvard University and finished when the author was an NSF postdoctoral fellow at MIT.

Notation

$\mathbb{N} = \{1, 2, \dots\}$

$\mathbb{Z}_+ = \{0, 1, 2, \dots\}$

$[n] = \{1, 2, \dots, n\}$

$\begin{bmatrix} n \\ k \end{bmatrix}$ - k-subsets of an n-set $[n]$

$n! = 1 \cdot 2 \cdot \ldots \cdot n, \; n \in \mathbb{N}$

$\binom{n}{k} = \frac{n!}{k!\,(n-k)!} = \left| \begin{bmatrix} n \\ k \end{bmatrix} \right|$

$(n)_q = 1 + q + q^2 + \dots + q^{n-1}, \; n \in \mathbb{N}$

$(n!)_q = (1)_q \cdot (2)_q \cdot \ldots \cdot (n)_q, \; n \in \mathbb{N}$

$\binom{n}{k}_q = \frac{(n!)_q}{(k!)_q\,(n-k)!_q}$

$\mathfrak{h}_n = 1 + \frac{1}{2} + \frac{1}{3} + \dots + \frac{1}{n}$ - partial harmonic sum

$\mathbb{Z}$ - group of integers

$\mathbb{Z}_m \simeq \mathbb{Z}/m\mathbb{Z}$ - cyclic group with m elements

$\mathbb{F}_q$ - finite field with q elements

$GL(n, \mathbb{F}_q)$ - full linear group over the field $\mathbb{F}_q$

$B(n, \mathbb{F}_q)$ - Borel subgroup over the field $\mathbb{F}_q$

$U(n, \mathbb{F}_q)$ - unipotent group over the field $\mathbb{F}_q$

$Fl(n, \mathbb{F}_q)$ - set of complete flags over the field $\mathbb{F}_q$

$Gr(n, k; \mathbb{F}_q)$ - set of k-subspaces of $\mathbb{F}_q^n$

S_n - group of permutations of n elements

A_n - alternating group of permutations of n elements

$B_n = S_n \ltimes \mathbb{Z}_2^n$ - group of symmetries of an n-dimensional cube

$\beth(p)$ - coin with probability p of heads, $0 \le p \le 1$.

1. How many are there?

1.1 Permutations

By the *symmetric group* S_n we mean a group of all permutations of an n-set $[n] = \{1, 2, \dots, n\}$. The number of all permutations i.e. the order of S_n is $n!$. Indeed, there are n ways to choose position of 1, there are $n-1$ ways to choose position of 2 among remaining places, ... , finally there is only one remaining place left for n.

We use the standard notation to denote permutations: $(3, 2, 1) \in S_3$ is a permutation $1 \to 3$, $2 \to 2$ and $3 \to 1$.

1.2 k-subsets of n-set

By the $\begin{bmatrix} n \\ k \end{bmatrix}$ we denote the set of all k-subsets of n-set $[n]$. The number of k-subsets of an $[n]$ is given by the binomial coefficient:

$$\left| \begin{bmatrix} n \\ k \end{bmatrix} \right| = \binom{n}{k} = \frac{n!}{k!(n-k)!}$$

Indeed, consider a surjective map $\varphi : S_n \to \begin{bmatrix} n \\ k \end{bmatrix}$ which maps a permutation $\sigma \in S_n$ onto the set of its first k elements. Observe that the preimage of a k-subset $A \in \begin{bmatrix} n \\ k \end{bmatrix}$ consists of all permutations $\sigma \in S_n$ with given sets of the first k and last $n-k$ elements regardless of their relative order. Therefore $|\varphi^{-1}(A)| = k!(n-k)!$ and

$$\left| \begin{bmatrix} n \\ k \end{bmatrix} \right| = \frac{|S_n|}{|\varphi^{-1}(A)|} = \binom{n}{k}$$

In other words, the elements in $\begin{bmatrix} n \\ k \end{bmatrix}$ are in a bijection with cosets in G/B, where $G = S_n$, $B = S_k \times S_{n-k}$.

1.3 Nonsingular matrices

By $GL(n, \mathbb{F}_q)$ we denote the group of nonsingular matrices over the finite field $\mathbb{F}_q$ with q elements. We claim that

$$|GL(n, \mathbb{F}_q)| = q^{\binom{n}{2}}(q^n - 1)(q^{n-1} - 1) \cdot \dots \cdot (q - 1)$$

Indeed, the number of ways to choose the first row vector in a nonsingular matrix is $(q^n - 1)$: all but a zero row; the number of ways to choose the second row vector is $(q^n - q)$: all vectors that are not proportional to the first one, etc. This immediately yields the result.

1.4 Complete flags over $\mathbb{F}_q$

By $Fl(n, \mathbb{F}_q)$ we denote the set of all *complete flags* in $\mathbb{F}_q^n$, i.e. the sequences $\emptyset = V^0 \subset V^1 \subset \cdots \subset V^n = \mathbb{F}_q^n$ where $\dim(V^i) = i$, $i = 1, \ldots, n$. We claim that the number of these flags is given by the formula

$$|Fl(n, \mathbb{F}_q)| = (n!)_q = \prod_{i=1}^{n} (1 + q + \cdots + q^{i-1})$$

Before we prove it, note that $(n!)_1 = n!$, which means that we can think of the complete flags as a *q-analog* of permutations, even though there is no such thing as a finite field with 1 element.

To count the number of complete flags, define a surjective map $\psi : GL(n, \mathbb{F}_q) \to Fl(n, \mathbb{F}_q)$. Take a matrix M with row vectors $v_1, \ldots, v_n$ and define a vector space $V^i = \langle v_1, \ldots, v_i \rangle$, $i = 1, \ldots, n$. As before, let us compute the preimage of a fixed flag $\emptyset = V^0 \subset V^1 \subset \cdots \subset V^n = \mathbb{F}_q^n$. Consider the *Borel subgroup* $B(n, \mathbb{F}_q) \subset GL(n, \mathbb{F}_q)$ of upper triangular matrices $B = (b_{i,j})$, $1 \le i \le j \le n$, $b_{i,j} \in \mathbb{F}_q$, $b_{i,i} \ne 0$. Denote row vectors of $B^T M$ by $\widetilde{v}_1, \ldots, \widetilde{v}_n$ and the corresponding flag $\widetilde{V}^0 \subset \widetilde{V}^1 \subset \cdots \subset \widetilde{V}^n$. By definition,

$$\widetilde{v}_i = b_{1,i} v_1 + \cdots + b_{i-1,i} v_{i-1} + b_{i,i} v_i$$

Observe that since $b_{i,i} \ne 0$, the linear span of V^{i-1} and $\widetilde{v}^i$ is exactly V^i. Therefore by induction we have $\widetilde{V}^i = V^i$ for all $i = 1, \ldots, n$. This means that the whole orbit of B^T lies in the preimage of ψ. Similarly one can show that there is nothing but orbit of B^T in the preimage and

$$|Fl(n, \mathbb{F}_q)| = \frac{|GL(n, \mathbb{F}_q)|}{|B(n, \mathbb{F}_q)|} = \frac{q^{\binom{n}{2}} (q^n - 1)(q^{n-1} - 1) \cdot \ldots \cdot (q - 1)}{q^{\binom{n}{2}} (q-1)^n} = (n!)_q$$

In other words, the complete flags in $Fl(n, \mathbb{F}_q)$ are in the bijection with cosets in G/B, where $G = GL(n, \mathbb{F}_q)$ and $B = B(n, \mathbb{F}_q)$ as above.

1.5 k-subspaces of $\mathbb{F}_q^n$

Define a *Grassmanian* $Gr(n, k, \mathbb{F}_q)$ to be the set of all k-dimensional subspaces $V^k \in \mathbb{F}_q^n$. We will show that

$$|Gr(n, k, \mathbb{F}_q)| = \binom{n}{k}_q = \frac{(n!)_q}{(k!)_q (n-k)!_q}$$

To prove that, define a map $\varphi : Fl(n, \mathbb{F}_q) \to Gr(n, k, \mathbb{F}_q)$ by forgetting all subspaces of a flag but the k-dimensional one. We have :

$$|Gr(n, k, \mathbb{F}_q)| = \frac{|Fl(n, \mathbb{F}_q)|}{|Fl(k, \mathbb{F}_q)| \cdot |Fl(n-k, \mathbb{F}_q)|} = \binom{n}{k}_q$$

Again, from here k-subspaces in $Gr(n, k, \mathbb{F}_q)$ are in the bijection with cosets in G/B, where $G = GL(n, \mathbb{F}_q)$ and $B = (b_{i,j}) \subset GL(n, \mathbb{F}_q)$, such that $b_{i,j} = 0$ for all $1 \le j < k < i \le n$.

1.6 Remark

As the reader may have noticed, everything in this chapter is very well known and can be generalized in many directions (see e.g. [GR, GJ, H1, H2, NW, P, S1, St]). For example, one may look at other root systems, such as C_n. The analogs of S_n for other root systems are called *Weyl group*, which in the case of C_n is the group of symmetries of the n-dimensional cube. Also, in this case the analog of $GL(n)$ is the *symplectic group* $Sp(2n)$ (see [H1, St, W]).

Note also that if we look at incomplete flags with a fixed sequence of dimensions we get q-multinomial coefficients as their cardinalities (see [GR, GJ, S1]).

2. How do you generate them? (First part)

The mathematical formulation of the question is the following: how do you choose an element of the classical set with equal probability?

2.1 Permutations

2.1.1 Here is a simple algorithm. Take an element 1 and put it on a line. Take 2. We have two possible places to put it: either to the left or to the right of 1. Do it with equal probability. Keep going. At the end we have n places to put n. Pick one at random and place n there. Stop.

2.1.2 There are thousands of others, more or less analogous ways to generate a random permutation. Here is a bit unusual way to do that. We are going to be marking elements of the permutation according to some rule. The rule is designed in such a way that each time we will have a random permutation of the marked elements with the rest of the elements remaining in their initial places.

Start with the identity permutation $id = (1, \ldots, n) \in S_n$ and mark the last element. We will mark a new element at each step. Suppose at step m we have m marked elements. Pick an unmarked element (it doesn't have to be a random choice; say it is the smallest unmarked element). Mark it. Exchange it with a randomly chosen marked element (including itself, in which case the element stays where it is). Stop after $(n-1)$ steps.

Checking by induction that we indeed get a random permutation is easy. Indeed, after m-th step we get a permutation with the new marked element be the smallest, the second smallest, etc., all with equal probability $\frac{1}{m+1}$. Therefore by induction it's a random permutation of these $m+1$ elements. This finishes the proof.

2.1.3 Define $T(i)$, $1 < i \leq n$ to be the following operation on a permutation $\sigma \in S_n$:

- Flip a coin $\beth\left(\frac{i-1}{i}\right)$. If heads, exchange $\sigma(i)$ and $\sigma(i-1)$.

Here by $\beth(p)$ we denote a coin with probability p of heads. Now, denote by $R(i) = T(i) \circ T(i-1) \circ \ldots \circ T(2)$ (means we apply $T(i)$ first, then $T(i-1)$, etc.).

Consider the following procedure. Starting with identity permutation and apply $R(2) \circ R(3) \circ \ldots \circ R(n)$. We claim that this gives us an algorithm for generating a random permutation. Indeed, after $R(2)$ we get either id or $(1,2)$ with equal probability. Suppose after $R(2) \circ R(3) \circ \ldots \circ R(k)$ every permutation of $[k]$ occurs with equal probability. After the first operation $T(k+1)$ the element $(k+1)$ stays with probability $\frac{1}{k+1}$ and the operations $T(k), \ldots, T(2)$ do not move it. Analogously the element $(k+1)$ gets to the place k with probability $\frac{k}{k+1} \cdot \frac{1}{k} = \frac{1}{k+1}$, etc. Thus after $R(k+1)$ the element $(k+1)$ gets to every place with equal probability. Recall that we started with a random permutation of $[k]$. Now use induction. This proves the claim.

2.1.4 Remarks

The algorithm in §2.1.2 was used by the author in [P] to bound the rate of convergence of random walks generated by transpositions (see also section 6 of this paper). The main advantage of this algorithm is that it can be easily generalized in many directions. Some of them can be found below.

The algorithm in §2.1.3 was introduced by Diaconis in [D2]. It is based on the following identity:

$$\overrightarrow{\prod_{k=n-1\ldots 1}} \overrightarrow{\prod_{i=1\ldots k}} \left(\frac{1}{i+1} e + \frac{i}{i+1} (i, i+1) \right) = \frac{1}{n!} \sum_{\sigma \in S_n} \sigma$$

where both sides are considered to be elements in group algebra $\mathbb{C}[S_n]$. Similar identities were obtained earlier by Jucys and Nazarov (see [J, N]) and in a less explicit context for any root system by Demazure (see [De]).

2.2 k-subsets of a n-set

2.2.1 There are also great many ways you can approach this problem. The most trivial way is simply to reduce the problem to the previous one by taking the first k elements of the permutation. Use §1.2 to prove that this works. Unfortunately when k is relatively small compared to n, this algorithm is very excessive.

2.2.2 One can try to use another simple idea. Start with an empty set. Pick an element $a_1 \in [n]$ in random and add it to your set. Pick an element $a_2 \in [n] \setminus \{a_1\}$ etc. until we get a_k. Stop. The disadvantage of this method is that it requires too much memory. It also outputs an unsorted k-subset so if we want it them to be sorted, we need to use an additional sorting at the end. This can be avoided by the tricked we present in the next subsection.

2.2.3 The algorithm works the following way. We will go over all numbers from 1 to n and for each we will decide whether we want to take it in a k-set or not. The only information we need to keep is the number l of elements we still need to choose ($l = k$ at the beginning) and the number m of elements left ($m = n$ at the beginning). When we accept or reject some element, we simply output it or not, change numbers l and m appropriately and move on to the next element. We claim that each time the probability of accepting a number is exactly $P(l, m) = \frac{l}{m}$. Indeed, this probability can be computed as follows.

$$P(l, m) = \frac{\left| \left\{ A \in \begin{bmatrix} m \\ l \end{bmatrix} : 1 \in A \right\} \right|}{\left| \begin{bmatrix} m \\ l \end{bmatrix} \right|} = \frac{\binom{m-1}{l-1}}{\binom{m}{l}} = \frac{l}{m}$$

This gives us a probabilistic rule for accepting/rejecting the next element. Stop when $l = 0$, i.e. when no elements are yet to be accepted.

Proof is clear.

2.2.4 Finally, we would like to present a version of the algorithm in §2.1.2 suited for this particular case. Start with the subset $A = \{1, \ldots, k\}$. At each step we will be marking one element in A, while making some changes in A. Now, suppose after m steps we have a set A with k elements of which exactly m are marked. Choose any unmarked element $a \in A$ (say, the smallest one). Flip a coin $\beth\left(\frac{m+1}{n-k+m+1}\right)$. If heads, mark the element and leave it there. If tails, choose randomly an element $b \in [n] \setminus A$, replace a by b and mark the latter. Stop after k-th step.

In order to show that this algorithm works the way it is supposed to, use the following trick. We can think that at the beginning all the elements $k+1, \ldots, n$ are marked and written to the right of A in some order. Then what we do each time is pick an unmarked element, mark it and exchange it with an unmarked one. What happens is that there is no need to exchange the element with the other one that are in the set A. That's where the number $\frac{m+1}{n-k+m+1}$ comes from. Hence we reduced the algorithm to the one considered in §2.1.2.

2.3 Nonsingular matrices over the finite field $\mathbb{F}_q$

2.3.1 The simplest known way to generate a random nonsingular matrix is just by choosing a random matrix, check whether it's nonsingular and if it is singular, try again. We will show that on average we wouldn't need many tries. Still, this algorithm takes about $O(n^3)$ steps for checking whether the matrix is nonsingular using the standard *Gaussian elimination algorithm.*

The first problem with this algorithm is that it uses on average more random bits than are really necessary. The second problem with this algorithm is that it is unclear how to generalize it for other algebraic groups such as the symplectic or orthogonal group over a finite field etc. (They are very sparsely distributed among all matrices). Later on we will explain the "right" way to generate nonsingular matrices which can be easily generalized in many directions.

2.3.2 **Lemma** *A randomly chosen matrix over $\mathbb{F}_q$ is nonsingular with probability greater than $1/4$ if $q = 2$ and $1/2$ if $q \geq 3$.*

Proof. The desired probability is given by

$$\begin{aligned} p(n,q) = \frac{|GL(n,\mathbb{F}_q)|}{q^{n^2}} &= \frac{q^n-1}{q^n}\,\frac{q^n-q}{q^n}\cdots\frac{q^n-q^{n-1}}{q^n} \\ &= \left(1-\frac{1}{q}\right)\left(1-\frac{1}{q^2}\right)\cdots\left(1-\frac{1}{q^n}\right) \\ &> \prod_{i=1}^{\infty}\left(1-\frac{1}{q^i}\right) = 1-\frac{1}{q}-\frac{1}{q^2}+\frac{1}{q^5}+\frac{1}{q^7}-\ldots \end{aligned}$$

The last product can be expanded into a sum using Euler's pentagonal theorem

$$\prod_{i=1}^{\infty}\left(1-\frac{1}{q^i}\right) = \sum_{m=0}^{\infty}\frac{(-1)^m}{q^{m(3m\pm1)/2}}$$

Therefore the desired probability can be bounded as

$$p(n,q) > 1 - \frac{1}{q} - \frac{1}{q^2} + \frac{1}{q^5} - \left(\frac{1}{q^6} + \frac{1}{q^7} + \frac{1}{q^8} + \dots\right) =$$

$$= 1 - \frac{1}{q} - \frac{1}{q^2} + \frac{1}{q^5} - \frac{1}{q^6}\frac{1}{1-\frac{1}{q}} = 1 - \frac{1}{q} - \frac{1}{q^2} + \frac{1}{q^5}\left(1 - \frac{1}{q-1}\right)$$

Finally we get

$$p(n,q) > 1 - \frac{1}{q} - \frac{1}{q^2}$$

In particular,

$$p(n,2) > 1 - 1/2 - 1/4 = 1/4$$

Also for all $q \geq 3$ we have

$$p(n,q) \geq p(n,3) > 1 - 1/3 - 1/9 = 5/9 > 1/2$$

This proves the lemma. □

We decided to include the proof of the lemma for the sake of completeness. From here one can see that the algorithm in §2.3.1 does indeed work very fast.

2.4 Complete flags over $\mathbb{F}_q$

2.4.1 We have a small problem here. Namely, we need to decide how we are going to encode the flags. By analogy with the idea described in §1.4 we will simply encode them by the sequence of vectors, which forms a basis $(v_1, \dots, v_n)$, and such that the vector subspaces are given by $V^i = \langle v_1, \dots, v_i \rangle$, $i = 1, \dots, n$.

2.4.2 Now we are ready to reduce the problem of generating a random complete flag to the problem of generating a random nonsingular matrix. Indeed, using the argument of §1.4 we can simply take row vectors of our randomly produced matrix and we are done. The big advantage of this algorithm is that it is very easy to code (assuming that the Gaussian elimination algorithm is available in some kind of mathematical package). The only problem is that one can actually generate a random complete flag much faster, in about $O(n^2)$ steps. You can't beat this bound since $O(n^2)$ is the number of steps required just to write the vectors down. We will come back to this question later in §5.1, where we present generalizations of the algorithm in §2.1.1.

2.4.3 If we were working over $\mathbb{C}$ we also could require our basis to be orthonormal, which would make this basis uniquely defined for a given flag. The only difference is that at the end we will now have to apply orthonormalization algorithm. Generally speaking we can do the same over finite fields, but some of the technical details become more complicated and we decided to omit this case.

2.5 k-subspaces of $\mathbb{F}_q^n$

2.5.1 Again we have a similar problem. We can try to encode k-dimensional subspaces by the k linearly independent vectors and try to generate them as in §2.2.1 by taking the first k vectors or a random matrix. In fact, we can pick a random $k \times n$ matrix and check whether it has the maximal rank (which it obviously does at least as often as is predicted by Lemma 2.3.2). Now use Gaussian elimination algorithm, which takes $O(k^2n)$ steps in this case.

2.5.2 To conclude, we would like to show how the algorithm in §2.2.4 can be modified to apply in this case.

Start with some k-dimensional linear subspace $V \in \mathbb{F}_q^n$. Mark it. We will be gradually unmarking parts of V, reducing the dimension of the marked subspace by one after each step. We use induction to get k-subspaces which are chosen at random conditioned on having contain the marked subspace. In other words, at the beginning our marked subspace is all of V so there is only one possible k-dimensional subspace containing the marked subspace. After one step we get a $k-1$-dimensional marked subspace and a random k-dimensional subspace containing it. Keep on decreasing the dimension of the marked subspace. At the end we don't have anything marked, so we have a randomly chosen k-dimensional subspace.

Suppose we are given a k-dimensional subspace V with the marked subspace $W \subset V$, $\dim(W) = m$. Choose *any* $(m-1)$-dimensional subspace $W' \subset W$. Unmark everything in $W \setminus W'$. Flip a coin $\daleth(P)$, where the probability $P = P(m,k,n)$ of heads is given by

$$P(m,k,n) = \frac{(k-m+1)_q}{(n-m+1)_q} = \frac{1+q+\cdots+q^{k-m}}{1+q+\cdots+q^{n-m}}$$

If it's heads, stay. If it's tails, choose a random $V' \supset W'$, $\dim(V') = k$, $\dim(V' \cap V) = 1$. Go to the next step. Stop when $W = \{0\}$.

To make this algorithm rigorous we need to define how we are going to choose subspaces V' and W' (see §5.2.3), explain how we are going to keep track of the spaces, and prove the correctness of the algorithm (see below).

2.5.3 To show why the above algorithm works, compare it with the algorithm in §2.2.4. Note that the idea is completely analogous except for the technical difference of *unmarking* the "random part" instead of marking it.

We prove the correctness of the algorithm by induction. We claim that after step i we have a random uniformly chosen k-subspace $V \in \mathbb{F}_q^n$ conditioned V contains the marked subspace W, where $\dim(W) = n-i$. The claim is clear at the beginning when $V = W = \mathbb{F}_q^n$. Define the probabilities

$$P'(m,k,n) = \frac{|\{V \in Gr(n,k,\mathbb{F}_q) : V \supset W\}|}{|\{V \in Gr(n,k,\mathbb{F}_q) : V \supset W'\}|}$$

If we assume that $P'(m,k,n) = P(m,k,n)$, then the step of induction is clear. Therefore modulo proof of the identity above, we prove the claim. Thus after k steps we indeed get uniform k-subspaces.

2.5.4 Now in order to finish the proof we need to show that the probabilities

$$P'(m,k,n) = P(m,k,n)$$

for all $0 \le m \le k \le n$.

Define $FL(m,k,n,\mathbb{F}_q)$ to be the set of sequences $\emptyset = V^0 \subset V^m \subset V^k \subset V^n = \mathbb{F}_q^n$, where $\dim(V^i) = i$. It is easy to see that

$$|FL(m,k,n,\mathbb{F}_q)| = \binom{n}{k}_q \binom{k}{m}_q$$

since the number of ways to choose the sequence $V^m \subset V^k \subset V^n$ can be thought as the number of ways to choose $V^k \subset V^n$ times the number of ways to choose $V^m \subset V^k$.

We have a group $GL(n,\mathbb{F}_q)$ acting transitively on $Fl(m,k,n,\mathbb{F}_q)$, which implies that the number of ways to choose a k-dimensional subspace which contain a given m-space is

$$c(m,k,n;q) = \frac{\binom{n}{k}_q \binom{k}{m}_q}{\binom{n}{m}_q} = \binom{n-m}{k-m}_q$$

From here we have

$$P'(m,k,n) = \frac{c(m,k,n;q)}{c(m-1,k,n;q)} = \frac{\binom{n-m}{k-m}_q}{\binom{n-(m-1)}{k-(m-1)}_q} = \frac{1+q+\cdots+q^{k-m}}{1+q+\cdots+q^{n-m}}$$

which finishes the proof.

We will get back to this algorithm in §5.2.3 with an explanation on how we can actually choose the subspaces.

2.6 Remarks

More about generation of permutations and k-subsets can be found in [K]. Various algorithms for generation of other combinatorial objects can be also found in [NW]. See also [DSh2] for the general treatment of subgroup algorithms.

The result of Lemma 2.3.2 is known, although use of the Euler's identity is not standard. Ironically Euler invented this identity also for computational purposes. He needed to compute the number $p(n)$ of partitions of n for large n ([An, Eu]). Indeed, the identity can be presented as follows

$$p(n) = p(n-1) + p(n-2) - p(n-5) - p(n-7) + p(n-12) + p(n-15) - \dots$$

which gives us a fast recursive algorithm.

Algorithm 2.2.3 can be found in [MR]. Another randomized algorithm was recently found in [Ra]. Algorithm 2.2.4 and its q-version given in §2.5.2 are probably new but not terribly interesting. They will be important to us when we get to the random walks in section 6.

3. What if you like some better than others?

3.1 Permutations

3.1.1 Statistics on permutations

There are great many different statistics on permutations and each of them can be transformed into a question about sampling from a nonuniform distribution associated with this statistics. Here is the definition.

Statistic on a finite set X is a function $\gamma : X \to \mathbb{Z}_+$. Fix some positive real number q. Define $\mathbf{S}(X, \gamma, q) = \sum_{x \in X} q^{\gamma(x)}$. The probability distribution on X associated with γ is defined as

$$P_{\gamma,q}(x) = \frac{q^{\gamma(x)}}{\mathbf{S}(X, \gamma, q)}$$

We needed to normalize it by $\mathbf{S}(X, \gamma, q)$ so that the total sum adds up to 1.

Now the question is: Given q and statistic γ, how can we sample from the distribution $P_{\gamma,q}$?

3.1.2 Inversions

Inversion in a permutation $\sigma \in S_n$ is a pair (i, j), $1 \le i < j \le n$ such that $\sigma(i) > \sigma(j)$. Define *inv* to be a statistic on S_n by the number of inversions. In other words,

$$inv(\sigma) = |\{(i, j) : \sigma(i) > \sigma(j), 1 \le i < j \le n\}|$$

Let us prove that

$$\mathbf{S}(S_n, inv, q) = \prod_{i=1}^{n} (1 + q + \cdots + q^{i-1}) = (n!)_q$$

Indeed, use the algorithm described in §2.1.1 and let t be a formal variable. We are going to compute the polynomial

$$f_n(t) = \sum_{\sigma \in S_n} t^{inv(\sigma)}$$

and then evaluate it at $t = q$.

Recall the algorithm in §2.1.1 for generating a permutation. Let us run this algorithm and keep track of the number of inversions it generates. There is only one way to put 1 on a line so $f_1 = 1$. There are two ways to put 2 : one to the right of 1 without any inversions and one to the left of 1 with one inversion. Therefore $f_2(t) = f_1(t)(1 + t)$ etc. There are n ways to put n : one with 0, 1, ... , $n - 1$ additional inversions. By induction we have that

$$f_n(t) = \prod_{i=1}^{n} (1 + t + \cdots + t^{i-1})$$

and by definition $\mathbf{S}(S_n, inv, q) = f_n(q)$. This completes the proof.

We would like to mention the important distinction here between the *polynomial* $f_n(t)$ and its *value* $f_n(q)$. Note also that $f_n(q) = (n!)_q$ is the number of complete flags over $\mathbb{F}_q$, where q is a power of a prime.

3.1.3 Standard sampling

Observe that the above stated idea can be reduced to an algorithm which generates a permutation σ with probability $q^{inv(\sigma)}/f_n(q)$. Indeed, when placing i simply place it in the right-most position with probability $1/(i)_q$, in the second position from the right with probability $q/(i)_q$, etc., in the left-most position i-th probability $q^{i-1}/(i)_q$, where $(i)_q = (1 + q + \cdots + q^{i-1})$. By definition we get exactly the probability distribution associated with statistic *inv* on S_n (i.e. by the number of inversions). Use induction to finish the proof.

Note also that if were trying to generalize an algorithm in §2.1.2, we wouldn't be able to prove the multiplicative formula for $f_n(t)$ since sometimes when inserting a new number we get a multiple of a different form. For example, if we had a permutation $(1, 2, 3) \in S_3$ with elements $1, 3$ marked, then three permutation occur after we exchanged 2 are $(2, 1, 3)$, $(1, 2, 3)$ and $(1, 3, 2)$ with 1, 0 and 1 inversions respectively, which is different from the needed 0, 1 and 2 distribution.

3.1.4 Product sampling

By analogy with §2.1.3, define $T(i; q)$, $1 < i \leq n$ to be the following operation on a permutation $\sigma \in S_n$.

- Flip a coin $\beth\left(\frac{(i-1)_q}{(i)_q}\right)$. If heads, exchange $\sigma(i)$ and $\sigma(i-1)$.

Denote $R(i; q) = T(i; q) \circ T(i-1; q) \circ \ldots \circ T(2; q)$. The algorithm works as follows. We start with a permutation $(n, n-1, \ldots, 1)$ and apply $R(2; q) \circ \ldots \circ R(n; q)$. We claim that this algorithm outputs a permutation σ with probability $P(\sigma; q) = \frac{q^{inv(\sigma)}}{(n!)_q}$. The proof goes through verbatim of the proof in §2.1.3 with the substitution of q-numbers in appropriate places. Note also that if we start with identity permutation $id \in S_n$ we get a permutation σ with probability $P(\sigma; \frac{1}{q})$.

Now observe that result in this subsection is equivalent to the following identity in $\mathbb{C}[S_n]$ which generalizes an identity presented in §2.1.4:

$$\overrightarrow{\prod_{k=n-1\ldots1}} \overrightarrow{\prod_{i=1\ldots k}} \left(\frac{q^i}{(i+1)_q} e + \frac{(i)_q}{(i+1)_q} (i, i+1) \right) = \frac{1}{(n!)_q} \sum_{\sigma \in S_n} \sigma \cdot q^{\binom{n}{2} - inv(\sigma)}$$

3.1.5 Metric on S_n

Define a *Coxeter graph* to be a graph with vertices in S_n and two permutation σ and σ' connected by an edge if and only if $\sigma = \sigma' \cdot (i, i+1)$ for some $i = 1, \ldots, n-1$. Call a *distance* $d(\sigma, \omega)$ the length of the minimal path between two permutation σ and ω. We call it distance because it is obviously nonnegative, symmetric, and satisfies the triangle inequality. We claim that

$$d(\sigma, \omega) = inv(\sigma^{-1}\omega)$$

First, observe that by the definition of a distance we have

$$d(\sigma, \omega) = d(\tau\sigma, \tau\omega)$$

for any $\sigma, \omega, \tau \in S_n$. Indeed, if we had a path $(\sigma, \sigma', \sigma'', \ldots, \omega)$, we would also have a path $(\tau\sigma, \tau\sigma', \tau\sigma'', \ldots, \tau\omega)$, and vice versa. Therefore it is enough to prove that

$$d(\sigma, id) = inv(\sigma)$$

for all σ (to see that, take $\tau = \sigma^{-1}$). We prove it by induction on the length of σ. Suppose the shortest path from σ to id starts with σ' and we know that $d(\sigma', id) = inv(\sigma') = l$. Hence $d(\sigma, id) = l + 1$ and we are left to prove that $inv(\sigma) = l + 1$.

By definition, there exists some i, such that $\sigma = \sigma' \cdot (i, i+1)$. It is easy to see that when we exchange i and $i+1$ in σ' we can either increase or decrease the number of inversions by exactly one depending on whether $(i, i+1)$ is an inversion or not. By induction it can't be $l - 1$. Therefore $inv(\sigma) = l + 1$ and we proved the result.

3.1.6 Braid Arrangements and Permutohedra

Consider a configuration of hyperplanes $\mathcal{H}_n = \bigcup H_{i,j} \subset \mathbb{R}^n$, where $H_{i,j}$ is defined as $(z_i - z_j = 0)$, $1 \leq i < j \leq n$. This configuration is usually called *braid arrangement.* The complement $(\mathbb{R}^n \backslash \mathcal{H}_n)$ is a disjoint set of regions which is in a natural bijection with the set of permutations. Indeed, it is not hard to see that each region is a set of a form $x_{\sigma(1)} > x_{\sigma(2)} > \cdots > x_{\sigma(n)}$ which makes the bijection obvious.

Construct a graph with vertices to be these regions and two vertices connected by an edge whenever we can cross only one hyperplane in order to get from one region to another. A little less obvious is the result that this graph is exactly a Coxeter graph after taking inverses of the permutations. Indeed, one can see that the hyperplanes adjacent to the region $x_{\sigma(1)} > x_{\sigma(2)} > \cdots > x_{\sigma(n)}$ are of the form $x_{\sigma(i)} = x_{\sigma(i+1)}$, $i = 1, \ldots n-1$ and therefore after crossing this hyperplane we end up at the region $x_{\sigma(1)} > \cdots > x_{\sigma(i)} > x_{\sigma(i+1)} > \cdots > x_{\sigma(n)}$ i.e. the new region corresponds to a permutation $\sigma' = (i, i+1)\sigma$ which proves the result.

Define a *permutahedron* to be a polytope in $\mathbb{R}^n$ with vertices $\big(\sigma(1), \ldots, \sigma(n)\big)$. This polytope is a translation of the Minkowsky sum of the vectors $e_i - e_j$, $1 \leq i < j \leq n$ and its 1-skeleton is again a Coxeter graph. The reason is that this polytope is dual to the braid arrangement which follows right from the definition. Another description of this polytope is a convex hull of its set of vertices which form a general orbit of a symmetric group acting on $\mathbb{R}^n$ by permutating coordinates. This is where the name comes from.

3.2 k-subsets of a n-set

3.2.1 Inversions

Define *inversion* in a k-subset $A \in \begin{bmatrix} n \\ k \end{bmatrix}$ to be a pair (i, j) such that $i \in A$, $j \in [n] \setminus A$, and $j > i$. Denote $inv(A)$ to be the number of inversions in A. We can think of inv as statistic on k-subsets of n-set. Here is a somewhat expected result (by analogy with §3.1.2)

$$\mathbf{S}(S_n, inv, q) = \binom{n}{k}_q$$

To show this formula, let us define a polynomial

$$f_{n,k}(t) = \sum_{A \in \begin{bmatrix} n \\ k \end{bmatrix}} t^{inv(A)}$$

We need to prove that

$$f_{n,k} = \frac{f_n}{f_k f_{n-k}}$$

We prove it using bijection φ defined in §1.2. Think about k-subsets of $[n]$ as permutations with increasing sequences of the first k and the last $n-k$ elements. For example, $A = \{1,4,6\} \in \begin{bmatrix} 6 \\ 3 \end{bmatrix}$ corresponds to $\sigma = (1,4,6,2,3,5) \in S_6$. Clearly $\varphi(\sigma) = A$ and by definition $inv(\sigma) = inv(A)$. On the other hand, when permuting first k and the last $n-k$ elements, we can only add the number of inversions in each part. Therefore

$$\sum_{\omega \in \varphi^{-1}(A)} t^{inv(\omega)} = t^{inv(A)} f_k(t) f_{n-k}(t)$$

Summing over all $A \in \begin{bmatrix} n \\ k \end{bmatrix}$ and dividing both sides by $f_k(t)f_{n-k}(t)$ we get the result.

3.2.2 Sampling

Suppose now we would like to sample an element $A \in \begin{bmatrix} n \\ k \end{bmatrix}$ from a probability distribution $P(\cdot)$:

$$P(A) = \frac{q^{inv(A)}}{\binom{n}{k}_q}$$

for some $q \in \mathbb{R}$, $q > 0$. As before, we can sample a permutation in the same manner as it is done in §3.1.3, and then take the first k elements (i.e. use φ). Of course one can do much better than that which we will show in the next subsection.

3.2.3 An algorithm

As in §2.2.3 we go over all numbers from 1 to n and for each of them decide whether we want to take it in a k-set or not. The only information we need to keep is the number l of elements we still need to choose and the number m of elements left, ($l = k$ and $m = n$ at the beginning).

We take an element $(n - m + 1)$ and decide whether we would like to take it in a k-set or not. To do that we simply flip a coin $\beth(P)$, where the probability $P = P(l,m)$ is given by the formula

$$P(l,m) = \frac{1 + q + \ldots q^{l-1}}{1 + q + \cdots + q^{m-1}}$$

If heads, take $(n - m + 1)$ in a k-set, and if tails skip this number and move to the next available. Stop when $l = 0$.

We claim that this algorithm does indeed produce a random k-subset A with probability proportional to the $q^{inv(A)}$. The algorithm itself is completely analogous to the one in §2.2.3. We just need to show that the probabilities $P(l,m)$ are chosen in the right way.

Indeed, let us compute the probabilities $P(l,m)$:

$$P(l,m) = \frac{\left|\left\{A \in \begin{bmatrix} m \\ l \end{bmatrix} : 1 \in A\right\}\right|}{\left|\begin{bmatrix} m \\ l \end{bmatrix}\right|} = \frac{\binom{m-1}{l-1}_q}{\binom{m}{l}_q} = \frac{(l)_q}{(m)_q}$$

Now, when we take a new elements into the k-subset we don't force any new inversions in the future since all the elements that can possibly make an inversion with $(n-m+1)$ must be smaller. On the other hand, the probability

$$Q(l,m) = 1 - P(l,m) = \frac{q^l(1+q+\dots q^{m-l-1})}{1+q+\dots+q^{m-1}}$$

has a multiple q^l because if we don't take an element $(n-m+1)$, by doing so we guarantee exactly l future inversions with all the numbers that *will be* added to our k-subset. This finishes the proof.

3.2.4 Metric on $\begin{bmatrix} n \\ k \end{bmatrix}$

There are several ways to define a distance on k-subsets. We briefly mention three of them.

1) Define a graph Γ_1 with all k-subsets as vertices and edges (A, B), where $A, B \in \begin{bmatrix} n \\ k \end{bmatrix}$ such that $|A \cap B| = k-1$. Define $d_1(A,B)$ to be a distance between A and B in this graph. It is easy to see that

$$d_1(A,B) = |A \cap B|$$

This is an interesting notion of a distance and it is symmetric under the natural action of S_n. Unfortunately this distance has nothing to do with inversions. In §6.3 we shall study nearest neighbor random walk on a graph Γ_1.

Here is another place where Γ_1 occurs. Define a *hypersimplex* to be the convex hull of characteristic functions of all k-subsets. It is an $(n-1)$-dimensional polytope which is a simplex when $k=1$. Another way to define this polytope is to take an orbit of the point $(1,\dots,1,0,\dots,0)$ (k ones) of S_n acting on $\mathbb{R}^n$ by permuting coordinates. An easy theorem says that Γ_1 corresponds to 1-skeleton of this polytope (see [EKK, Z]).

2) Define a graph Γ_2 on $\begin{bmatrix} n \\ k \end{bmatrix}$ as follows. Two k-subsets A and B are connected by an edge if and only if $|A \cap B| = k-1$, and the unique elements in $B \setminus A$, $A \setminus B$ are consecutive integers.

By definition Γ_2 is a subgraph of Γ_1, but it is no longer regular and has fewer symmetries. We can still define a distance $d_2(A,B)$ as a distance in this graph. One can easily see now that

$$d_2(J,A) = inv(A)$$

where $J = \{1,\dots,k\}$.

There is one more interpretation of Γ_2. Consider a set of all sequences $(a) = (a_1, \ldots, a_k)$ such that $n-k \geq a_1 \geq a_2 \geq \cdots \geq a_k \geq 0$. We say that two sequences are *similar* if they differ by 1 at exactly one place. Define a graph on these sequences by connecting pairs of similar sequences by edges. This graph appears to be isomorphic to Γ_2. To prove that we define an intermediate set of combinatorial objects which are easier to deal with and construct two simple bijections between our sets and this new set.

Consider set of all lattice paths in a rectangular $k \times (n-k)$, which start from the southwest corner, go either right or up, and end at the northeast corner. We can think of the area above the path as a diagram (which is called *Young diagram*) with the number of squares in each row corresponding to elements in a sequence. This gives the first bijection.

On the other hand, we can move along the path starting from the southwestern corner and ending at the northeastern corner. We need exactly n moves. Now we can either take an element into a k-subset or not depending whether our move was vertical or horizontal. It gives another bijection and establishes the result.

Although d_2 has some nice properties, we are not going to use this notion of a distance.

3) Denote the $\mathcal{INV}(A)$ to be a set of inversions in a k-subset A (see §3.2.1). Define a new distance as follows:

$$d_3(A, B) = |\mathcal{INV}(B) \setminus \mathcal{INV}(A)|$$

Although it is not true that $d_3(A, B) = d_3(B, A)$, we still have the triangle inequality:

$$d_3(A, B) + d_3(B, C) \geq d_3(A, C)$$

On the other hand there is an obvious connection with inversions:

$$d_3(J, A) = inv(A)$$

where $J = \{1, \ldots, k\}$. Moreover, if $\mathcal{INV}(A) \subset \mathcal{INV}(B)$, then

$$d_3(A, B) = d_2(A, B)$$

Therefore it makes sense to think of d_3 as some kind of "oriented distance".

As far as we know d_3 does not correspond to any distance in an oriented graph. It occurs in combinatorics of k-dimensional subspaces. We shall come back to it later in the paper.

3.3 Remarks

The number of inversions in a permutation can be generalized for any Weyl group where it becomes the *length* of an element (see [Bou, H2, W]).

The q-identities similar to the one given in §3.1.4 for all root systems were found in [Ch].

Inversions in permutations and k-subsets have been heavily studied in Combinatorics (see e. g. [GJ, S1, SW]). They will reappear in connection with flags and k-dimensional subspaces. There are many other important combinatorial statistics.

Braid arrangements are very important by themselves. Considered over $\mathbb{C}$ their complement is an Eilenberg-MacLain space $K(\pi, 1)$, where π is a colored braid group (see [Ar, Ca]).

Permutohedron as an abstract polytope is simply a Coxeter complex of S_n and Coxeter generators (see [CM]). Translations of permutohedron can tile the whole $(n-1)$-dimensional space. This is obvious once you consider Coxeter complex of affine Weyl group $\hat{A}_n$.

Both permutahedron and hypersimplex are examples of the *orbit polytopes*, which can be obtained by taking a convex hull of an S_n-orbit of a particular point in $\mathbb{R}^n$.

The distance d_1 is well known and has been studied in numerous papers (see e.g. [Be, DSh3]). Distance d_2 in a different form was also considered in various places (see e.g. [GJ, S1, SW]). Distance d_3 is probably new.

4. What do these flags look like, anyway?

The idea is to break a set of flags into many pieces (cells) with each of them looking like finite vector spaces. Then we can try to work with each of the cells individually.

4.1 Complete flags

4.1.1 Schubert cells

Let us fix a basis $E = (e_1, \ldots, e_n)$ in a vector space $V = \mathbb{F}_q^n$ and fix a complete flag $\emptyset = V^0 \subset V^1 \subset \cdots \subset V^n = V$ associated with E, where $V^i = \langle e_1, \ldots, e_i \rangle$, $i = 1, \ldots, n$ (cf. §1.4). Now with each flag $\emptyset = W^0 \subset W^1 \subset \cdots \subset W^n = V$ associate set of numbers

$$\alpha_{i,j} = \dim(V^i \cap W^j)$$

Define *Schubert cells* to be unions of flags with the same set of numbers $\alpha_{i,j}$, $1 \leq i, j < n$. Of course not all sets of numbers can appear and we will give a complete description of them in the next few subsections.

4.1.2 Unipotent group and its action on flags

Define the *unipotent group* $U(n, \mathbb{F}_q) \subset B(n, \mathbb{F}_q)$ to be the group of upper triangular matrices with 1's on the diagonal. Recall that complete flags correspond to cosets $B^T(n, \mathbb{F}_q) \setminus GL(n, \mathbb{F}_q)$ where $B(n, \mathbb{F}_q)$ is a Borel subgroup of all upper triangular matrices. We claim that Schubert cells correspond to double cosets $B^T(n, \mathbb{F}_q) \setminus GL(n, \mathbb{F}_q)/U(n, \mathbb{F}_q)$. In other words Schubert cells are the orbits of the unipotent group acting on flags through its action on matrices.

First, notice that by associativity of matrix multiplication it doesn't matter whether we first multiply a matrix from the right and then from the left or vice versa. Hence the action of $U(n, \mathbb{F}_q)$ on complete flags is indeed well-defined. Now, take a flag $W^0 \subset W^1 \subset \cdots \subset W^n = V$ associated with the basis $(w_1, \ldots, w_n)$, where w_i is an i-th row vector of a matrix M. Take an upper triangular matrix $U \in U(n, \mathbb{F}_q)$. We can think of the complete flag corresponding to $M \cdot U$ as *the same* flag written in different coordinates. Observe that since U is upper triangular, the original basis flag $V^0 \subset V^1 \subset \cdots \subset V^n$ wouldn't change either (see §1.4) and therefore all the dimensions of intersections will remain the same. It proves that the orbits of the unipotent group do indeed lie inside the Schubert cells. The other direction, which claims that the numbers $r_{i,j}$ actually distinguish between the orbits, will be proven later using a combinatorial description of the orbits.

4.1.3 Permutation flags

Since we haven't yet fully established the connection between Schubert cells and orbits of the unipotent group, in this subsection we will continue to analyze the latter.

With each permutation σ associate a *permutation flag* V_σ corresponding to a basis $(e_{\sigma(1)}, \dots, e_{\sigma(n)})$. We claim that each orbit of the unipotent group $U(n, \mathbb{F}_q)$ acting on the complete flags $Fl(n, \mathbb{F}_q)$ contains exactly one permutation flag.

First of all, no two flags can lie in the same orbit since otherwise they would have to be in the same Schubert cell (by the result in §4.1.2) and have the same set of numbers $\alpha_{i,j}$, which is clearly false. Therefore if we prove that each orbit contains at least one permutation flag, we would instantly prove not only this claim but also the result about Schubert cells being the orbits of the unipotent group.

Fix a permutation $\sigma \in S_n$ and consider the orbit $O(\sigma, \mathbb{F}_q)$ of the unipotent group containing the permutation flag V_σ, $V_\sigma \in O(\sigma, \mathbb{F}_q)$. Recall that in a basis E we can think of $O(\sigma, \mathbb{F}_q)$ as an image of the map ψ (see §1.4) of an orbit of the $U(n, \mathbb{F}_q)$ acting on $GL(n, \mathbb{F}_q)$ by the right multiplication. Now we are ready to give an explicit combinatorial description of the orbit $O(\sigma, \mathbb{F}_q)$ (and, as we will see in a moment, of Schubert cells).

4.1.4 Combinatorial construction

By definition a flag V_σ corresponds to a matrix $D(\sigma) = (d_{i,j})$, where $d_{i,j} = 1$, $j = \sigma(i)$ and $d_{i,j} = 0$ otherwise. By definition an orbit of $U(n, \mathbb{F}_q)$ is the set of all matrices that can be obtained by adding columns to the columns with bigger numbers. Since all the ones are in different columns in $D(\sigma)$, it is easy to see that we get a set of matrices $\mathcal{R}(\sigma)$ such that $R = (r_{i,j}) \in \mathcal{R}(\sigma)$ if and only if $r_{i,\sigma(i)} = 1$ and $r_{i,j} = 0$, $j < \sigma(i)$. Indeed, both conditions are necessary since we can't possibly change $r_{i,\sigma(i)}$ and any numbers to the left of it (in standard matrix notations). Suppose now we have a matrix $R \in \mathcal{R}(\sigma)$. There is nothing to the right of $(\sigma^{-1}(n), n)$. There is exactly one matrix element to the right of $(\sigma^{-1}(n-1), n-1)$ and we can make it zero by subtracting the $(n-1)$-th column from n-th column appropriate number of times. Proceed analogously until we are left with matrix $D(\sigma)$. Therefore $\mathcal{R}(\sigma)$ is indeed an orbit of $U(n, \mathbb{F}_q)$ in $GL(n, \mathbb{F}_q)$.

Now let us find the equivalence classes of matrices $R \in \mathcal{R}(\sigma)$ under the action of $B^T(n, \mathbb{F}_q)$ by multiplication from the left. In other words, we would like to find a representative in a preimage $\psi^{-1}(F)$ for each of the $F \in O(\sigma, \mathbb{F}_q)$. As we know (see §1.4) we can add a row of the matrix $R \in \mathcal{R}(\sigma)$ to a row with a bigger number. By an argument similar to the one above we can get all the elements below $(i, \sigma(i))$ to be zeros.

Note that in this construction new zeros are placed in those positions (i, j) which correspond to the inversions in σ (see §3.1.2). Indeed, if we have a matrix element to the right of $(i, \sigma(i))$ and below $(\sigma^{-1}(j), j)$, then we have $i > \sigma^{-1}(j)$ and $\sigma(i) < j$ i.e. pair $(i, \sigma^{-1}(j))$ is an inversion in σ.

Define $\mathcal{S}(\sigma, \mathbb{F}_q)$ to be a set of matrices $M = (m_{i,j})$, $M \in \mathcal{R}$, such that $r_{i,\sigma(i)} = 1$, and $r_{i,j} = 0$ if $j < \sigma(i)$ or $\sigma^{-1}(j) < i$. We just showed that for each $F \in O(\sigma, \mathbb{F}_q)$ there exist exactly one matrix $M \in \mathcal{S}(\sigma, \mathbb{F}_q)$ such that $F = \psi(M)$.

4.1.5 Example

Let $n = 5$, $\sigma = (2,4,3,1,5) \in S_5$. All matrices $R \in \mathcal{R}(\sigma, \mathbb{F}_q)$ and $M \in \mathcal{S}(\sigma, \mathbb{F}_q)$ look as follows.

$$R = \begin{pmatrix} 0 & 1 & * & * & * \\ 0 & 0 & 0 & 1 & * \\ 0 & 0 & 1 & * & * \\ 1 & * & * & * & * \\ 0 & 0 & 0 & 0 & 1 \end{pmatrix}, \quad M = \begin{pmatrix} 0 & 1 & * & * & * \\ 0 & 0 & 0 & 1 & * \\ 0 & 0 & 1 & 0 & * \\ 1 & 0 & 0 & 0 & * \\ 0 & 0 & 0 & 0 & 1 \end{pmatrix}$$

where by "$*$" we mean any number $x \in \mathbb{F}_q$. Four new zeros in M at places $(3,4)$, $(4,2)$, $(4,3)$ and $(4,4)$ correspond to four inversions in $\sigma = (2,4,3,1,5)$, namely to $(2,3)$, $(1,4)$, $(3,4)$ and $(2,4)$ respectively.

4.1.6 End of the proof

Recall that in §4.1.3 we showed that all we have left to prove is that each orbit of the unipotent group contains at least one permutation flag. We already know all permutation flags lie in different orbits. Now what we can do is compute the number of flags in orbits containing a permutation flag and then compare this number with the total number of complete flags.

By the §4.1.4 we have

$$|O(\sigma, \mathbb{F}_q)| = |\mathcal{S}(\sigma, \mathbb{F}_q)|$$

Recall that matrices $M \in \mathcal{M}(\sigma, \mathbb{F}_q)$ have exactly n ones, $0 + 1 + \cdots + (n-1) = \binom{n}{2}$ zeros to the left of ones and an additional $inv(\sigma)$ number of zeros which are below and to right of ones. Therefore the total number of "stars" in M (see example above) is $\left(n^2 - n - \binom{n}{2} - inv(\sigma)\right) = \binom{n}{2} - inv(\sigma)$ i.e.

$$|\mathcal{S}(\sigma, \mathbb{F}_q)| = q^{\binom{n}{2} - inv(\sigma)}$$

Summing over all permutations we get

$$\sum_{\sigma \in S_n} q^{\binom{n}{2} - inv(\sigma)} = q^{\binom{n}{2}} f_n\left(\frac{1}{q}\right) = f_n(q)$$

where $f_n(q) = (n!)_q = |Fl(n, \mathbb{F}_q)|$ is the total number of the complete flags of length n over $\mathbb{F}_q$ (see §1.4 and §3.2.1).

Let us summarize what we did. We proved that all the orbits of the unipotent group containing at least one permutation flag do not overlap and cover the whole set of complete flags. This implies that all orbits must contain a permutation flag which respectively implies that these orbits are indeed the Schubert cells. This completes the proof of the Schubert cell description.

4.2 k-subspaces of $\mathbb{F}_q^n$

The case of k-subspaces of $\mathbb{F}_q^n$ is almost exactly similar to the case of complete flags. So we will just state what Schubert cells look like in this case and skip much of the proof.

4.2.1 Schubert cells

As in §4.1.1, fix a basis $E = (e_1, \dots, e_n)$ in a space $V = \mathbb{F}_q^n$ and an associated with E complete flag $\emptyset = V^0 \subset V^1 \subset \cdots \subset V^n = V$ defined as $V^i = \langle e_1, \dots, e_i \rangle$, $i = 1, \dots, n$.

Now with each k-dimensional subspace $W \subset V$ we associate a set of numbers

$$\beta_i = \dim(V^i \cap W)$$

Define *Schubert cells* to be unions of subspaces with the same set of numbers β_i, $1 \le i < n$.

Rather than the whole unitary group in this case one should consider its subgroup $U(n, k, \mathbb{F}_q) \subset U(n, \mathbb{F}_q)$ such that $U = (u_{i,j}) \in U(n, k, \mathbb{F}_q)$ if and only if $u_{i,j} = 0$, $1 \le i < j \le k$ or $k < i < j \le n$.

The role of permutation flags would play *basis k-subspaces* i.e. subspaces generated by any subset of k basis vectors. Basis k-subspaces are in bijection with k-subsets $A \in \begin{bmatrix} n \\ k \end{bmatrix}$, $A = \{i_1, i_2, \dots, i_k\}$, $1 \le i_1 < i_2 < \cdots < i_k \le n$. They correspond to $k \times n$ matrices $D(A, \mathbb{F}_q)$ which have ones in places $(1, i_1)$, $(2, i_2)$, ..., (k, i_k), and zeros elsewhere.

As in the case of complete flags, we claim that each Schubert cell contains exactly one basis k-subspace and can be obtained as an orbit of $U(n, k, \mathbb{F}_q)$ acting on $Gr(n, k, \mathbb{F}_q)$. These orbits $O(A, \mathbb{F}_q)$ can be lifted up to a set of matrices $\mathcal{S}(A, \mathbb{F}_q)$ which can be obtained from matrices $D(A, \mathbb{F}_q)$ by putting "stars" on places to the right of ones and with no 1 in the same column. One can observe that

$$|O(A, \mathbb{F}_q)| = |\mathcal{S}(A, \mathbb{F}_q)| = q^{k \cdot (n-k) - inv(A)}$$

which by analogy with §4.1.6 proves the claim.

4.2.2 Example

Let $n = 9$, $k = 4$, $A = \{1, 4, 6, 7\} \in \begin{bmatrix} 9 \\ 4 \end{bmatrix}$. All matrices $M \in \mathcal{S}(A, \mathbb{F}_q)$ look as follows:

$$M = \begin{pmatrix} 1 & * & * & 0 & * & 0 & 0 & * & * \\ 0 & 0 & 0 & 1 & * & 0 & 0 & * & * \\ 0 & 0 & 0 & 0 & 0 & 1 & 0 & * & * \\ 0 & 0 & 0 & 0 & 0 & 0 & 1 & * & * \end{pmatrix}$$

The number of inversions $inv(A) = 8$. Here we have

$$|\mathcal{S}(A, \mathbb{F}_q)| = q^{4 \cdot (9-4) - 8} = q^{12}$$

which is easy to see since M has exactly 12 "stars" in it.

4.3 Remarks

Although the results in this section are known, our elementary approach is not standard. The careful treatment of Bruhat decomposition for all root systems can be found in [Bou, H1]. Lie groups over the finite field are called *Chevalley groups* (see e.g. [Bor, St]). The group $U(n, k, \mathbb{F}_q)$ introduced in §4.2 is an example of a *parabolic subgroup.* These parabolic subgroups can be used in a similar way to study incomplete flag varieties.

Schubert cells by themselves represent a major development. They are heavily used in different branches of mathematics. There is even a *Schubert calculus*, which is a part of Algebraic Geometry (see e.g. [GH, M2]). Quite recently people in Algebraic Combinatorics have also invaded this area (see [LS, M2, S2]).

5. How do you generate them? (Second part).

The idea is to generate flags or subspaces in two steps. First, choose a Schubert cell with probability proportional to the number of elements it contains. Second, choose an element in the chosen cell.

As we already know these Schubert cells look like vector spaces over $\mathbb{F}_q$ of dimension equal to $\binom{n}{2}$ minus the number of inversions. So the first step is basically done in §3 while the second is easy since we have a complete combinatorial description of these cells. Here is a careful explanation of this idea.

5.1 Complete flags

Let q_0 be a power of a prime, and let $q = \frac{1}{q_0}$. We would like to sample complete flags over $\mathbb{F}_{q_0}$ from the uniform distribution. Following §2.4.1 the output must be of a form $(v_1, \ldots, v_n)$.

5.1.1 The idea stated above can be reformulated the following way. By §4.1.4 and §4.1.6 choosing a complete flag is equivalent to first choosing the permutation σ and then a matrix $M \in \mathcal{S}(\sigma, \mathbb{F}_q)$. In the notation of §3.1.1 we therefore need to sample σ from the distribution $P_{inv,q}$, which can be done by the algorithm described in §3.1.3. Note that we need to use $q = \frac{1}{q_0}$ since by the result in §4.1.6 we have

$$|\mathcal{S}(\sigma, \mathbb{F}_q)| = q^{\binom{n}{2} - inv(\sigma)}$$

which forces us to take an inverse scale.

The second step is sampling a matrix $M \in \mathcal{S}(\sigma, \mathbb{F}_q)$. But this is easy since the set $\mathcal{S}(\sigma, \mathbb{F}_q)$ has an explicit description (see §4.1.4). Basically all we need to do is replace all the "stars" by the randomly chosen numbers in $\mathbb{F}_{q_0}$. Here is the final version of the algorithm.

5.1.2 Algorithm

1)Input n, q_0; $q := \frac{1}{q_0}$;
2)Sample $\sigma \in S_n$ from the distribution $P_{inv,q}$ using the Algorithm in §3.1.3;
3)Sample $M \in \mathcal{S}(\sigma, \mathbb{F}_q)$ uniformly;
4)Output M.

5.2 k-subspaces of $\mathbb{F}_{q_0}^n$

5.2.1

Again let q_0 be a power of a prime, $q = \frac{1}{q_0}$. We can work out the generating algorithm of the k-dimensional subspaces by analogy with flags. Following §2.4.1 the output must be of form k independent vectors $(v_1, \ldots, v_k)$. Now we can use a very efficient algorithm described in 3.2.3 to generate a k-subset from the nonuniform distribution associated with statistic *inv* and a constant q. Here is the new algorithm.

5.2.2 Algorithm

1)Input n, k, q_0; $q := \frac{1}{q_0}$;
2)Sample $A \in \begin{bmatrix} n \\ k \end{bmatrix}$ from the distribution $P_{inv,q}$ using the Algorithm in §3.2.3;
3)Sample $M \in \mathcal{S}(A, \mathbb{F}_q)$ uniformly;
4)Output M.

5.2.3 Now we can also formalize the idea of an algorithm in §2.5.2. As usual we will encode the k-dimensional subspace V by k linearly independent vectors $v_1, \dots, v_k$ In addition to that we will also keep in memory $n-k$ vectors $v_{k+1}, \dots, v_n$ which form a basis together with $v_1, \dots, v_k$. Since we are allowed to choose marked m-dimensional subspace W in any way we like, we will do it by making it to be a span of $v_1, \dots, v_m$, $0 \le m \le k \le n$.

Now let us describe what we do at each step. At step $k - m + 1$ first we need to choose an $m - 1$-dimensional subspace W' which is going to be a span of $v_1, \dots, v_{m-1}$. Then we need to choose a new k-dimensional subspace $V' \supset W'$. We flip a coin $\beth(P)$, where $P = P(m, k, n) = \frac{(k-m+1)_q}{(n-m+1)_q}$. If it's heads, let $V' = V$ and keep all the vectors.

If it's tails, we replace v_m by a vector $v'_m \notin V$ in such a way that a subspace $V' = \langle v_1, \dots, v'_m, \dots, v_k \rangle$ is randomly chosen k-dimensional subspace in condition $\dim(V' \cap V) = 1$ and $V' \supset W'$.

Let a vector v'_m to be of a form $a_1 v_m + \dots + a_{k-m+1} v_k + b_1 v_{k+1} + \dots + b_{n-k} v_n$ Choose a number $1 \le i \le n - k$ with probability

$$p(i) = \frac{q^{i-1}}{1 + q + \dots + q^{n-m}}$$

Set values $b_{i+1}, \dots, b_n = 0$, $b_i = 1$ and choose random values $a_1 \dots, a_{k-m+1}$ and $b_1, \dots, b_{i-1} \in \mathbb{F}_q$. Now have the subroutine output vectors $v_1, \dots, v'_m, \dots, v_k$.

What we basically did here was show how to choose a line in a $(n-k)$-dimensional space by choosing first a *Schubert cell* and then its random element. This establishes a rigorous procedure for the algorithm in §2.5.2.

5.3 Nonsingular matrices over $\mathbb{F}_{q_0}^n$

5.3.1 We can use the algorithm described in §5.1.2 in order to generate random nonsingular matrices. Indeed, as we showed in §1.4, complete flags correspond to the cosets $B^T(n, \mathbb{F}_q) \backslash GL(n, \mathbb{F}_q)$, so all we need to do is choose a random complete flag and multiply it by a random lower triangular matrix. Here is an algorithm formalizing the idea.

5.3.2 Algorithm

1)Input n, q_0; $q := \frac{1}{q_0}$;
2)Sample $\sigma \in S_n$ from the distribution $P_{inv,q}$ using the Algorithm in §3.1.3;

3)Sample $M \in \mathcal{S}(\sigma, \mathbb{F}_q)$ uniformly;
4)Sample $B \in B(n, \mathbb{F}_q)$ uniformly;
5)Compute $M' = B^T \cdot M$;
6)Output M'.

5.4 "Lazy programmer" solution

Sometimes you can see "lazy programmers" who are capable of sacrificing the speed of an algorithm for the sake of writing less of a computer code. We can actually help these people.

5.4.1 Complete flags

Observe that instead of sampling a random matrix $M \in \mathcal{S}(\sigma, \mathbb{F}_q)$ we could have sampled a random unipotent matrix $U \in U(n, \mathbb{F}_q)$ which is easy to do since these are upper triangular matrices with ones on diagonal and then output the product $D(\sigma, \mathbb{F}_q) \cdot U$, where $D(\sigma, \mathbb{F}_q)$ is a permutation matrix corresponding to σ.

Sampling a random unipotent matrix might be considered easier to code especially if it's already done in some kind of math. package. By using it we still get a uniform distribution of corresponding complete flags which is exactly what we need. It may take more time to sample a unipotent matrix since it may have many more random entries to be chosen. On average however it is off only by a factor of 2 since the average number of inversions in a permutation is $\frac{\binom{n}{k}}{2}$, simply because $f_n(t) = t^{\binom{n}{k}} f_n(t^{-1})$.

The constant 2 is so small that the employer of the "lazy programmer" can probably live with it. Here is an algorithm which realizes this idea.

Algorithm

1)Input n, q_0; $q := \frac{1}{q_0}$;
2)Sample $\sigma \in S_n$ from the distribution $P_{inv,q}$ using the Algorithm in §3.1.3;
3)Sample $U \in U(\sigma, \mathbb{F}_q)$ uniformly; $M := D(\sigma, \mathbb{F}_q) \cdot U$;
4)Output M.

5.4.2 k-subspaces of $\mathbb{F}_{q_0}^n$

Exactly the same thing happens in here. One can come up with the following algorithm.

Algorithm

1)Input n, k, q_0; $q := \frac{1}{q_0}$;
2)Sample $A \in \begin{bmatrix} n \\ k \end{bmatrix}$ from the distribution $P_{inv,q}$ using the Algorithm in §3.2.3;
3)Sample $U \in U(n, k, \mathbb{F}_q)$ uniformly; $M := D(A, \mathbb{F}_q) \cdot U$;
4)Output M.

5.4.3 Nonsingular matrices over $\mathbb{F}_{q_0}^n$

Finally in the case of nonsingular matrices we get an amazingly simple algorithm by choosing two upper triangular matrices. Even High School students can learn this algorithm but only few of them will really understand why it works.

Algorithm

1)Input n, q_0; $q := \frac{1}{q_0}$;
2)Sample $\sigma \in S_n$ from the distribution $P_{inv,q}$ using the Algorithm in §3.1.3;
3)Sample $B_1, B_2 \in B(n, \mathbb{F}_q)$ uniformly;
4)Compute $M = B_1^T \cdot D(\sigma, \mathbb{F}_q) \cdot B_2$;
5)Output M.

5.5 Product sampling

Here we present an analog of the algorithm in §2.1.3 for generating nonsingular matrices over the finite field $\mathbb{F}_q$. The idea is again based on the identity in the group algebra $\mathbb{C}[GL(n, \mathbb{F}_q)]$ (see §2.1.4).

Recall that by $D(\sigma)$, $\sigma \in S_n$ we denote a matrix $D(\sigma) = (d_{i,j})$, where $d_{i,j} = 1$, $j = \sigma(i)$ and $d_{i,j} = 0$ otherwise. Denote by $E(i, j; a)$, where $1 \le i, j \le n$, a matrix with 1-s on diagonal, $a \in \mathbb{F}_q$ in place (i, j) and 0 elsewhere. Define

$$\mathcal{E}(i,j) = \sum_{a \in \mathbb{F}_q} E(i,j;a)\,, \qquad \text{where } i \neq j$$

$$\mathcal{H}(i) = \sum_{a \in \mathbb{F}_q, a \neq 0} E(i,i;a)$$

$$\mathcal{E}_- = \overrightarrow{\prod_{i=2\ldots n}} \; \overrightarrow{\prod_{j=1\ldots i-1}} E(i,j)$$

$$\mathcal{E}_+ = \overrightarrow{\prod_{i=1\ldots n-1}} \; \overrightarrow{\prod_{j=i+1\ldots n}} E(i,j)$$

$$\mathcal{H} = \prod_{i=1\ldots n} \mathcal{H}(i)$$

An easy argument shows that

$$\mathcal{E}_- = \sum_{M \in U^T(n, \mathbb{F}_q)} M$$

and

$$\mathcal{H} \circ \mathcal{E}_+ = \sum_{M \in B(n, \mathbb{F}_q)} M$$

Now consider a product

$$\mathcal{D} = \overrightarrow{\prod_{k=n-1\ldots 1}} \; \overrightarrow{\prod_{i=1\ldots k}} \left(\frac{q^i}{(i+1)_q}\, id + \frac{(i)_q}{(i+1)_q}\, D(i, i+1) \right)$$

Recall that by §3.1.4 we have

$$\mathcal{D} = \frac{1}{(n!)_q} \sum_{\sigma \in S_n} q^{\binom{n}{2} - inv(\sigma)} \cdot D(\sigma)$$

Thus the result in the previous subsection implies that

$$\mathcal{E}_- \circ \mathcal{D} \circ \mathcal{H} \circ \mathcal{E}_+ = \sum_{M \in GL(n,\mathbb{F}_q)} M$$

which leads to another sampling algorithm, too sluggish to be useful.

5.6 Gaussian elimination algorithm

We would like to conclude this section by establishing the connection between Gaussian elimination algorithm and the Bruhat decomposition

$$GL(n) = \coprod_{\omega \in S_n} B(n)wB(n)$$

Indeed, it's a common mistake to think that all the Gaussian algorithm does is invert matrices. The other common mistake is to think that it breaks a matrix into the product of an upper triangular and a lower triangular ones. What it really does, is present a matrix as a product of a lower triangular, permutation matrix and an upper triangular one, which is exactly what the Bruhat decomposition does in case of the $GL(n)$.

To see that, let us recall how Gaussian elimination algorithm works. If the first upper left corner is non-zero, we subtract the first row from each of the rows below it with some coefficients to place zeros in the first column everywhere but in the upper left corner. Then we subtract the second row from each of the consecutive rows, etc. This way we get an upper triangular matrix as a result of the subtractions and a lower triangular matrix indicating what were the coefficients. Note however that if the upper left corner is zero, the conventional wisdom says that it's not a problem since all we need to do is find a row with a nonzero entry in the first column. Let it be the first nonzero entry. The same thing is repeated for the second row etc. and at the end we also get a permutation matrix.

In other words what we just did was show how to break the set of all matrices into cells using the Gaussian algorithm. One can proceed with this idea and prove all the results contained in §4.1. We just think it easier to understand as described here.

5.7 Remarks

Observe that the number of steps needed for algorithm 5.3.2 can be bounded by the number of steps one needs to multiply two matrices. The number of random bits required by the algorithm is about n^2 i.e. the same as in the case of the algorithm in §3.1.2.

Algorithm 5.2.2 originates in [CW]. Unfortunately the authors omitted many of the details.

Algorithms 5.1.2 and 5.3.2 seem to be new. Algorithm 5.4.3 is brand new and should be patented by somebody. Although known, the connection between Gaussian elimination algorithm and the Bruhat decomposition became a part of mathematical folklore nowhere to be found in print.

The subsection §5.1.5 is of theoretical impotence and is meant to answer a question of Diaconis whether there are analogs of the identity in §2.1.4 for the full linear group.

6. How about taking a random walk?

This is another approach commonly used in practice for the purposes of random generation. The idea is to start from somewhere and walk on the objects you wish to generate at random. One may think that after some large number of steps the walk is at a random state. It is actually true indeed and can be made precise and proved in large generality. In this section we will study several interesting random walks and show that up to a logarithmic factor they are almost as fast as the direct generating algorithms (see §2, §5).

6.1 Definitions

6.1.1 Random walks

Define the *nearest neighbor random walk* as follows. Start at any vertex x and flip a fair coin. If it's heads, stay there. If it's tails, move to a randomly chosen vertex adjacent to x. This defines a reversible aperiodic *Markov chain*. Denote $P^k(x)$ to be the probability of a walk being at a vertex $x \in \Gamma$ after k steps. If Γ is regular, the walk has a uniform stationary distribution, i.e.

$$P^k(x) \to \frac{1}{N} \quad \text{as } k \to \infty \quad \text{for all } x \in \Gamma$$

where $N = |\Gamma|$ is the number of vertices in Γ. For the rest of the paper we will consider only regular graphs.

Let G be a finite group, S be its set of generators. We say that S is symmetric if $s \in S$ implies $s^{-1} \in S$. Define a *Cayley graph* $\Gamma(G, S)$ to be a graph with group elements $g \in G$ as vertices and pairs $(g, g \cdot s)$ as edges, $s \in S$.

6.1.2 Separation distance and mixing times

There are several ways to measure how fast the random walk mixes, most notably the *total variation distance*

$$tv(k) = \frac{1}{2} \sum_{x \in \Gamma} \left| P^k(x) - \frac{1}{N} \right|$$

and the *separation distance*

$$s(k) = N \cdot \max_{x \in \Gamma} \left(\frac{1}{N} - P^k(x) \right)$$

For more general definitions, properties of these distances and references see e.g. [AF, D1, P].

Although the total variation distance is used more often than the separation distance, the latter will be better suited for our purposes. Here are a few properties of the separation distance:

$$s(0) = 1$$

$$s(m+k) \le s(m) \cdot s(k)$$

$$s(k) \le tv(k)$$

for all $m, k > 0$.

It is useful sometimes to quantify the rate of convergence by a single number, which is called *mixing time.* Again, several definitions are known, namely

$$n_{1/2} = \min\{i : s(i) \leq \frac{1}{2}\} = \min\{i : P^i(x) \geq \frac{1}{2N} \text{ for all } x \in \Gamma\}$$

and

$$\xi = 1 + s(1) + s(2) + \cdots < \infty$$

The latter quantity is called *the total separation.* It is always finite and is of the same order of magnitude as $n_{1/2}$:

$$\xi \leq n_{1/2} \leq 2\xi$$

(see [P] for proofs and rèferences). Note also that if the starting state of the walk is x_0, then $\xi \geq diam(\Gamma, x_0)$, where $d = diam(\Gamma, x_0)$ is the maximal distance in graph Γ between x_0 and other vertices. Indeed, we have $P^1(x_0) = \cdots = P^{d-1}(x_0) = 0$ so $s(1) = \cdots = s(d-1) = 1$ and therefore $\xi \geq d$.

For the rest of the paper by the mixing time we will mean the total separation. Finally, note that the submultiplicativity of the separation distance and the inequality above gives us

$$s(k) \leq 2^{-\lfloor k/2\xi \rfloor}$$

which is usually a sharp upper bound.

6.1.3 Strong uniform times

Since the probability distribution P^k can be very complicated, it is usually very hard to compute the separation distance directly. Here we briefly describe how to use the strong uniform time technique to get the sharp estimates.

Define a *stopping time* by a stopping rule which observes the walk and tells us when to stop based on the previous observations and additional randomness if necessary. We think of the stopping time of the walk as a random variable which will be denoted T.

The stopping time T is called *uniform* if the probability of stopping at each state is uniform. The stopping time T is called *strong uniform time* if the above probability is uniform and independent on the time the walk stopped:

$$P(x_t = z \,|\, T = t) = \frac{1}{N}$$

Heuristically, even if we know the time when the walk stopped, we still don't have any information about *where* it stopped.

Let us consider an example. Suppose we walk on C_4, a circle with four vertices. Start at state 1. Define the stopping time as follows: choose a random state and walk until we get there, then stop. It is easy to see that this is not a strong uniform time because if it stopped after one step, it couldn't possibly be the vertex 3, which is opposed to the starting one. On the other hand here is a right strong uniform time: walk till we hit either 2 or 4. Do one more step and then stop.

The theorem due to Aldous and Diaconis (see [AD2,D1]) gives a bound for the separation distance given the strong uniform time:

$$s(k) \le P(T > k)$$

From here for the total separation we have (see [P])

$$\xi \le E(T)$$

A strong uniform time for which the above inequality becomes an equality is called *perfect time.* It is known that for every random walk there exist a perfect time (see [AD2, D1]).

Given the random walk and a stopping time, define a *halting state* to be a vertex x_0 such that whenever the walks hits x_0, it always stops there. In general, there can be no halting states or there can be many of them. It is known (see [DF, P]) that if a strong uniform time has at least one halting state then it is perfect.

For example, recall the random walk on C_4 and a construction of the strong uniform time defined above. Observe that vertex 3 is a halting state, which means the walk never leaves it. Therefore our stopping time is perfect and the total separation is given by $\xi = E(T) = 2 + 1 = 3$ (it takes on average two steps to hit either 1 or 4).

6.1.4 Remarks

While very fast strong uniform times are known to exist in theory, they are usually very difficult to construct in practice. To generate a graph vertex uniformly it is sufficient to consider just uniform stopping times. Thus constructing strong uniform times is generally a harder problem. It turns out, that in the classical cases this problem can be solved in a way similar to generation algorithms.

The method of strong uniform times is originated in works by Aldous and Diaconis (see [AD1, AD2]) and was later developed in [DF, P, AP]. The notion of the separation distance is also due to Aldous and Diaconis, while the total separation was introduced by the author (see [AF, D1, P]). Halting elements were defined by Diaconis and Fill in [DF] (see also [AF, P]).

6.2 Permutations

6.2.1 Random walk

Let $R = \{(1,2), \ldots, (1,n)\} \in S_n$ be the set of *star transpositions* in S_n. Denote by $\Gamma(S_n, R)$ the corresponding Cayley graph. Rather than taking the nearest neighbor random walk, take the probability of staying to be $1/n$. We can think of each walk steps as follows:

- Uniformly choose i, $1 \le i \le n$. Exchange elements in places 1 and i.

In other words, multiply a permutation by a random transposition $(1, i)$ where $1 \le i \le n$.

The main result of this subsection is the computation of the total separation ξ for this random walk:

$$\xi < 2n \ln n + \gamma n + \frac{1}{2}$$

where $\gamma \approx 0.5772156649$ is the Euler-Mascheroni constant (see e.g. [BE], §1.1; [WW], §12.1). We use an explicit construction of the strong uniform time to prove this claim.

6.2.2 Strong uniform time construction

Start with the identity permutation. The stopping rule consists of marking permutation elements according to walk movements. Once the element is marked, it stays marked forever. We stop when all the elements are marked. Here is how we mark elements:

- Mark the element n at the beginning.

Suppose the walk is at the state $\sigma \in S_n$ and is moving to $\sigma \cdot (1, i)$, $1 \leq i \leq n$.

- Mark $\sigma(1)$ if it is unmarked and $i = 1$ or $\sigma(i)$ is marked.

Stop when all n elements are marked.

6.2.3 Proof

Let us prove that the algorithm actually works which means that it indeed defines a strong uniform time. Note the similarity of this algorithm with the one in §2.1.2.

We will use the induction on the number k of marked elements. The inductive assumption is that all marked elements are in random order with respect to each other. In other words, even if we know the stopping time, which elements are marked and which k-subset of places they occupy, we still don't have any knowledge of their relative order. When $k = 1$ the condition is trivial.

Now suppose we have a k-subset A of marked elements. If the unmarked element $\sigma(1)$ is either exchanged with any of the marked ones or stays (as in §2.1.2) we can mark this new element. Indeed, the element $\sigma(1)$ after the walk step now has an equal chance to be the first, the second, dots, the last in the relative order of $A \cup \{\sigma(1)\}$.

Suppose now we exchange either two marked or two unmarked elements. In these cases we do not mark a new element. But then, the relative order of marked elements is still equally likely to be any given linear order on A, so we come back to the previous situation.

Recall that we stop when all the elements are marked. This means that the probability of stopping at any permutation conditioned the stopping time is uniform. This proves the result.

6.2.4 Mixing time

In contrast with the algorithm in §2.1.2 the random walk takes longer than just n steps, but only by a logarithmic factor. Indeed, let us compute the average stopping time under this stopping rule.

Observe that once we get an unmarked element in the first place, the only way to get a marked element there is by marking a new element. If we have k marked elements, $1 \leq k < n$, this will take on average $\frac{n}{k+1}$ steps.

On the other hand, after we mark a new marked element we need to wait till we get an unmarked element in the first place. If we have k marked elements, $2 \leq k < n$, this will take on average $\frac{n}{n-k}$ steps. Therefore we can break the stopping time into $2n - 2$ periods described above. For the total separation we have

$$\xi \le E(T) = \sum_{k=1}^{n} \frac{n}{k+1} + \sum_{k=2}^{n-1} \frac{n}{n-k} < 2n\,\mathfrak{h}(n) = 2n\,\ln n + O(n)$$

where $\mathfrak{h}(n) = 1 + \frac{1}{2} + \frac{1}{3} + \cdots + \frac{1}{n}$. Now the estimate for ξ given in §6.2.1 follows form the asymptotic expansion for $\mathfrak{h}_n$ (see e.g. [BE, WW]):

$$\mathfrak{h}(n) = \ln n + \gamma + \frac{1}{2n} + \ldots$$

An easy argument shows also that up to a constant this is the best one can get from this random walk. Indeed, it takes on average $n \log n$ steps just to *touch* all the elements. See [D1, P] on how to make this argument precise.

6.3 k-subsets of n-set

6.3.1 Random walk

Consider a graph Γ with all k-subsets of n-set as vertices. Connect two subsets A, A' by an edge if $|A \cap A'| = k - 1$. We can think of the nearest neighbor random walk on Γ as follows. Suppose the walk is at state $A \in \begin{bmatrix} n \\ k \end{bmatrix}$. Randomly choose a pair (i, j), such that $i \in A$, $j \notin A$. Flip a fair coin. If it's heads, replace i by j. If it's tails, stay.

Here we will present a simple construction of a strong uniform time for this walk. It will give us the exact value of the total separation

$$\xi = E(T) = 2k\,\frac{n-k}{n+1}\,\big(\mathfrak{h}(k) + \mathfrak{h}(n) - \mathfrak{h}(n-k)\big)$$

where $\mathfrak{h}(m) = 1 + \frac{1}{2} + \frac{1}{3} + \cdots + \frac{1}{m}$.

6.3.2 Perfect time construction

Start with some k-subset A. Mark all the elements in $[n] \setminus A$. While walking on Γ we will be marking some of its elements in A according to the following rule. Suppose the walk is at a k-subset B with exactly m marked elements. Assume by induction that all elements in $[n] \setminus B$ are marked. This means we have a total of $m + n - k$ marked elements.

As in §6.3.1 we think of our walk in the following way. Randomly choose a pair (i, j), such that $i \in B$, $j \notin B$. Flip a fair coin.

- If heads, exchange i and j. If i is unmarked, mark it.
- If tails, stay and if i is unmarked, mark it with probability $P = \frac{m+1}{n-k}$.
- Go to the next step.

As before, we stop when all elements are marked.

We claim that this defines a perfect time.

6.3.3 Proof

First let us proof that we have a strong uniform time. The proof is very much similar to the one in §6.2.3 and is based on the same idea as §2.2.4. Some of the numbers are different however.

Our inductive assumption is that at each time the m-set of marked elements in our k-subset is drawn uniformly from the set of all $m+n-k$ marked elements. In other words, we want to prove that all the probabilities of having in our k-subset a particular m-subset of an $(m+n-k)$-set of marked elements are equal even conditioned that we know *which* elements are marked and time of the observation. Of course, when $m = k$ this condition is equivalent to the definition of a strong uniform time.

The base of induction is obvious: we have no marked elements in our k-subset. Consider what happens when the walk is at a k-subset B with m marked elements. In the next step we can either increase the number of marked elements by one or keep it the same. In the latter case we either stay at B or exchange two marked elements i.e. inductive assumption still holds.

The case when we increase the number of marked elements is more complicated. First of all it could happen either when the walk stays or when we exchange a marked and an unmarked element. We use the same trick as in §2.2.4 thinking that what we have is actually a permutation with elements in B placed on the left in their natural order, and elements in $[n] \setminus B$ are on the right. Following the logic of §2.4.4 we can think of our step as picking a random unmarked element and exchanging it with a randomly chosen marked element. Depending on whether this marked element is in our set B or not, we either stay or exchange these two elements (if both chosen elements are in B there is no point in exchanging them). The conditional probability of the marked element being not in B (given the first element is unmarked) is $Q_1 = \frac{n-k}{m+n-k+1}$.

The conditional probability of staying, whether because we want our new element to stay or because we need to exchange it with the marked element in B, is $P_1 = 1 - Q_1 \frac{m+1}{m+n-k+1}$.

Therefore in order to compute the conditional probability of staying and marking an unmarked element i we need to choose the probability of staying and marking to be

$$P = \frac{P_1}{Q_1} = \frac{m+1}{n-k}$$

Let us repeat what we just did. We showed that conditioned we mark a new element, the probability P of staying is chosen in such way that now we have in our k-subset exactly $m+1$ marked elements, which are chosen randomly from the set of all $m+n-k+1$ marked elements. This implies that our time is strong uniform.

We have yet to prove that our time is perfect. Suppose the walk is started at a k-subset A_0. Consider any k-subset $B \in [n] \setminus A_0$. Recall that $k \leq n-k$ so $B \cap A_0 = \emptyset$. If the walk gets to B, it must have all the elements in $[n]$ marked. Thus the walks always stops at B and B is a halting state. Therefore our strong uniform time is perfect. This completes the proof.

6.3.4 Mixing time

Let us now compute the expected number of steps for this algorithm to stop. We assume that $k \leq \frac{n}{2}$, otherwise we can simply switch the set with its complement.

Suppose we currently have m marked elements in our k-subset. Observe that the probability of marking a new element is equal to

$$\frac{k-m}{k}\left(\frac{1}{2}+P\right)=\frac{k-m}{k}\frac{m+n-k+1}{2(n-k)}$$

Therefore the expected stopping time of this algorithm is equal to

$$\begin{aligned}
E(T) &= (2\,k\,(n-k))\left(\frac{1}{1\cdot n}+\cdots+\frac{1}{k\,(n-k+1)}\right)\\
&= \frac{2\,k\,(n-k)}{n+1}\left(\frac{1}{1}+\frac{1}{2}+\cdots+\frac{1}{k}+\frac{1}{n-k+1}+\cdots+\frac{1}{n}\right)\\
&= 2\,k\,\frac{n-k}{n+1}(\mathfrak{h}(k)+\mathfrak{h}(n)-\mathfrak{h}(n-k)) \;=\; 2\,k\,\frac{n-k}{n+1}\left(\ln\frac{k\,n}{n-k}+O(1)\right)
\end{aligned}$$

Since our stopping time T is perfect we have

$$\xi=E(T)=2\,k\,\frac{n-k}{n+1}\left(\mathfrak{h}(k)+\mathfrak{h}(n)-\mathfrak{h}(n-k)\right)$$

Note that when the ratio $\frac{k}{n}$ is fixed and k grows, we have $\xi=O(k\log(k))$.

6.4 k-subspaces of $\mathbb{F}_q^n$

6.4.1 Random walk

Consider a graph Γ with all k-dimensional subspaces of $\mathbb{F}_q^n$ as vertices. Vertices corresponding to subspaces V and V' are connected by an edge if and only if $\dim(V\cap V')=k-1$. We can think of the nearest neighbor random walk on Γ as follows. Choose a random $V'\in G(n,k,\mathbb{F}_q)$, $\dim(V\cap V')=k-1$. Flip a fair coin. If heads, stay. If tails, move to V'.

We assume that $2\,k\le n$ since otherwise one can use a duality to walk on $(n-k)$-dimensional subspaces. As before, we would like to analyze this walk by use of an explicit construction of a perfect time.

6.4.2 Perfect time construction

As in §2.5.2, at the beginning we mark our k-dimensional subspace. Following the walk we will be unmarking this subspace, leaving marked a subspace of a smaller dimension.

Suppose the walk is at the k-dimensional space V, and $W\subseteq V$ is its m-dimensional marked subspace. Choose a random $V'\in G(n,k,\mathbb{F}_q)$, $\dim(V\cap V')=k-1$. Denote $W'=W\cap V'$. It is easy to see that the dimension $d=\dim(W')$ can be either m or $m-1$. The rules are different depending on which one of them is the actual value of d. Flip a fair coin.

- If heads and $d=m$, stay.
- If tails and $d=m$, move to V'.
- If heads and $d=m-1$, stay. Unmark everything but W' with probability $P(m,k,n)$ given by

$$P(m,k,n)=\frac{1+q+\cdots+q^{k-m}}{1+q+\cdots+q^{n-m}}$$

- If tails and $d = m - 1$, move to V'. Unmark everything but W'.

Stop when $m = 0$ i.e. everything but 0 is unmarked.

6.4.3 Proof

Let us first prove that the constructed stopping time T is indeed strong uniform. We need to recall what happened in §2.5.2. There we made the assumption that at each time our k-subspace V is uniformly chosen given that it contains marked m-subspace W. Moreover, it is uniformly chosen even conditioned we know W and time of the event. The main difference is that in §2.5.2 we were able to decrease m at each step, so it took us only k steps to get $m = 0$. Here we break the set of all events into several distinct possibilities, some of them exactly like in §2.5.2 and the others are different.

Start with the case when we get tails and $d = m$. It is clear, that if $V \supset W$ is randomly chosen, then so is $V' \supset W$. This means that the assumption remains valid. Similarly to the cases when heads and we don't unmark, we simply stay and nothing changes.

At last, we have two cases when we do unmark and therefore decrease m by 1. We claim that these are exactly the cases for which everything worked in §2.5.2. Indeed, the only difference is that we restrict our attention to the cases when $m = d - 1$. Therefore the inductive assumption should still work, and T is strong uniform.

To prove that our stopping time is perfect we shall exhibit a halting state. Suppose we start at a k-dimensional subspace V. Recall that $2\,k \leq n$. Consider any k-dimensional subspace V' such that $V \cap V' = \emptyset$ By analogy with §6.3.3, whenever the walk gets into V' it must always stop there. Therefore T is perfect and this completes the proof.

6.4.4 Mixing time

Since T is perfect, we need to find the expected stopping time. First, compute the total number of k-dimensional subspaces V', such that $\dim(V \cap V') = k - 1$. In other words we would like to compute the *degree* $deg(\Gamma)$, i.e. the number of edges going out of each vertex in Γ.

Denote $U = V \cap V'$. We can think of choosing V' as a two step procedure of choosing $U \subset V$ and then choosing $V' \supset U$, $V' \neq V$. The number of ways to choose U is $\binom{k}{k-1}_q = (1 + q + \cdots + q^{k-1})$ (see §1.5). To compute the number of ways to choose V' given U we can use the result in §2.5.4, which gives us $c(k-1, k, n; q) = \binom{n-k+1}{1}_q = (1+q+\cdots+q^{n-k})$. We also need to subtract 1 from the last expression to satisfy the condition $V' \neq V$. Finally, we get

$$|\{V' \in Gr(n, k, \mathbb{F}_q) : \dim(V \cap V') = k - 1\}| = q\,(k)_q (n-k)_q$$
$$= q\,\left(1 + q + \cdots + q^{k-1}\right)\,\left(1 + q + \cdots + q^{n-k-1}\right)$$

Let us fix the m-dimensional marked subspace W. We need to compute the probability of unmarking at the next step. Before we can do that we need to compute the probability that the dimension $d = \dim(W \cap V')$ is equal to m.

Observe that $W \cap V' = W \cap U$ since $U = V \cap V'$. The number of ways to choose U, $W \subset U \subset V$ is equal to

$$c(m, k-1, k; q) = \binom{k-m}{k-m-1}_q = (1 + q + \cdots + q^{k-m-1})$$

Therefore the desired probability is given by the formula

$$P(d = m-1) = 1 - \frac{1+q+\cdots+q^{k-m-1}}{1+q+\cdots+q^{k-1}} = q^{k-m}\,\frac{1+q+\cdots+q^{m-1}}{1+q+\cdots+q^{k-1}}$$

Now, according to §6.4.2, the actual decrease in m happens with probability more than 1/2 given $d = m-1$. Indeed, it happens always when tails and with probability $P(m,k,n)$ when heads. To simplify the calculations we estimate it by 1/2 from the below. This will give us an upper bound. Analogously the lower bound would be less than the upper bound by a factor of 2. We get

$$E(T) < 2\sum_{m=1}^{k} \frac{1+q+\cdots+q^{k-1}}{q^{k-m}\,(1+q+\cdots+q^{m-1})}$$

which finally gives us

$$\xi = E(T) < 2\,\frac{1+q+\cdots+q^{k-1}}{q^k}\sum_{m=1}^{k}\frac{q^m}{(1+q+\cdots+q^{m-1})}$$

The sum on the right hand side can be (and sometimes is, see remarks below) considered as a q-analog of logarithm. When $q = 1$ we get the bound $2\,k\,\log(k)$ which is the upper bound found in §6.3.3. This is not a coincidence. In §6.5 we present a rigorous argument for that.

Now consider what happens when $q \geq 2$, which is very common among powers of primes. First, take $q = 2$. We get

$$\xi < 2\,\frac{2^k-1}{2^k}\sum_{m=1}^{k}\frac{2^m}{2^m-1} < 4\,k$$

On the other hand $\xi \geq k$ since $k = diam(\Gamma, V)$. Thus the total separation $\xi = O(k)$ where the constant varies between 1 and 4 and depends on q, n and k.

Note also that for fixed n and k the mixing time ξ is decreasing as q is increasing.

6.5 Back to k-subsets of n-set

6.5.1 We would like to conclude this section by explaining why the random walk on k-subspaces of an n-space behaves like a q-version of a random walk on k-subsets of an n-set. It turns out that there is a formal argument for that.

Suppose that our random walk on k-subspaces starts at V_0. Notice that by symmetry the only invariant of a k-subspace V is the dimension $V_0 \cap V$. Observe also that this dimension can either remain the same or change by at most 1 at each step. This means that we basically can make this walk into a birth and death chain by simply projecting the walk. This simplifies the problem significantly.

Here is a different way of doing the same kind of procedure. Recall that each k-subspace lies in a Schubert cell corresponding to exactly one of the k-subsets. Suppose our basis $E = (e_1, \ldots, e_n)$ was chosen in such a way that the starting k-subspace V_0 was a span of $(e_1, \ldots, e_k)$. Then by definition all the elements of the Schubert cell should have the same dimension $\beta_k = \dim(V_0 \cap V)$. It implies that at each time the probability of the walk being at V is constant on all V in the same

Schubert cell. Therefore we can project our walk on a set of k-subsets of an n-set. It would not be a random walk of course, but rather some new Markov chain, which can be described.

6.5.2 First observe that since the stationary distribution for the random walk on k-subspaces is uniform, the stationary distribution of the projected chain must be proportional to the size of the Schubert cell. Recall the definitions and results in §4.2. We have the stationary distribution:

$$\pi(A) = \frac{q^{k\,(n-k)-inv(A)}}{\binom{n}{k}_q}$$

Observe also that we can move from any k-subspace in the Schubert cell corresponding to A to at least one k-subspace in the Schubert cell corresponding to A' if and only if $|A \cap A'| = k - 1$. One can actually compute the transition probabilities of this Markov chain:

$$p(A \to A') = \frac{q^{d_3(A,B)}}{q\,\left(1+q+\cdots+q^{k-1}\right)\,\left(1+q+\cdots+q^{n-k-1}\right)}$$

where $d_3(A, B)$ is defined as in §3.2.4, 3). From here one can get a combinatorial interpretation for d_3 in terms of the number of k-dimensional subspaces with certain properties.

This gives us a complete description of the projected Markov chain.

6.5.3 Now we can use the following trick. We can now extend the definition of the Markov chain described in §6.5.2 to *any* $q > 0$. Fix n and k and consider $\begin{bmatrix} n \\ k \end{bmatrix}$ as the set of states of the chain. Define transition probabilities using formulas in §6.5.2.

First we need to show that they indeed define transition probabilities for *any* $q > 0$, not just powers of primes. In other words we need to show that they indeed add up to 1 in each state. Multiply these probabilities by the denominator $g(n, k; q) = q\,\left(1+q+\cdots+q^{k-1}\right)\,\left(1+q+\cdots+q^{n-k-1}\right)$. We know that these corrected probabilities are polynomial in q and add up to $g(n, k; q)$ for the infinite number of values of q. Since the degree of polynomials is bounded by n, we immediately get that these polynomials are *equal* to $g(n, k; q)$ and therefore their ratio is 1 for all $q > 0$. Therefore the Markov chain is indeed correctly defined.

Using a similar argument we can prove that the stationary distribution is given by

$$\pi(A) = \frac{q^{k\,(n-k)-inv(A)}}{\binom{n}{k}_q}$$

for any $q > 0$. Moreover, given the right definition of separation distance in case of non-uniform stationary distribution, strong stationary time, etc. (see [AF, DF]), one can show that the bounds found in §6.4.3 will work in this case as well. The reason for that is that the stopping time like everything else in this section is "polynomial in q" which means it can be defined and proved for any q.

6.5.4 Now we can finally formalize the connection between random walk on k-subsets and on k-subspaces announced earlier.

Clearly when $q = 1$ all transition probabilities in the Markov chain defined in §6.5.2 become equal. Also the stationary distribution becomes uniform. In fact our Markov chain becomes a random walk on k-subsets of an n-set (see §6.3.1). But by the argument given in §6.5.3 this means that we can get "for free" both stopping time and bounds for this random walk (see §6.3.2, 3). By simply substituting $q = 1$ in all the formulas we get in §6.4.2, 3. Moreover, one can show that we get precisely *the same* stopping time and bounds. The details are left to the reader.

6.6 Other random walks and remarks

6.6.1 The theory of Markov chains and random walks in particular is very extensive. See [F] for an introduction into the field and [AF] for an up to the date study of the random walks on graphs. For the total variation distance there exist a coupling technique which is another probabilistic argument for working with general Markov chains See [AF, D1] for the definition references and applications.

6.6.2 The case of the random walk on S_n generated by *all* transpositions has been heavily studied. The mixing time $\xi = O(n \log n)$ in this case. See [D1, D3] for the references and history of the subject. In this case there are several known strong uniform time constructions (see [D1, Ma, P]). The walk by star transpositions was earlier studied in [FOW]. The construction in §6.2.2 first appeared in [P] in a more general context. See also [DSa, P] for other random walks on S_n.

Random walk on k-subsets was introduced and analyzed in [DSh3] (see also [Be, D1, Gr]). The strong uniform time construction given here is the only known and was given in [P]. Note that the upper bound found in [DSh3] is asymptotically tight only when $k = \Omega(n)$.

Random walk on k-dimensional subspaces was studied in [Ari]. Again we follow [P] in our treatment.

There are various other random walks on $GL(n, \mathbb{F}_q)$ which have a property of rapid mixing. Most notably, the random walk on trasvections which generate $SL(n, \mathbb{F}_q)$ was analyzed in [Hi]. By transvection we mean the linear transformation that fixes a hyperplane. It turns out that the random walk is rapidly mixes after $O(n)$ steps. Since generation of a transvection is fast and multiplication of a matrix by a transvection takes $O(n^2)$ multiplications, this gives us a random matrix generating algorithm in $O(n^3)$ steps. See [Gl] for walks on other linear groups generated by conjugacy classes.

6.6.3 Although much of the technical details were omitted, we believe that the section §5.5 has a philosophical importance. We showed how to construct a Markov chain which would simultaneously generalize both random walk on k-subsets and on k-dimensional spaces over the finite field. This explains the similarities one can observe throughout the paper. One can also make the same kind of argument for permutations and complete flags.

7. Black box groups

7.1 Definitions

Suppose the elements of a group are given as binary stings of a fixed length, say N, and a "black box", also called an *oracle* performs group operation: multiplication, taking an inverse, and recognition of the identity element. Such a group G

is called a *black box group*. Note that the encoding need not be bijective: different strings may correspond to the same group element while some strings may not correspond to group elements at all. Note also that $|G| \leq 2^N$.

Suppose we are also given a set S of generators of a black group G. The first problem is to find an algorithm in such a generality for generation of the uniform group elements. In general, this must take an exponential time (in N), since the problem is harder than computing the order $|G|$ and the latter was proved to be exponentially hard in [Ba1]. However, if we weaken the problem to finding *nearly-uniform* group elements the problem can be resolved in a polynomial time. This is the result of Babai we describe in §7.2. Thus in a sense the situation is similar to the approximate counting problem.

By nearly-uniform group elements we mean that all group elements are generated with probability at least $1/2|G|$. Of course, if we repeat the algorithm several times and take a product of the resulting group element, we can get as close to the uniform distribution as we need. Formally if we achieve near-uniformity after N steps, then after $k\,N$ steps we get the probability of each element at least $(1 - 1/2^k)/|G|$ (see §6.1.2 above).

The second problem is the *black group recognition problem*. Given a black box group G we would like to test whether it is isomorphic to a fixed finite group H. Here in §7.3, 4 we describe algorithms for recognition of S_n and $SL(n, \mathbb{F}_q)$ given as lack box groups.

Unfortunately, black group recognition problem in general is again too hard to have a polynomial solution. It is useful sometimes to assume the existence of an oracle which computes the order of group elements. Amazingly, one can tell a lot just by the order of the group elements. We call G a *gray box group* then.

We give a constructive solution for the recognition problem. This means that me not only prove that $G \simeq H$, but also construct such an isomorphism $\phi : G \to H$ explicitly. This approach is called *constructive recognition*.

About the algorithms. It is important to distinguish algorithms that are correct most of the time from algorithms that are correct all the time. The latter are called *Las Vegas* algorithms while various formulations of the former are called *Monte Carlo* algorithms. When appropriate we will specify the type of our algorithms.

Everywhere in this section μ is the time necessary to multiply two group elements. For simplicity, we will suppress most of the constants using notation $O(f(n))$ instead. The right constants can be found in the literature.

7.2 Generating random group elements

7.2.1 Reducing the number of generators

Let G be a finite black box group and let S be its set of generators. At this stage we construct another set of generators S_1 such that $|S_1| = O(\log |G|)$. Of course, any increasing subgroup chain of G must have length at most $l = \log_2 |G|$ so having more than l generators would be redundant. Here is how the this can be done in practice.

Let $S = \{s_1, \ldots, s_m\}$. By a *random subproduct* denote a product $w = s_1^{\epsilon_1} \cdot \ldots \cdot s_m^{\epsilon_m}$ where $\epsilon_i \in \{0, 1\}$ are chosen by independent flips of a fair coin.

Lemma Let S generate group G and let $H \subset G$ be an arbitrary proper subgroup of G. Then a random subproduct w on S is not a member of H with probability at least $1/2$.

Proof Consider a largest i such that $s_i \notin H$. Then w can be decomposed into the form $u\, s_i^{\epsilon_i}\, v$, where $v \in H$.

We have two cases. If $u \in H$ then with probability $1/2$, $\epsilon_i = 1$ and $w = u\, s_i\, v \notin H$. If $u \notin H$ then with probability $1/2$, $\epsilon_i = 0$ and $w = u\, v \notin H$. This proves the lemma. □

Now one can show that with a large probability $O(\log |G|)$ random subproducts will generate the whole group G. This will be our new set of generators S_1. See [BCFLS,CF] for details and further references.

Note that the complexity of our subroutine is $O(\mu\, |S| \log |G|)$. We assume that S is reasonably small (e.g. $|S| = O(\log^3 |G|)$) which is often the case in practice. Otherwise even reading the set S might take time longer than the rest of the algorithm.

7.2.2 Constructing fast generators

For a finite group G and its set of generators S define the *length* of an element $l = l_S(g)$ to be the minimum length of a decomposition $g = s_{i_1} s_{i_2} \dots s_{i_l}$. The *diameter* $D = diam(S)$ is defined as the maximum length $l_S(g)$ among all group elements $g \in G$.

Here we will present an algorithm which will construct a set of generators S_2 such that $|S_2| = O(\log |G|)$ and $D = diam(S_2) = O(\log |G|)$. The idea is to start with set of generators S_1 and repeatedly take a random walk on g and at the end of each walk add the last element into our set of generators.

For a symmetric set of generators $S = \{s_1, \dots, s_m\}$ (see §6.1) denote by $g \leftarrow rand(G, S, t)$ the state of the nearest neighbor random walk on a Cayley graph $\Gamma(G, S)$ after t steps.

Algorithm

1)Input L, N, $S := S_1 \cup S_1^{-1}$;
2)For $j = 1$ to N Do:
- Choose $t \in [L]$ uniformly; Let $g \leftarrow rand(G, S)$, $S \leftarrow S \cup \{g, g^{-1}\}$;

End Do;
3)Output $S_2 := S$.

Now take $N = O(\log |G|)$ and $L = O(\log^3 |G|)$. A hard and tedious analysis of random walks on finite groups shows that we do get a generating set S_2 such that the diameter $diam(S_2) = O(\log |G|)$ with large probability (see [Ba1]).

Note that the complexity of the algorithm is $O(\mu \log^4 |G|)$

7.2.3 Final random walk

Now we have a generating set S_2, $|S_2| = O(\log |G|)$ for which we have an upper bound on diameter $D = diam(S_2) = O(\log |G|)$ (with high probability). Again consider a random walk on Cayley graph $\Gamma(G, S_2)$. It is a known result in the theory of random walks (see e.g. [DSa,Ba1]) that the total separation

$$\xi \le 16\, |S_2|\, D^2 \log |G| = O(\log^4 |G|)$$

Recall that $n_{\frac{1}{2}} \le 2\xi$ (see §6.1.2). Then after $O(\log^4 |G|)$ steps of the walk we get each group element with probability at least $1/2$.

To summarize, we constructed an algorithm which with large probability will produce the nearly-uniform group elements taking $O(\log^4 |G|)$ group multiplication on each. Also, assuming $|S| = O(\log^3 |G|)$ the preprocessing stage will also take $O(\log^4 |G|)$ group multiplications.

7.2.4 An Erdös-Rényi machine

Here is the last trick one can employ when there is a large number of random group elements needed to be generated. We will show that when $O(\mu \log^5 |G|)$ time used for preprocessing, then one can produce nearly-uniform group elements in time $O(\mu \log |G|)$.

Let $k = O(\log |G|)$ and using the random walk on $\Gamma(G, S_2)$ compute k nearly-uniform group elements $g_1, \dots, g_k$.

Lemma Random subproducts $u = g_1^{\epsilon_1} \cdot \dots \cdot g_k^{\epsilon_k}$ are nearly-uniformly distributed in G.

The Lemma is due to Erdös and Rényi. It implies that having spent $(\mu \log^5 |G|)$ time for finding k random group elements, we get a algorithm for producing nearly-uniform group elements at a cost of k multiplications.

7.2.5 Remarks

the first two parts of the algorithm above are Monte Carlo. We had no way of checking whether it does what it is supposed to. Put it this way, after §7.2.2 the algorithm outputs the fast generators with probability $1 - \delta$ for some δ which can be made as small as desired. On the other hand our objective was not a uniform distribution but a near-uniform distribution. So after part §7.2.2 we get *some* symmetric generating set which is fast with probability $> 1 - \delta$. Thus combined with the part §7.2.2 we *always* get near-uniform random group elements at the end. Thus our algorithm is Las Vegas.

Situation with Erdös-Rényi machine is different. Our machine is either works or it doesn't (say, the random elements we get could generate only a subgroup of G). Thus this is a Monte Carlo algorithm.

In our short presentation of the algorithm above we followed the pioneer work [Ba1] of Babai. The random subproducts were invented by Erdös and Rényi in [ER] where they also discovered the machine. The result of §7.2 originated in [BCFLS, CF].

It might be worth noting that the algorithm takes too much time to be practical. Recently there has been an attempt in [CLMNO] to find a fast alternative approach. See [Ba3, CG] for more on this algorithm.

7.3 Recognition of the black box group isomorphic to S_n

Suppose we have a black box group G. Let μ be the time necessary to multiply two elements, and let ρ be the time to generate nearly-uniform random elements of G. Assume that it takes ν time to find an order of an element on G. In the worst case we can always keep multiplying until get an identity (see also §7.3.3 below). Here we present a constructive recognition algorithm which compares G with S_n.

7.3.1 Goldbach elements

Fix the number n. Observe that if n is a prime, then every element of order n is a long cycle. On the other hand, every n-th element of S_n is a long cycle (so has an order n) so in this case we can get a long cycle just by checking the order of an element. For general n this idea wouldn't work, so we must find an alternative.

Let n be any even number and suppose $n = p_1 + p_2$ where p_1, p_2 are distinct odd primes. Then every permutation $\sigma \in S_n$ of the order $p_1 \times p_2$ is must have cycles of length p_1 and p_2. On the other hand the probability that a random permutation is of this type is equal to $\frac{1}{p_1 p_2} > \frac{4}{n^2}$, so there must be many of such elements in a group. We call them *Goldbach elements.*

Now, let us estimate the number of Goldbach elements in S_n, where n is even. Unfortunately even their existence is an unsolved problem. The Goldbach conjecture states that there must be some, but heuristically it is clear that there must be many. Indeed, if there are about $\frac{n}{\ln n}$ primes less than n they must form $\frac{n^2}{\ln^2 n}$ pairs and therefore if sums of pairs were "uniformly distributed", we would have $Const\left(\frac{n}{\ln^2 n}\right)$ solutions of the equation $n = p_1 + p_2$. Sylvester was the first to observe that this must be the case(see [Sy]). Later Hardy and Littlewood proved that this formula is correct for almost all n (see [HL]), so there is a reason to believe that it is true in general. In [BP] the authors computed that the number of such solutions is $> \frac{n}{3 \ln^2 n}$ for all even n, $20 < n < 10^6$. We will assume here that this is generally the case.

From the estimates above we conclude that a permutation is a Goldbach element with probability at least $\left(\frac{4/3}{n \ln^2 n}\right)$.

7.3.2 Constructing an isomorphism

Here is a general description of the algorithm.

Suppose n is even and $G \simeq S_n$. Pick a random element x and compute its order $d(x)$. Repeat the process until $d(x) = p_1\, p_2$ where $p_1 + p_2 = n$. Then x is a Goldbach element with cycles of lengths p_1 and p_2. Compute these cycles: $x_1 = x^{p_1}$, $x_2 = x^{p_2}$.

Repeat picking random group elements y until $d(y) = 2\, p_3\, p_4$ where $p_3 + p_4 + 2 = n$. Then y^2 is a Goldbach element in S_{n-2} with cycles of lengths p_1 and p_2. Thus $z = y^{p_3 p_4}$ is a transposition.

Observe that for random $g \in G$ a conjugate gzg^{-1} is a random transposition. Repeat picking random elements g until $b = gzg^{-1}$ does not commute with neither of the cycles x_1 and x_2.

Now observe that $a = x_1 b x_2$ is a long cycle. Then a map $\phi : a \to (1, 2, \dots)$ and $\phi : b \to (1, 2)$ defines an isomorphism.

Compute the average expected time of this algorithm. Since finding a transposition which does not commute with x_1 and x_2 takes on average at most $O(\rho n + \mu n)$ steps, the expected time is bounded by a constant times the expected time to get a Goldbach element. Thus the total expected time is $O\big((\rho + \mu + \nu)\, n \log^2 n\big)$ steps.

7.3.3 Gray $\to$ black

Let us present an argument here that we having G presented as a gray box group was not unreasonable. In other words, we will show that there is a fast way to compute the order of black box group elements.

First, observe that never in the algorithm we needed to compute the order d which was larger than n^2. Therefore a simple idea would be simply keep multiplying until we get the identity and if we don't get it before n^2, move to the next element. There is a trick, however, how one can achieve $O(n)$ multiplications instead of n^2.

Given a black box group element a, compute two sets of elements a^n, a^{2n}, ... , a^{n^2} and id, a^{-1}, a^{-2}, ... , a^{-n}. Now if $a^{kn+i} = id$ then we would have two *identical* elements in both sets. But comparing lists of elements can be done very fast since we do not have to use all the information but just a few randomly chosen bits. If they coincide, we can check them in full then. This hashing idea is now commonly used in Computer Science (see e.g. [MR]). We conclude that in the algorithm one can assume that $\nu = O(\mu n)$.

7.3.4 Miscellanea

Our algorithm is defined when n is even. When n is odd, one can find a Goldbach element and two transpositions that connect the cycles with a fixed point. The details are given in [BP].

Note that we have yet to check that we do indeed have an isomorphism. Simply compute the Coxeter transpositions and check whether the relations are satisfied. There are only $O(n)$ relations. This proves only that $S_n \subset G$.

Now let us find the action of the original generators. First reconstruct the underlying set $[n]$ by acting on transpositions and checking whether they commute or not. Now if all the original generators are mapped to permutations, then $G \simeq S_n$. This makes the algorithm Las Vegas. The cost is $O(|S|\mu n^2)$.

One can also use Goldbach elements in trying to determine n if it is unknown at the beginning. Simply repeat picking elements and adding their prime divisors. After about $O(\rho n \ln^2 n)$ steps we shall get either n or $n-1$ because of the large number of Goldbach elements in S_n. Distinguishing between the two possibilities is easy again. Thus if n is unknown but we have an upper bound N on n (say, the length of the binary string that presents black box group elements) we can get a Monte Carlo algorithm for finding n.

7.4 Recognition of the black box group isomorphic to $SL(n, \mathbb{F}_q)$

Suppose we have a black box group G. As before, we assume that it is isomorphic to $SL(n, \mathbb{F}_q)$. Then, when we find such an isomorphism, check whether our assumption was valid.

Let μ be the time necessary to multiply two elements, and let ρ be the time to generate nearly-uniform random elements of G. Assume that it takes ν time to find an order of an element on G. In [CL] authors present an algorithm for finding the order of elements in linear groups. Their algorithm works in time $\log|G| \log\log|G|$ group multiplications. Thus in our case $|SL(n, \mathbb{F}_q)| \sim q^{n^2}$ and we have $\nu = O(n^2 \log n \log q \log\log q)$ which is sufficient for our purposes.

7.4.1 Finding a transvection and coinciding primitive element

A *transvection* is a nontrivial linear transformation of $\mathbb{F}_q^n$ that fixes an $(n-1)$-dimensional subspace (a hyperplane). The transvections play a role in linear groups that is similar to the role of transpositions in S_n. In this section we will find a transvection in a gray box $G \simeq SL(n, \mathbb{F}_q)$.

For a pair q and m define a *primitive prime divisor* (*ppd*) to be a prime p such that $p|q^m - 1$ and $p \nmid q^i - 1$ for $1 \le i \le m-1$. A known result of Zsigmondy shows that such ppd exist for all pairs $(q, m) \neq (2, 6)$.

Lemma If $x \in SL(n, \mathbb{F}_q)$, $q = p^l$ and the order $d(x) = p \cdot p' \cdot r\, p'$ is a primitive prime divisor for a pair q and $n-2$, then $x^{r\,p'}$ is a transvection. Moreover, the probability that an element of $SL(n, \mathbb{F}_q)$ satisfies the conditions above is $O\left(\frac{1}{n\,q}\right)$.

The proof which is a combination of number theoretic arguments and some linear algebra results (see [BCFL]).

Now the algorithm for finding a transvection follows directly from Lemma. Choose a random element of a gray box group G. Compute the order. Repeat the process until an element x which satisfies conditions of the lemma is found. Output $x^{d(x)/p}$.

By the second part of the lemma the running time of the algorithm is $O((\nu + \rho)\, n\, q)$ steps. We assume here that the decomposition into prime is relatively fast compared to the finding a random group element. The reason lies in a special structure of the numbers $q^m - 1$ which have been studied heavily in computational number theory.

Consider a *coinciding primitive element* $z = x^p$. We will need it in the next section for the final construction. Heuristically it plays the same role as an $(n-2)$-cycle in the case when $(n-2)$ is a prime.

7.4.2 Finding elementary transvections

The idea is to construct an isomorphism which maps certain gray box group elements into elementary transvections. By elementary transvections we mean matrices in $SL(n, \mathbb{F}_q)$ with ones on diagonal and only one nonzero element elsewhere.

The technique presented in [BCFL] is nice but rather laborious so we will just give a short overview.

First, authors observe that by conjugating transvection x with by random element one gets random transvections. Second, a commutator $x\,y\,x^{-1}y^{-1}$ of x with a random transvection with probability about $1/q$ is a transvection which fixes the same hyperplane as x. The last property is also not hard to check. Now, after such a transvection y is found taking commutators of y with powers of the coinciding primitive element z will lead to $n-1$ "linearly independent transvections" that all fix the same hyperplane. The authors in [BCFL] then define a linear algebra technique for working with transvections.

Now the process is repeated for n other hyperplanes and finally authors employ a version of orthogolization algorithm to get the desired elementary transvections. This is just the rough idea of the technique used.

The overall algorithm takes $O(n^3\epsilon + n^2 q\mu + \rho\mu)$ where ϵ is the time for field operations in $\mathbb{F}_q$. The way the algorithm is presented, it is a Monte Carlo algorithm. It can be made Las Vegas by checking the relations at the end. This can be done at low cost. The minimal set of relations for $SL(n, \mathbb{F}_q)$ can be found in [CM].

7.5 Remarks

The recognition problem of the black box groups has been recently studied in a number of papers (see e.g. [CFL,NP1,KS]).

In our presentation of the recognition algorithm of the gray box group isomorphic to S_n we follow the paper [BP]. A similar idea can be used to find a recognition in case of the alternating group A_n.

The sketch of the recognition algorithm of the gray box group isomorphic to $SL(n, \mathbb{F}_q)$ follows the paper [BCFL]. The paper generalizes the approach started in [CFL]. The alternative version of the recognition algorithm for all Chevalley groups was developed in [KS]. Their result is quite general but in case of $SL(n, \mathbb{F}_q)$ the algorithm is significantly slower.

REFERENCES

[A1] D. Aldous, *Random walks on finite groups and rapidly mixing Markov chains*, Lecture Notes in Mathematics **986** (1983).

[A2] D. Aldous, *Random walks on finite groups and rapidly mixing Markov chains*, J. Theor. Probability **2** (1980), 91–100.

[AD1] D. Aldous, P. Diaconis, *Shuffling cards and stopping times*, Amer. Math. Monthly **93** (1986), 333–348.

[AD2] D. Aldous, P. Diaconis, *Strong uniform times and finite random walks*, Advances in Applied Math. **8** (1987), 69–97.

[AF] D. Aldous, J. Fill, *Reversible Markov Chains and Random Walks on Graphs*, monograph in preparation, 1996.

[An] G. Andrews, *The Theory of Partitions*, Addison-Wesley, New York, 1976.

[Ari] A. D'Aristotile, *The nearest neighbor random walk on subspaces of a vector space and rate of convergence*, J. Theoretical Probability **8** (1995), 321–346.

[Arn] V. I. Arnol'd, *The cohomology ring of the group of dyed braids*, Mat. Zametki **5** (1969), 227–231.

[AP] A. Astashkevich, I. Pak, *Random walks on nilpotent and supersolvable groups*, in preparation (1997).

[Ba1] L. Babai, *Local expansion of vertex-transitive graphs and random geneartion in finite groups*, in Proc 23rd ACM STOC (1991), 164–174.

[Ba2] L. Babai, *Automorphism groups, isomorphism, reconstruction*, in Handbook of Combinatorics (R. L. Graham, M. Groetschel, and L. Lovasz, eds.) (1996), Elsevier.

[Ba3] L. Babai, *Randomization in group algorithms: Conceptual questions*, in Groups and Computation II (L. Finkelstein, W.M. Kantor, eds.) DIMACS Workshops on Groups and Computation (1997), AMS, Providence.

[BCFLS] L. Babai, G. Cooperman, L. Finkelstein, E.M. Lucks, and A. Seress, *Fast Monte Carlo algorithms for permutation groups*, Proc 23-rd ACM STOC (1991), 90–100.

[BE] H. Bateman, A. Erdélyi, *Higher Transcedential Functions, Volume 1*, Mc Graw-Hill, New York, NY, 1953.

[Be] E. Belsley, *Random walks on distance regular graphs* (1996), in preparation.

[Bor] A. Borel et al, *Seminar on algebraic groups and related finite groups*, Springer, Berlin, 1970.

[Bou] N. Bourbaki, *Groupes et algèbres de Lie, Ch I, IV, V, VI*, Hermann, Paris, 1960.

[BP] S. Bratus, I. Pak, *Fast constructive recognition of a gray box group isomorphic to S_n or A_n using Goldbach's Conjecture* (1997), preprint.

[BCFL] S. Bratus, G. Cooperman, L. Finkelstein, and S. Linton, *Constructive recognition of black box groups isomorphic to $SL(n,q)$ or $PSL(n,q)$*. (1997), preprint.

[CW] E. Calabi, H. Wilf, *On the sequential and random selection of subspaces over a finite field*, J. Comb. Theory (A) **22** (1977), 107–109.

[Ca] P. Cartier, *Lecture Notes in Math., 901*, vol. **1980/81**, Springer, Berlin, pp. 1–22.

[CL] F. Celler, C.R. Leedham-Green, *Calculating the Order of an Invertible Matrix*, in Groups and Computation II (L. Finkelstein, W.M. Kantor, eds.) DIMACS Workshops on Groups and Computation (1997), AMS, Providence.

[CLMNO] F. Celler, C.R. Leedham-Green, S. Murray, A. Niemeyer, and E.A. Obrien, *Generating random elements of a finite group*, Comm. Alg. **23** (1995), 4931–4948.

[Che] I.V. Cherednik, *A new interpretation of Gelfand-Tzetlin bases*, Duke Math. J. **54** (1987), 563–571.

[CG] F.R.K. Chung, R.L. Graham, *Random walks on generating sets for finite groups*, The Electronic J. of Comb. **4 No 2.** (1997), #R7.

[CF] G. Cooperman, L. Finkelstein, *Combinatorial tools for computational group theory*, in Groups and Computation I (L. Finkelstein, W.M. Kantor, eds.) DIMACS Workshops on Groups and Computation (1993), AMS, Providence.

[CFL] G. Cooperman, L. Finkelstein, and S. Linton, *Constructive recognition of a black box group isomorphic to GL(n , 2)*, in Groups and Computation II (L. Finkelstein, W.M. Kantor, eds.) DIMACS Workshops on Groups and Computation (1997), AMS, Providence.

[CM] H.S.M. Coxeter, W.O.J. Moser, *Generators and relations for discrete groups (third edition)*, Springer, Berlin, 1972.

[D1] P. Diaconis, *Group Representations in Probability and Statistics*, IMS, Hayward, California, 1988.

[D2] P. Diaconis, *A group theoretic interpretation of the records to cycles map* (1995), unpublished manuscript.

[D3] P. Diaconis, *The cutoff phenomenon in finite Markov chains*, Proc. Nat. Acad. Sci. U.S.A. **93** (1996), 1659–1664.

[DF] P. Diaconis, J. Fill, *Strong stationary times via a new form of duality*, Ann. Prob. **18** (1990), 1483–1522.

[DSa] P. Diaconis, L. Saloff–Coste, *Comarison techniques for random walk on finite groups*, The Annals of Probability **21** (1993), 2131–2156.

[DSh1] P. Diaconis, M. Shahshahani, *Generating a random permutation with random transpositions*, Z. Wahr. verw. Gebiete **57** (1981), 159–179.

[DSh2] P. Diaconis, M. Shahshahani, *The subgroup algorithm for generating uniform random variables*, Prob. in Eng. and Info. Sci. **1** (1987), 15–32.

[DSh3] P. Diaconis, M. Shahshahani, *Time to reach stationarity in the Bernoulli–Laplace diffusion model*, SIAM J. Math'l Analysis **18** (1987), 208–218.

[Dem] M. Demazure, *Désingularisation des variétes de Schubert généralisées*, Ann. Sci. Éc. Norm. Sup. **7** (1974), 53–88.

[EKK] V.A. Emelichev, M.M. Kovalev, M.K. Kravtsov, *Polytopes, Graphs and Optimization*, Cambridge University Press, New York, 1984.

[ER] P. Erdös, A. Rényi, *Probabilistic methods in group theory*, J. d'Analyse Math. **14** (1965), 127–138.

[Eu] L. Eulero, *Introductio in Analysin Infitorum*, Marcum-Michaelem Bousquet et Socios, Lausanne, Switzerland, 1748.

[F] W. Feller, *An introduction to Probability theory and its applications (third edition)*, John Wiley, New York, 1970.

[FOW] L. Flatto, A. M. Odlyzko, D. B. Wales, *Random shuffles and group representations*, Ann. Prob. **13** (1985), 155–178.

[Gl] D. Gluck, *Characters and random walks on finite classical groups*, Advances in Mathematics, to appear (1995).

[GR] J. Goldman, G.-C. Rota, *Recent Progress in Combinatorics*, Acad. Press, New York, 1969, pp. 75–83.

[GJ] I. P. Goulden, D. M. Jackson, *Combinatorial enumeration*, John Wiley, New York, 1983.

[Gr] A. Greenhalgh, *Random walks on groups with subgroup invariance properties, Ph. D. dissertation*, Stanford U., 1987.

[GH] D. Griffiths, J. Harris, *Principles of Algebraic Geometry*, John Wiley, New York, 1978.

[HL] G.H. Hardy, J.E. Littlewood, *Some problems of 'Partitio Numerorum'; III: On the expression of a number as a sum of primes*, Acta Math. **13** (1944), 1–70.

[Hi] M. Hilderbrand, *Generating random elements in $SL_n(\mathbb{F}_q)$ by random transvections*, J. Alg. Combinatorics **1** (1992), 133–150.

[H1] J. Humphreys, *Linear algebraic groups*, Springer, Berlin, 1975.

[H2] J. Humphreys, *Reflection groups and Coxeter groups*, Cambridge University Press, Cambridge, UK, 1990.

[J] A. Jucys, *On the Young operators of the symmetric group* (in Russian), Lietuvos Fizikos Rinkinys **6** (1966), 163–180.

[KS] W. Kantor, A. Seress, *Black box classical groups*, preprint (1997).

[K] D. E. Knuth, *The Art of Computer Programming,* Vol. 2, Seminumerical Algorithms, Addison-Wesley, 1969.

[LS] A. Lascoux, M.-P. Schützenberger, *Polynômes de Schubert*, C. R. Acad. Sci. Paris **294** (1982), 447–450.

[M1] I. G. Macdonald, *Symmetric Functions and Hall Polynomials*, Oxford University Press, London, 1979.

[M2] I. G. Macdonald, *Notes on Schubert Polynomials*, Université du Québec, Montréal, 1991.

[Ma] P. Matthews, *A strong uniform time for random transpositions*, J. Theoretical Probability **1** (1988), 411–423.

[MR] R. Motwani and P. Raghavan, *Randomized Algorithms*, Cambridge University Press, Cambridge, UK, 1995.

[N] M. Nazarov, *Young's Symmetrizers for Projective Representations of the Symmetric Group*, RIMS Preprint No 900, Kyoto University, Kyoto, Japan, 1992.

[NP1] P. Neumann, C. Praeger, *A recognition algorithms for special linear groups*, Proc. London Math. Soc. **65** (1992), 555 – 603.

[NP2] P. Neumann, C. Praeger, *Cyclic matrices over finite fields*, J. London Math. Soc. (2) **52** (1995), 263–284.

[NW] A. Nijenhuis, H. Wilf, *Combinatorial Algorithms*, Academic Press, New York, 1978.

[P] I. Pak, *Random walks on permutation: strong uniform time approach*, Ph.D. Thesis, Harvard U., 1997.

[R] D. Randall, *Efficient random generation of nonsingular matrices*, Random Structures and Algorithms **4** (1993), 111–118.

[RND] E. Reingold, J. Nievergelt, N. Deo, *Combinatorial Algorithms*, Prentice-Hall, Englewood Cliffs, New Jersey, 1977.

[S1] R. P. Stanley, *Enumerative Combinatorics,* Vol. 1, Wadsworth & Brooks/Cole, California, 1986.

[S2] R. P. Stanley, *Some combinatorial aspects of the Schubert calculus, in*, Lecture Notes in Math. **579** (1971), Springer, Berlin, 225–259.

[SW] D. Stanton, D. White, *Constructive Combinatorics,* Undegraduate Texts in Mathematics, Springer–Verlag, Berlin, 1986.

[St] R. Steinberg, *Lectures on Chevalley groups*, Yale U., 1967.

[Sy] J.J. Sylvester, *On the partition of an even number into two primes*, Proc. London Math. Soc. (Series 1) **4** (1871), 4–6.

[W] H. Weyl, *Classical Groups*, Princeton University Press, Princeton, NJ, 1939.

[WW] E. T. Whittaker, G. N. Watson, *A Course of Modern Analysis* (Fourth Edition), Cambridge University Press, Cambridge, UK, 1927.

[Z] G. Ziegler, *Lectures on Polytopes, Graduate Texts in Mathematics 152*, Springer, New York, 1995.

Department of Mathematics, Yale University, New Haven CT 06520
E-mail address : **paki@math.yale.edu**

DIMACS Series in Discrete Mathematics
and Theoretical Computer Science
Volume **43**, 1999

Computing On Optical Models

Sanguthevar Rajasekaran

ABSTRACT. Optical models of computing have several attractive features absent in conventional models of computing. In this paper we consider three optical models of computing, namely, the Arrays with Reconfigurable Optical Buses (AROBs), Optical Transpose Interconnection Systems (OTISs), and the Optical Communication Parallel Computers (OCPCs). We provide a survey of algorithms that have been proposed on these models for fundamental problems. Problems of our interest are prefix computation, packet routing, selection, sorting, and matrix operations.

1. Introduction

Models of computing based on optical technology offer superior power and speed. Several optical computing models have been proposed and studied in the literature in the past several years. In this paper we consider three models that have been widely explored: the Optical Communication Parallel Computer (OCPC), the Array with Reconfigurable Optical Buses (AROB), and the Optical Transpose Interconnection System (OTIS). We survey fundamental algorithms that have been developed for these models.

To begin with we give an introduction to the models of computing. There are two versions of these models, namely 1D and 2D. The 1D version has a $1 \times n$ configuration and the 2D version has a $\sqrt{n} \times \sqrt{n}$ configuration.

Arrays with Reconfigurable Optical Buses. An AROB [**24, 18**] is an $m \times n$ reconfigurable mesh [**11**] wherein the buses are realized using optical technology. This model has been extensively studied.

A reconfigurable mesh is a conventional mesh in which local switches in each processor can be dynamically set to form buses connecting contiguous processors. Many closely related variants of the reconfigurable mesh with optical buses have been proposed in the literature (see e.g. [**24, 18**]). In the model assumed by this paper the switch settings of the processors permitted are the same as those in the RN model of [**2**]. Any bus link connects two adjacent processors. Two wave guides are employed, one for left to right transmission and the other for right to left transmission. Local bus links can be set to form disjoint buses. The length of

---

1991 *Mathematics Subject Classification.* Primary 68Q10, 68P10, 68Q22; Secondary 68Q20, 68Q25.

This research was supported in part by an NSF Award CCR-95-03-007 and an EPA Grant R-825-293-01-0.

a bus is the number of links on that bus. The time taken by a message to traverse the bus once is called a cycle.

A cycle can be thought of as consisting of *slots* of duration τ. Each slot carries a different optical signal. The time taken by an optical pulse to move down a bus link is τ. In particular, τ time is sufficient to send a b-bit message, where each bit is a light pulse with a w second duration, for suitable values of $b \geq 1$ and w. It is argued that (see e.g., [**24**]) for reasonable size meshes (say up to 1000×1000), the number of slots in a cycle may be assumed to be n for an $n \times n$ mesh. Also, it is reasonable to assume that the duration of a cycle is constant and comparable to the time for a CPU operation.

In an AROB each processing element has an associated slot counter. These counters may be started at the beginning of a cycle. For each processor, a slot in the cycle is allocated for reading and writing. Any processor can write to or read from the bus only in the allocated slots. In cases of write conflicts, the last written message remains in the slot. A processor cannot poll the up to n light pulses moving through it in one cycle since it can read from or write to its bus during only one slot of a cycle. Another special feature of an AROB is the delay unit at each processor. A processor can introduce a delay of one time slot in the light pulses passing through it.

The Optical Transpose Interconnection System. This model of computing has been proposed in [**9, 17, 39**]. In an OTIS, processors are partitioned into groups where each group is realized as chips with electronic interprocessor connections. Free space optical links are used to realize connections among the groups. It has been shown that free space optical links provide superior speed and power when the connect distance is more than a few millimeters. This is the reason for the choice of optical links for connecting the groups.

Interconnections among the processors in any group can be realized as a mesh, a hypercube, or any other network. Accordingly the OTIS-Mesh, the OTIS hypercube, etc. will arise. If j is any processor in group i of an OTIS, this processor is connected to processor i of group j.

In an $n \times n$ OTIS-Mesh there are n groups where each group has n processors organized as a $\sqrt{n} \times \sqrt{n}$ mesh. It is known that the diameter of this mesh is $4\sqrt{n}-3$ [**34**]. Both SIMD and MIMD versions of the OTIS-Mesh can be conceived of.

The Optical Communication Parallel Computer. An OCPC consists of n processors, where any processor can communicate with any other processor in one step if there are no conflicts. If more than one processor sends a message to some processor π at the same time, then π does not receive any meaningful message. We can simulate each step of an EREW PRAM with n processors and n common memory cells on an OCPC with $2n$ processors in $O(1)$ time as follows. Employ n of the OCPC processors to simulate the PRAM processors. The other n processors are associated with the common memory cells one processor per cell. Each read or write step of the PRAM now can be simulated in $O(1)$ communication steps on the OCPC. The following Lemma is clear.

LEMMA 1.1. *We can simulate a single step of an N-processor EREW PRAM that uses M common memory cells in $O(1)$ steps on an OCPC with* $\max\{N, M\}$ *processors.*

2. Problems Definition

Given a sequence of numbers, say, $k_1, k_2, \ldots, k_n$, the problem of sorting is to rearrange them in nondecreasing order. The problem of selection is to identify the ith smallest number from out of n given numbers (where i is an input and $1 \leq i \leq n$).

Packet routing plays a vital role in any parallel machine. A single step of interprocessor communication in any interconnection network can be thought of as a packet routing task. Packet routing refers to the following problem. There is a packet of information at each node in the network that is destined for some other node. Send all the packets to their destinations as quickly as possible in such a way that no more than one packet crosses any edge at any time. We define the *run time* of any packet routing algorithm to be the time taken by the last packet to reach its destination. The *queue size* refers to the maximum number of packets that any processor has to store during the algorithm.

If at most one packet originates from any node and at most one packet is destined for any node, the problem of routing is called *partial permutation routing.* Likewise, if at most h packets originate from any node and at most h packets are destined for any node, then we have a *h-relations routing* or a *$h-h$ routing* problem [**37, 28**].

Matrix operations considered in this paper include multiplication and inverse.

3. Prefix Computation

Let $x_1, x_2, \ldots, x_n$ be any sequence of elements from the domain Σ. Let $\oplus$ be any binary, associative, and unit-time computable operator on Σ. The problem of prefix computation is to compute $x_1, x_1 \oplus x_2, x_1 \oplus x_2 \oplus x_3, \ldots, x_1 \oplus x_2 \oplus \cdots \oplus x_n$. This problem forms the basis for computing on any parallel model. Optimal algorithms are known for prefix computation on the three models of our interest.

AROB. The following Lemma is due to Pavel and Akl [**24**].

LEMMA 3.1. *Let L be a $1 \times n$-processor AROB. If each processor of L has a bit, then the prefix sums of these bits can be computed in $O(1)$ cycles* [**24**]. *Also, the addition of n $\log n$-bit numbers can be performed in $O(1)$ cycles on a $\log n \times n$ AROB.*

Using Lemma 3.1, we can get an $O(1)$-cycle algorithm for prefix sums of bits on a $\sqrt{n} \times \sqrt{n}$ AROB (see [**24, 31**]).

LEMMA 3.2. *If there is a bit at each node of a $\sqrt{n} \times \sqrt{n}$ AROB, we can compute the prefix sums of these bits in $O(1)$ cycles.*

The maximum bus length employed by [**24**]'s algorithm is $3\sqrt{n}$. The bus length is reduced to $\sqrt{n} + o(\sqrt{n})$ in [**31**].

The OTIS-Mesh. Consider an $n \times n$ OTIS-Mesh where there are n groups of processors each group being a $\sqrt{n} \times \sqrt{n}$ Mesh. Index the processors in any group using the snake-like row-major order. There is an element to begin with at each processor. The prefix indexing used is the lexicographical ordering of $(group, processor)$.

We can perform prefix computation on the OTIS-Mesh in $6\sqrt{n}+O(1)$ time using the following algorithm. Each group computes all the local prefixes in $2\sqrt{n}-2$ time.

At the end of this step, processor n in every group has the sum of all the elements in the group. Send the sum of group i to processor i of group n, for $1 \leq i \leq n$. This takes only one time unit using the transpose OTIS connections. After this, group n performs a prefix computation on the sums in $2\sqrt{n}-2$ time. The result is moved right by one position in one time unit. Now send out these prefixes using the transpose OTIS connections. After this, processor n in group i will have the sum of all the elements in groups 1 through $i-1$ (for $2 \leq i \leq n$). Finally, a broadcast local to groups is done in order to update the prefixes.

LEMMA 3.3. *We can do prefix computation on an $n \times n$ OTIS-Mesh in $6\sqrt{n}+O(1)$ time.*

OCPC. Since an OCPC with n processors can simulate a single step of an n-processor $O(n)$-memory EREW PRAM in $O(1)$ time (c.f. Lemma 1.1), the following Lemma results.

LEMMA 3.4. *Prefix computation on a sequence of length n can be computed in $O(\log n)$ time on an n-processor OCPC.*

4. Packet Routing

In this section we survey packet routing algorithms that have been developed for the AROB, OTIS, and OCPC models.

AROB. Proof of the following Lemma can be found in [**24**].

LEMMA 4.1. *In an AROB of size $1 \times n$ any permutation can be routed in $O(1)$ cycles* [**24**].

The basic idea behind the algorithm is the following. There is a fixed time slot assigned for each processor for reading from or writing into the bus. Any processor π chooses a time slot depending on the destination π' of the packet it has and writes in that time slot. This time slot will be such that the message will reach π' in the time slot allocated to it for reading.

The preceding Lemma can be strengthened as follows [**24, 31**]:

LEMMA 4.2. *Let L be an AROB of size $1 \times n$. Consider a routing problem where $O(1)$ packets originate from any node and $O(1)$ packets are destined for any node. This problem can also be solved in $O(1)$ cycles.*

LEMMA 4.3. *There are k elements arbitrarily distributed in a $\sqrt{n} \times \sqrt{n}$ AROB, with ≤ 1 element per processor. The problem is to collect them in the first $\lceil \frac{k}{\sqrt{n}} \rceil$ rows. This problem can be solved in $O(1)$ cycles.*

There is an $\widetilde{O}(\log\log n)$ time algorithm available for routing any partial permutation on a 2D OCPC. The same algorithm can be simulated on a 2D AROB to get the same run time. (We say a randomized algorithm uses $\widetilde{O}(f(n))$ amount of any resource (like time, space, etc.) if the amount of resource used is no more than $c\alpha f(n)$ with probability $\geq (1-n^{-\alpha})$ for any α, $c > 0$ being a constant.)

LEMMA 4.4. *Any partial permutation can be routed in $\widetilde{O}(\log\log n)$ time on a $\sqrt{n} \times \sqrt{n}$ AROB.*

Now we consider the problem of h-relations routing. This problem has been studied well on the conventional mesh as well as the OCPC. Rajasekaran and Sahni have given a randomized algorithm for this problem that runs in $\widetilde{O}(h)$ cycles on a 1D AROB [**31**]. On the other hand, the best known h-relations routing algorithm on the OCPC has a run time of $\widetilde{O}(h + \log\log n)$ [**8**].

The OCPC and the AROB differ in two critical ways. In an OCPC if two or more processors try to send a message to the same processor at the same time, then no message reaches the destination. But in an AROB, under the same conditions, an arbitrary packet reaches the destination. In addition, prefix sums of bits can be computed in $O(1)$ cycles on the AROB model and there is no such algorithm known for the OCPC.

[**31**]'s algorithm runs in *stages*. A stage consists of many *basic steps*. During a basic step, a processor chooses one of its packets at random and sends it in the bus. The chosen packet π may not reach its destination in cases of conflicts.

At the end of every stage a *load balancing* operation is performed. The load balancing operation distributes unsuccessful packets equally among all the processors. If the maximum number of packets in any processor is k, then the load balancing operation takes only $O(k)$ time.

If k_i is the maximum number of packets in any node at the beginning of stage i, then the value of k_i decreases by a constant factor at the end of the stage with high probability. The total time spent in any stage is $\widetilde{O}(k_i)$. Since $k_1 = h$, the run time of the algorithm is $\widetilde{O}(h)$.

LEMMA 4.5. *Any h-relation can be routed in $\widetilde{O}(h)$ cycles on a $\sqrt{n}\times\sqrt{n}$ AROB.*

The above algorithm can be extended [**31**] to a $\sqrt{n}\times\sqrt{n}$ AROB. The run time of the extension is $\widetilde{O}(h+\log\log n)$ cycles. This result can be contrasted with the run time of $\widetilde{O}(h+\log n/\log\log n)$ that the algorithm of [**33**] takes for the same problem on the OCPC model. The basic idea is to employ the two phase routing algorithm of Valiant. In the first phase the packets are routed to random destinations and in the second phase they are routed to their actual destinations.

The OCPC algorithm of [**33**] consists of a sequence of stages. The algorithm of [**31**] also consists of stages except that they perform a load balancing operation after every stage.

THEOREM 4.6. *Any h-relation can be routed in $\widetilde{O}(h + \log\log n)$ cycles on a $\sqrt{n}\times\sqrt{n}$ AROB.*

A deterministic algorithm for routing has also been given in [**31**].

LEMMA 4.7. *We can route any partial permutation deterministically in $O(\log n)$ cycles on a $\sqrt{n}\times\sqrt{n}$ AROB. Any h-relation can be deterministically routed in $O(h\log n)$ cycles on a $\sqrt{n}\times\sqrt{n}$ AROB.*

OTIS. It is known that a 4D mesh can be simulated efficiently on the OTIS-Mesh. Thus it is possible to import 4D mesh algorithms to the OTIS-Mesh. However, the underlying constants in the time bounds will be large. Thus it is essential to develop algorithms specific to the OTIS-Mesh.

Rajasekaran and Sahni [**30**] show that any partial permutation can be routed in $4\sqrt{n}+\widetilde{o}(\sqrt{n})$ time on an $n\times n$ OTIS-Mesh. The algorithm is an adaptation of [**32**]'s algorithm. In the algorithm of [**32**], a packet chooses a random node within a restricted space around its origin and goes there. From then on it greedily progresses toward its destination traveling the dimensions one at a time. The resultant Lemma is:

LEMMA 4.8. *Any partial permutation can be routed on an $n \times n$ OTIS-Mesh in $4\sqrt{n} + \widetilde{o}(\sqrt{n})$ time, the queue sizes being $\widetilde{O}(1)$.*

The problem of h-relations routing can be solved by applying the algorithm given in [**28**].

LEMMA 4.9. *Any h-relation can be routed on an $n \times n$ OTIS-Mesh in $\widetilde{O}(h\sqrt{n})$ time, the queue sizes being $\widetilde{O}(h)$.*

OCPC. On a one-dimensional OCPC, any partial permutation can be routed in one unit of time. Rao and Tsantilas [**33**] have given a partial permutation routing algorithm for a $\sqrt{n} \times \sqrt{n}$ OCPC that runs in time $\widetilde{O}(\log \log n)$. Any packet π traverses to a random node in its row of origin and from there it goes to its destination using greedy paths. The time complexity of the algorithm is proven by showing that if there are $\alpha\sqrt{n}$ packets at any given time in any row, then after a constant amount of time the row will have no more than $\alpha^2\sqrt{n}$ packets, with high probability.

LEMMA 4.10. *Any partial permutation can be routed in $\widetilde{O}(\log \log n)$ time on a $\sqrt{n} \times \sqrt{n}$ OCPC.*

The h-relations routing problem has been studied by many. Anderson and Miller's algorithm can route any $\log n$-relation on an n-node 1D OCPC in $\widetilde{O}(\log n)$ time [**1**]. Valiant's algorithm works for any h-relation with a run time of $\widetilde{O}(h + \log n)$ [**37**]. Geréb-Graus and Tsantilas [**7**]'s algorithm has a run time of $\widetilde{O}(h + \log n \log \log n)$ for h-relations. An algorithm for arbitrary h-relations with a run time of $\widetilde{O}(h + \log \log n)$ has been given by Goldberg, Jerrum, Leighton, and Rao [**8**].

The algorithm of Goldberg, Jerrum, Leighton, and Rao is very complex. There are four stages in the algorithm each taking $\widetilde{O}(h + \log \log n)$ time. 1) In every step of the first stage a processor picks one of its remaining packets and tries to send it. At the completion of the first stage there will be only $\widetilde{O}\left(\frac{n}{h \log \log n}\right)$ unsent packets. 2) A load balancing operation is conducted in the second stage so that each processor ends up with at most one packet. Group the processors so that each group has $\log^c n$ processors, for some constant c. 3) Packets are sent to some processors in their target groups in the third stage. 4) Finally, the packets are routed within their groups.

LEMMA 4.11. *On an n-node 1D OCPC any h-relation can be routed within $\widetilde{O}(h + \log \log n)$ communication steps.*

The following Lemma can also be found in [**33**].

LEMMA 4.12. *An arbitrary h-relation can be realized in $\widetilde{O}\left(h + \frac{\log n}{\log \log n}\right)$ communication steps on a $\sqrt{n} \times \sqrt{n}$ OCPC.*

5. Selection

Given a sequence of n numbers $k_1, k_2, \ldots, k_n$ and an $i \leq n$, the problem of selection is to identify the ith smallest of the n numbers. Optimal sequential algorithms are known for selection (see e.g., [**10**]). A simple linear time randomized algorithm for sequential selection has been given by Floyd and Rivest [**6**].

Selection has also been studied widely on various parallel models of computing (see e.g., [**27**]). Many of the parallel selection algorithms (both deterministic and randomized) employ the technique of sampling. In this section we survey algorithms that have been developed for optical computers.

AROB. Pan's algorithm [**23**] for the 1D AROB selects a random input element k, partitions the input around k, determines which part has the key to be selected, and recursively performs a selection on this part. Each basic step of choosing a k and partitioning the input can be expected to reduce the input size by a constant factor.

LEMMA 5.1. *We can perform selection from out of n given keys in an expected time of $O(\frac{n}{p} \log n)$ cycles on an AROB of size $1 \times p$. The worst case run time is $O(n)$ cycles.*

Several $O(1)$ time algorithms are known for sorting on the reconfigurable mesh (e.g., [**11**], [**21**], [**16**]). Rajasekaran and Sahni [**31**]'s algorithm can select from out of n numbers in $\widetilde{O}(1)$ cycles on an AROB of size $\sqrt{n} \times \sqrt{n}$. Each processor is input a number. The underlying theme is the following: 1) Pick a random sample S of size $s = o(n)$; 2) Identify two elements ℓ_1 and ℓ_2 from the sample whose ranks in S are $i\frac{s}{n} - \delta$ and $i\frac{s}{n} + \delta$ for some appropriate δ. These elements will 'bracket' the element to be selected with high probability; 3) Eliminate those input keys that fall outside the range $[\ell_1, \ell_2]$; 4) Perform an appropriate selection from out of the remaining keys.

LEMMA 5.2. *We can perform selection from out of n keys in $\widetilde{O}(1)$ cycles on a $\sqrt{n} \times \sqrt{n}$ AROB.*

OTIS. Sahni and Wang's deterministic algorithm on an $n \times n$ SIMD OTIS-Mesh [**35**] has a run time of $22\sqrt{n} + o(\sqrt{n})$ which is the same as the run time of their sorting algorithm.

Rajasekaran and Sahni [**30**]'s selection algorithm on the OTIS-Mesh has the same theme as the one used for the AROB. A random sample is used to pick two very likely bracketing elements. Those input keys that can not possibly be the element to be selected are eliminated. An appropriate selection is performed in the set of remaining keys. Though this general idea has been applied over a variety of parallel models, the challenge is in coming up with the best implementation of this idea on the model under concern. In general this may not be a trivial task.

LEMMA 5.3. *Selection from out of n elements can be performed in $6\sqrt{n} + \widetilde{o}(\sqrt{n})$ time on an $n \times n$ OTIS-Mesh.*

OCPC. If the number of input keys is n, then at least n common memory cells will be used by any PRAM algorithm. In order to apply Lemma 1.1, at least n OCPC processors will be needed. Therefore, we can simulate any of the EREW PRAM sorting algorithms on the OCPC to perform selection. For instance, Cole's parallel merge sort [**5**] runs in $O(\log n)$ time using n processors.

LEMMA 5.4. *We can perform selection from out of n keys in $O(\log n)$ time on an n-processor 1D OCPC.*

6. Sorting

AROB. The lower bound result in [**3**] indicates that sorting of n numbers on a p-processor parallel comparison tree will need $\Omega\left(\frac{\log n}{\log(1+\frac{p}{n})}\right)$ time [**3**]. Since the parallel comparison tree is more powerful than the AROB, the same lower bound holds on the AROB as well. As a consequence, if sorting has to be done in $O(1)$ time, $\Omega(n^{1+\epsilon})$ processors will be needed, for some constant $\epsilon > 0$. Such an optimal algorithm has been given by Rajasekaran and Sahni [**31**].

LEMMA 6.1. *If there is one key per processor in the first row of an $n^\epsilon \times n$ AROB, then sorting of these n keys can be done in $O(1)$ time.*

They employ Leighton's column sort [**14**]. The same result can be obtained using the (l, m)-merge sort given in [**25**].

OTIS. A deterministic algorithm with a run time of $22\sqrt{n} + o(\sqrt{n})$ has been proposed by Sahni and Wang [**35**] for sorting on the SIMD OTIS-Mesh.

LEMMA 6.2. *Sorting and hence selection on an $n \times n$ SIMD OTIS-Mesh can be performed in $22\sqrt{n} + o(\sqrt{n})$ time.*

An $8\sqrt{n} + \tilde{o}(\sqrt{n})$ algorithm can be found in [**30**] for sorting on an $n \times n$ OTIS-Mesh. The main theme behind this algorithm is random sampling just like randomized sorting algorithms available for various other parallel models. The idea is to pick a random sample S with $s = o(n)$ keys, sort the sample, partition the input using the sample keys as *splitter* keys, and to sort the individual parts.

LEMMA 6.3. *Sorting on an $n \times n$ OTIS-Mesh takes $8\sqrt{n} + \tilde{o}(\sqrt{n})$ time.*

OCPC. Lemma 1.1 and Cole's parallel mergesort can be used to derive the following result.

LEMMA 6.4. *Sorting can be accomplished in $O(\log n)$ time on an n-processor OCPC.*

7. Matrix Operations

AROB. The results discussed in this section are due to Pavel and Akl [**24**].

The first algorithm deals with matrix multiplication and makes use of the fact that prefix sums can be computed in $O(1)$ cycles provided the numbers have $O(\log n)$ bits each (c.f. Lemma 3.1).

LEMMA 7.1. *We can multiply two $n \times n$ matrices in $O(r)$ cycles on an $n \times n \log n/r \times n$ AROB, for any $1 \le r \le \log n$, provided the numbers have $O(\log n)$ bits each.*

Their results on the problem of multiplying sparse matrices where each row (or column) has at most r nonzero elements can be summarized as follows.

LEMMA 7.2. *Let A be a matrix with at most r nonzero elements per row and B be a matrix with at most r nonzero elements per column. AB can be computed in $O(\log r)$ cycles on an $n \times n \times r$ AROB, in $O(r \log r)$ cycles on an $n \times n$ AROB, or in $O(1)$ cycles on an $n \times n \log n \times r$ AROB, where $r \le \frac{n}{\log n}$.*

In [**24**] an algorithm is also given for Gaussian elimination.

OTIS. Consider an $n \times n$ OTIS-Mesh. Let A and B be two given matrices of size $n \times n$ each that we are interested in multiplying. Let the first row of A or B be stored in the OTIS-mesh as follows. The first $\sqrt{n}$ elements are in the first row of group 1, the next $\sqrt{n}$ elements are in the first row of group 2, and so on. Similarly store the rest of the rows.

Now one could simulate the standard 'systolic' algorithm for matrix multiplication on the conventional Mesh (see e.g., [**13**]) and get the following Lemma.

LEMMA 7.3. *We can multiply two $n \times n$ matrices in $O(n)$ time on an $n \times n$ OTIS-Mesh.*

OCPC. The trivial algorithm for parallel matrix multiplication on the EREW PRAM can be simulated on the OCPC as follows.

LEMMA 7.4. *Two $n \times n$ matrices can be multiplied in $O(\log n)$ time using n^3 OCPC processors.*

Also sparse matrix multiplication algorithms known for the EREW PRAM can be transported to the OCPC.

8. Conclusions

We have considered the AROB, OTIS, and OCPC models in this paper. Algorithms known for fundamental problems on these models have been surveyed.

References

[1] R.J. Anderson and G.L. Miller, Optical Communication for Pointer Based Algorithms, Technical Report CRI-88-14, Computer Science Department, University of Southern California, 1988.

[2] Y. Ben-Asher, D. Peleg, R. Ramaswami, and A. Schuster, The Power of Reconfiguration, *Journal of Parallel and Distributed Computing*, 1991, pp. 139-153.

[3] R. Boppana, The Average-Case Parallel Complexity of Sorting, *Information Processing Letters* 33, 1989, 145-146.

[4] H. Chernoff, A Measure of Asymptotic Efficiency for Tests of a Hypothesis Based on the Sum of Observations, *Annals of Mathematical Statistics* 2, 1952, pp. 241-256.

[5] R. Cole, Parallel Merge Sort, *SIAM Journal on Computing* 17, 1988, pp. 770-785.

[6] R.W. Floyd, and R.L. Rivest, Expected Time Bounds for Selection, *Communications of the ACM* 18(3), 1975, pp. 165-172.

[7] M. Gereb-Graus and T. Tsantilas, Efficient Optical Communication in Parallel Computers, *Proc. ACM Symposium on Parallel Algorithms and Architectures*, 1992, pp. 41-48.

[8] L. Goldberg, M. Jerrum, T. Leighton, and S. Rao, A Doubly-Logarithmic Communication Algorithm for the Completely Connected Optical Communication Parallel Computer, *Proc. ACM Symposium on Parallel Algorithms and Architectures*, 1993, pp.300-309.

[9] W. Hendrick, O. Kibar, P. Marchand, C. Fan, D. V. Blerkom, F. McCormick, I. Cokgor, M. Hansen, and S. Esener, Modeling and Optimization of the Optical Transpose Interconnection System, in *Optoelectronic Technology Center, Program Review*, Cornell University, September 1995.

[10] E. Horowitz and S. Sahni and S. Rajasekaran, *Computer Algorithms*, Computer Science Press, 1998.

[11] J. Jang and V.K. Prasanna, An Optimal Sorting Algorithm on Reconfigurable Mesh, *Proc. International Parallel Processing Symposium*, 1992, pp. 130-137.

[12] J. Jenq and S. Sahni, Reconfigurable Mesh Algorithms for Image Shrinking, Expanding, Clustering, and Template Matching, *Proc. International Parallel Processing Symposium*, 1991, pp. 208-215.

[13] T. Leighton, *Introduction to Parallel Algorithms and Architectures: Arrays-Trees-Hypercubs*, Morgan-Kaufmann Publishers, 1992.

[14] T. Leighton, Tight Bounds on the Complexity of Parallel Sorting, *IEEE Transactions on Computers*, C-34(4), 1985, pp. 344-354.
[15] R. Lin, S. Olariu, Reconfigurable Buses with Shift Switching: Concepts and Applications, *IEEE Transactions on Parallel and Distributed Systems*, 6(1), 1995, pp. 93-102.
[16] R. Lin, S. Olariu, J. L. Schwine, and J. Zhang, Sorting in $O(1)$ Time on a Reconfigurable Mesh of Size $N \times N$, *Proc. European Workshop on Parallel Computing*, 1992, pp. 16-27.
[17] G. C. Marsden, P. J. Marchand, P. Harvey, and S. C. Esener, Optical Transpose Interconnection System Architectures, *Optic Letters*, 18(3):1083-1085, July 1993.
[18] R. G. Melhem, D. M. Chiarulli, and S. P. Levitan, Space Multiplexing of Waveguides in Optically Interconnected Multiprocessor Systems, *Computer Journal*, 32(4), 1989, pp. 362-369.
[19] R. Miller, V.K. Prasanna-Kumar, D. Reisis and Q.F. Stout, Meshes with Reconfigurable Buses, *IEEE Transactions on Computers* 42, 1993, pp. 678-692.
[20] D. Nassimi and S. Sahni, A Self-Routing Benes Network and Parallel Permutation Algorithms, *IEEE Transactions on Computers*, C-30(5), 1981, pp. 332-340.
[21] M. Nigam and S. Sahni, Sorting n Numbers on $n \times n$ Reconfigurable Meshes with Buses, *Proc. International Parallel Processing Symposium*, 1993, pp. 174-181.
[22] S. Olariu, J.L. Schwing and J. Zhang, Integer Problems on Reconfigurable Meshes, with Applications, *Proc. 1991 Allerton Conference* 4, 1991, pp. 821-830.
[23] Y. Pan, Order Statistics on Optically Interconnected Multiprocessor Systems, *Proc. First International Workshop on Massively Parallel Processing Using Optical Interconnections*, 1994, pp. 162-169.
[24] S. Pavel and S.G. Akl, Matrix Operations using Arrays with Reconfigurable Optical Buses, manuscript, 1995.
[25] S. Rajasekaran, A Framework For Simple Sorting Algorithms On Parallel Disk Systems, *Proc. 10th Annual ACM Symposium on Parallel Algorithms and Architectures*, 1998.
[26] S. Rajasekaran, Meshes with Fixed and Reconfigurable Buses: Packet Routing, Sorting and Selection, *Proc. First Annual European Symposium on Algorithms*, Springer-Verlag Lecture Notes in Computer Science 726, 1993, pp. 309-320. Also see *IEEE Transactions on Computers*, 45(5), 1996, pp.529-539.
[27] S. Rajasekaran, Sorting and Selection on Interconnection Networks, *DIMACS Series in Discrete Mathematics and Theoretical Computer Science* 21, 1995, pp. 275-296.
[28] S. Rajasekaran, $k-k$ Routing, $k-k$ Sorting, and Cut Through Routing on the Mesh, *Journal of Algorithms 19*, 1995, pp. 361-382.
[29] S. Rajasekaran and J.H. Reif, Derivation of Randomized Sorting and Selection Algorithms, in *Parallel Algorithm Derivation and Program Transformation*, Edited by R. Paige, J.H. Reif, and R. Wachter, Kluwer Academic Publishers, 1993, pp. 187-205.
[30] S. Rajasekaran and S. Sahni, Randomized Routing, Selection, and Sorting on the OTIS-Mesh, manuscript, 1997.
[31] S. Rajasekaran and S. Sahni, Sorting, Selection, and Routing on the Array with Reconfigurable Optical Buses, *IEEE Transactions on Parallel and Distributed Systems* 8(11), 1997.
[32] S. Rajasekaran and Th. Tsantilas, Optimal Routing Algorithms for Mesh Connected Processor Arrays, *Algorithmica* 8, 1992, pp. 21-38.
[33] S. Rao and T. Tsantilas, Optical Interprocessor Communication Protocols, *Proc. Workshop on Massively Parallel Processing Using Optical Interconnections*, 1994, pp. 266-274.
[34] S. Sahni and C. Wang, BPC Permutations on the OTIS-Mesh Optoelectronic Computer, *Proc. The Fourth International Conference on Massively Parallel Processing Using Optical Interconnections (MPPOI'97)*, 1997, pp. 130-135.
[35] S. Sahni and C. Wang, Basic Algorithms on the OTIS-Mesh Optoelectronic Computer, Manuscript, 1997.
[36] R.K. Thiruchelvan, J.L. Trahan, and R. Vaidyanathan, On the Power of Segmenting and Fusing Buses, *Proc. International Parallel Processing Symposium*, 1993, pp. 79-83.
[37] L.G. Valiant, General Purpose Parallel Architectures, in *Handbook of Theoretical Computer Science*: Vol. A (J. van Leeuwen, ed.), North Holland, 1990.
[38] L.G. Valiant and G.J. Brebner, Universal Schemes for Parallel Communication, *Proc. 13th Annual ACM Symposium on Theory of Computing*, 1981, pp. 263-277.
[39] F. Zane, P. Marchand, R. Paturi, S. Esener, Scalable Network Architectures Using the Optical Transpose Interconnection System (OTIS), *Proc. Third International Conference on*

Massively Parallel Processing Using Optical Interconnections (MPPOI'96), 1996, pp. 114-121.

DEPARTMENT OF COMPUTER AND INFORMATION SCIENCE AND ENGINEERING, UNIVERSITY OF FLORIDA, GAINESVILLE, FL 32611
E-mail address: raj@cise.ufl.edu

DIMACS Series in Discrete Mathematics
and Theoretical Computer Science
Volume **43**, 1999

Manipulating Statistical Difference

Amit Sahai and Salil Vadhan

ABSTRACT. We give several efficient transformations for manipulating the statistical difference (variation distance) between a pair of probability distributions. The effects achieved include increasing the statistical difference, decreasing the statistical difference, "polarizing" the statistical relationship, and "reversing" the statistical relationship. We also show that a boolean formula whose atoms are statements about statistical difference can be transformed into a single statement about statistical difference. All of these transformations can be performed in polynomial time, in the sense that, given circuits which sample from the input distributions, it only takes polynomial time to compute circuits which sample from the output distributions.

By our prior work (see FOCS 97), such transformations for manipulating statistical difference are closely connected to results about SZK, the class of languages possessing statistical zero-knowledge proofs. In particular, some of the transformations given in this paper are equivalent to the closure of SZK under complement and under certain types of Turing reductions. This connection is also discussed briefly in this paper.

1. Introduction

Statistical difference, also known as variation distance, is a fundamental measure of similarity between probability distributions. This measure is the most appropriate choice for many applications in algorithms and cryptography, so it is natural to seek efficient means of manipulating it. In this paper, we exhibit polynomial-time transformations mapping pairs of distributions to pairs of distributions which have the following effects:

A **(Increasing Statistical Difference)** "Noticeable" statistical difference is driven exponentially close to 1 (whereas negligible statistical difference remains negligible). This is a well-known technique — we simply repeat each distribution many times independently.

B **(Decreasing Statistical Difference)** Statistical difference that is bounded away from 1 is driven exponentially close to 0 (whereas statistical difference that is negligibly close to 1 remains as such). This is achieved by an XOR construction inspired by a technique of [**DDPY94**].

1991 *Mathematics Subject Classification.* Primary 68Q25, 60E05; Secondary 68Q15.

Both authors are supported by DOD/NDSEG Graduate Fellowships and partially by DARPA grant DABT63-96-C-0018.

A longer paper, containing the results of this paper along with those of [**SV97**], plus some additional results, will be made available at `http://www-math.mit.edu/~salil` and `http://theory.lcs.mit.edu/~amits`.

C **(Polarizing the Statistical Relationship)** If the original distributions have (moderately) large statistical difference, then the resulting distributions will have statistical difference exponentially close to 1, and if the original distributions have (moderately) small statistical difference, then the resulting ones will have statistical difference exponentially close to 0.

D **(Reversing the Statistical Relationship)** If the original distributions have small statistical difference, the resulting ones will have large statistical difference, and if the original ones have large statistical difference, the resulting ones will have small statistical difference.

Of these, Transformations C and D are the main new contributions of this work, though we also describe Transformations A and B in detail, as we make use of them for the former. The final result of this paper is an application of all of these techniques to show that a boolean formula whose atoms are statements about statistical difference can be efficiently transformed into a single statement about statistical difference. All of these results are discussed in more detail below.

Our initial motivation for addressing these questions arose from our recent work [**SV97**] showing a close relationship between statistical difference and statistical zero-knowledge (SZK) proofs. Specifically, the problem of distinguishing between pairs of (succinctly described) distributions with large statistical difference and pairs with small statistical difference was shown to be *complete* for SZK, the class of languages possessing statistical zero-knowledge proofs [**SV97**]. By that result, efficient transformations for manipulating statistical difference yield closure properties of SZK, and conversely. For example, one of the transformations given here is equivalent to the closure of SZK under complementation, and another is equivalent to the closure of SZK under certain types of Turing (or Cook) reductions. In fact, some of these transformations were developed in [**SV97**] in order to prove things about SZK, whereas others were obtained by extracted ideas which appeared in [**DDPY94, Oka96, SV97**] stated in terms of SZK.

Another reason for looking at statistical difference is that many computational problems of interest can be cast as statistical difference problems; examples include QUADRATIC RESIDUOSITY [**GMR89**], GRAPH ISOMORPHISM [**GMW91**], and approximate versions of the CLOSEST and SHORTEST VECTOR PROBLEMS [**GG98**]. Indeed, statistical zero-knowledge proofs are often constructed for such problems based on this observation.

1.1. Formal Setting. If X and Y are probability distributions (or random variables) on a discrete space D, the *statistical difference* between X and Y is defined to be

$$\|X - Y\| = \max_{S \subset D} |\Pr[X \in S] - \Pr[Y \in S]|.$$

In this paper, we focus on distributions Z which have a "succinct description" which enables them to be sampled efficiently. By "succinct description" we mean a circuit C which, when fed the uniform distribution, has output distribution Z. For example, if C has m input gates and n output gates, Z would be the distribution induced on $\{0,1\}^n$ when C is fed the uniform distribution on $\{0,1\}^m$. Thus, when we speak of "efficient transformations" of pairs of distributions, we mean that there is a polynomial-time computable function on pairs of circuits that achieves the desired transformation on the corresponding pair of distributions. For notational convenience, we write C for both the circuit and the distribution Z it defines.

1.2. Contrast with the "Standard" Setting. In the setting of randomized algorithms, there is typically a single probability space at hand, and one is interested in the probability of some event (such as the algorithm giving the correct answer.) To achieve various effects on the probability of this event (such as decreasing the error probability), a Chernoff bound argument is often all that is needed. In the setting of statistical difference, there are *two* probability spaces at hand, and we are interested in the maximum *difference* in probabilities, over all possible events. Because of this, we are limited in what we can achieve with Chernoff bounds alone. Chernoff bounds do, however, enable us to analyze the following *direct product* construction: Suppose we have a pair of distributions X_0 and X_1, and we let Y_0 and Y_1 consist of k independent copies of X_0 and X_1, respectively. Then a Chernoff bound argument tells us that

$$\|X_0 - X_1\| > \epsilon \Rightarrow \|Y_0 - Y_1\| > 1 - e^{-\Omega(k\epsilon^2)}. \tag{1.1}$$

Moreover, one can also show that

$$\|X_0 - X_1\| < \epsilon \Rightarrow \|Y_0 - Y_1\| < k\epsilon.$$

This gives us Transformation A — noticeable statistical differences are driven exponentially close to 1, whereas negligible statistical differences remain negligible.

1.3. Polarizing the Statistical Relationship. Our first main result is a transformation which "polarizes" the statistical relationship between two distributions. That is, pairs of distributions which are statistically close become much closer and pairs of distributions which are statistically far apart become much further apart. That is, we exhibit a polynomial-time computable transformation which takes a triple $(C_0, C_1, 1^k)$, where C_0 and C_1 are circuits and produces a new pair of circuits (D_0, D_1) such that

$$\begin{aligned} \|C_0 - C_1\| > 2/3 \quad &\Rightarrow \quad \|D_0 - D_1\| > 1 - 2^{-k} \\ \|C_0 - C_1\| < 1/3 \quad &\Rightarrow \quad \|D_0 - D_1\| < 2^{-k}. \end{aligned}$$

Note that this is *not* achieved by the direct product construction described above. Looking back at Equation 1.1, we see that the statistical difference will go to 1 in both cases, whereas we want the statistical difference to go to 1 in the first case and 0 in the second case. Thus, the Polarization Lemma[1] is achieved by carefully combining the direct product construction with another construction which decreases statistical difference.

To decrease statistical difference, we show that, given two pairs of distributions, we can efficiently produce a third pair of distributions whose statistical difference is exactly the *product* of the original two statistical differences. Iterating this drives statistical differences which are bounded away from 1 to 0 as desired, whereas statistical differences that are negligibly close to 1 remain as such. The construction is based on the intuition that the hardness of guessing the XOR of two bits should be the "product" of the hardnesses of the bits individually, and is inspired by [**DDPY94**].

[1]The Polarization Lemma stated here is called the Amplification Lemma in [**SV97**]. We change the name here to stress that the Polarization Lemma does not merely increase statistical difference.

1.4. Reversing the Statistical Relationship. Our second result is perhaps even more unexpected than the Polarization Lemma — we exhibit an efficient transformation that "reverses" the statistical relationship between a pair of distributions. That is, we show that there is polynomial-time computable function that maps pairs of circuits (C_0, C_1) to pairs of circuits (D_0, D_1) such that

$$\begin{aligned} \|C_0 - C_1\| < 1/3 &\Rightarrow \|D_0 - D_1\| > 2/3 \\ \|C_0 - C_1\| > 2/3 &\Rightarrow \|D_0 - D_1\| < 1/3 \end{aligned}$$

The techniques used in this transformation come from work on statistical zero-knowledge [**Oka96, SV97**], as discussed below.

1.5. Statistical Zero-Knowledge. Informally, zero-knowledge proofs [**GMR89**] are interactive proofs in which the verifier "learns nothing" other than the assertion being proven. A *statistical* zero-knowledge proof is a type of zero-knowledge proof in which the "learns nothing" condition is interpreted in a strong information-theoretic sense. Statistical zero-knowledge proofs are of interest both in cryptography and complexity, and the set of languages possessing such proofs, SZK, has been studied in a number of recent works.

In prior work [**SV97**], we related SZK to statistical difference by proving that the problem of deciding whether the statistical difference between two distributions is large or small is *complete* for SZK. This can be formally described as a "promise" problem[2] STATISTICAL DIFFERENCE (abbreviated SD) whose YES instances are pairs of circuits whose statistical difference is greater than 2/3 and whose NO instances are pairs whose statistical difference is less than 1/3:

$$\begin{aligned} \mathrm{SD}_Y &= \left\{(C_0, C_1) : \|C_0 - C_1\| > \frac{2}{3}\right\} \\ \mathrm{SD}_N &= \left\{(C_0, C_1) : \|C_0 - C_1\| < \frac{1}{3}\right\} \end{aligned}$$

The main theorem in relating SD to SZK is the following:

THEOREM 1.1 ([**SV97**]). SD *is complete for* SZK. *That is,* SD $\in$ SZK, *and every problem in* SZK *reduces to* SD *(via a many-one polynomial-time reduction).*

By Theorem 1.1, efficient transformations for manipulating statistical difference can yield closure properties of SZK, and conversely. For example, by Theorem 1.1, the Reversal Mapping described above is equivalent to the closure of SZK under complement. In fact, the existence of such a transformation was originally deduced from the fact that SZK is closed under complement [**Oka96, SV97**]. This result motivated our search for a more explicit description of such a mapping. By extracting ideas used in the transformations of statistical zero-knowledge proofs given in [**Oka96**] and [**SV97**], we obtained the description of this transformation given in this paper. The Polarization Lemma, on the other hand, was originally developed for the purpose of proving things about SZK [**SV97**], but now also serves as an essential tool in the construction of our other transformations on statistical difference.

[2] A *promise problem* is simply a decision problem in which some inputs are excluded [**ESY84**].

1.6. Boolean Closure. The final result of this paper is an application of all of the above transformations to show that a boolean formula whose atoms are statements about membership in SD can be efficiently transformed into a single statement about SD. This is a strengthening of a result of [**DDPY94**] who show how to do this for *monotone* formulae whose atoms are statements about *random self-reducible languages* (which can be reduced to an extreme case of SD in which the thresholds are 1 and 0).

This result is based on the observation that some of transformations described above can be interpreted as boolean operations on statistical difference. For example, suppose we let distribution Z_0 consist of a copy of of distribution X_0 followed by an independent copy of distribution Y_0, and similarly let Z_1 consist of X_1 followed by Y_1. Then we see that if either X_0 and X_1 or Y_0 and Y_1 are statistically far apart, then Z_0 and Z_1 will be statistically far apart. Similarly, if both pairs (X_0, X_1) and (Y_0, Y_1) are statistically very close, then Z_0 and Z_1 will be statistically close. Thus, this operation in some sense represents OR. Similarly, the XOR construction mentioned earlier represents AND, and the Reversal Mapping represents negation. Combining these operations with the Polarization Lemma, we see that, given a k-ary formula ϕ and k pairs of input distributions, we can produce a pair of distributions whose statistical difference indicates whether or not the formula is true when its variables are set according to whether the corresponding pairs of input distributions are statistically far or statistically close. How efficient is this procedure? We show that a careful implementation of this procedure, using these particular AND or OR operations, can be performed in time polynomial in the size of ϕ and the circuits describing the input distributions.

By Theorem 1.1, this implies a very strong boolean closure property of SZK, one that does not necessarily follow from the fact that SZK is closed under complement, union, and intersection. As explained in Section 6, this can also be viewed as closure under a weak form of polynomial-time Turing reductions, and a step towards determining whether SZK is closed under general polynomial-time Turing reductions.

2. Notation and Basic Facts

First, we introduce some notation that will be used throughout the paper. If X is a probability distribution (or random variable), we write $x \leftarrow X$ to indicate that x is a sample taken from X. If S is a set, we write $x \in_R S$ to indicate that x is uniformly selected from S. In this paper, we focus on probability distributions that have a "succinct description" which enables them to sampled efficiently. By "succinct distribution" we mean a circuit C which, when fed the uniform distribution, has output distribution Z. For example, if C has m input gates and n output gates, Z would be the distribution induced on $\{0,1\}^n$ by feeding C the uniform distribution on $\{0,1\}^m$. For notational convenience, we write C for both the circuit and the distribution Z it defines.

Recall the definition of statistical difference given in Section 1.1. For probability distributions (or random variables) X and Y on a discrete set D,

$$\|X - Y\| = \max_{S \subset D} |\Pr[X \in S] - \Pr[Y \in S]|.$$

This is often also called the *variation distance* between X and Y. There is an equivalent formulation of statistical difference in terms of the ℓ_1 norm $|\cdot|_1$ that will sometimes be more convenient for us. To every probability distribution X on a discrete set D, the *mass function* of X is a vector in $\mathbb{R}^D$ whose x'th coordinate is $\Pr[X = x]$. For the sake of elegance, we also denote this vector by X. With this notation, we can state the following well-known fact.

FACT 2.1. $\|X - Y\| = \frac{1}{2}|X - Y|_1$.

PROOF. For any set $S \subset D$,

$$\begin{aligned}
&2\left|\Pr[X \in S] - \Pr[Y \in S]\right| \\
&= \left|\Pr[X \in S] - \Pr[Y \in S]\right| + \left|\Pr[X \notin S] - \Pr[Y \notin S]\right| \\
&= \left|\sum_{x \in S} (\Pr[X = x] - \Pr[Y = x])\right| + \left|\sum_{x \notin S} (\Pr[X = x] - \Pr[Y = x])\right| \\
&\leq \sum_{x \in S} \left|\Pr[X = x] - \Pr[Y = x]\right| + \sum_{x \notin S} \left|\Pr[X = x] - \Pr[Y = x]\right| \\
&= |X - Y|_1 .
\end{aligned}$$

Equality is achieved by taking $S = \{x\colon \Pr[X = x] > \Pr[Y = x]\}$. □

It is immediate from this characterization of statistical difference that it is a metric (as long as we identify random variables that are identically distributed). In particular, it satisfies the Triangle Inequality.

FACT 2.2 (Triangle Inequality). *For any probability distributions X, Y, and Z, $\|X - Y\| \leq \|X - Z\| + \|Z - Y\|$.*

Recall that for any two vectors $v \in \mathbb{R}^m$ and $w \in \mathbb{R}^n$, their *tensor product* $v \otimes w$ is the vector in $\mathbb{R}^{nm}$, whose (i, j)'th component is $v_i w_j$. Now, if we have a pair of random variables (X, Y) (on the same probability space) taking values in $D \times E$, then X is independent from Y iff the corresponding mass functions satisfy $(X, Y) = X \otimes Y$, where we view the mass functions of X and Y as elements of $\mathbb{R}^D$ and $\mathbb{R}^E$, respectively. For this reason, if we have random variables X and Y taking values in sets D and E, respectively, we write $X \otimes Y$ for the random variable taking values in $D \times E$ which consists of independent samples of X and Y.

Now, for any two vectors v and w, $|v \otimes w|_1 = |v|_1 \cdot |w|_1$. In addition, for any mass function X, $|X|_1 = 1$. These facts enable one to show that the statistical difference behaves well with respect to independent random variables:

FACT 2.3. *Suppose X_1 and X_2 are independent random variables on one probability space and Y_1 and Y_2 are independent random variables on another probability space. Then,*

$$\|(X_1, X_2) - (Y_1, Y_2)\| \leq \|X_1 - Y_1\| + \|X_2 - Y_2\| .$$

PROOF.

$$\begin{aligned}
\|(X_1, X_2) - (Y_1, Y_2)\| &\leq \|(X_1, X_2) - (Y_1, X_2)\| + \|(Y_1, X_2) - (Y_1, Y_2)\| \\
&= \frac{1}{2}|X_1 \otimes X_2 - Y_1 \otimes X_2|_1 + \frac{1}{2}|Y_1 \otimes X_2 - Y_1 \otimes Y_2|_1 \\
&= \frac{1}{2}|(X_1 - Y_1) \otimes X_2|_1 + \frac{1}{2}|Y_1 \otimes (X_2 - Y_2)|_1 \\
&= \frac{1}{2}|X_1 - Y_1|_1 \cdot |X_2|_1 + \frac{1}{2}|Y_1|_1 \cdot |X_2 - Y_2|_1 \\
&= \|X_1 - Y_1\| + \|X_2 - Y_2\|
\end{aligned}$$

□

One basic fact about statistical difference is that it cannot be created out of nothing. That is, for any procedure A, even if it be randomized, the statistical difference between $A(X)$ and $A(Y)$ is no greater than the statistical difference betewen X and Y. Formally, if D is any set, a *randomized procedure* on D is a a pair $A = (f, R)$, where R is a probability distribution on some set E and f is a function from $D \times E$ to any set F. Think of the distribution R as providing a "random seed" to the procedure A. If X is a probability distribution on D, then $A(X)$ denotes the probability distribution on F obtained by sampling $X \otimes R$ and applying f to the result. Note that applying a *function* is a special case of applying a randomized procedure.

FACT 2.4. *If X and Y are random variables and A is any randomized procedure, then $\|A(X) - A(Y)\| \leq \|X - Y\|$.*

PROOF. Let $A = (f, R)$. Then, for any set $S \subset F$,

$$\begin{aligned}
&|\Pr[A(X) \in S] - \Pr[A(Y) \in S]| \\
&\quad = |\Pr[f(X \otimes R) \in S] - \Pr[f(Y \otimes R) \in S]| \\
&\quad = \left|\Pr\left[X \otimes R \in f^{-1}(S)\right] - \Pr\left[Y \otimes R \in f^{-1}(S)\right]\right| \\
&\quad \leq \|X \otimes R - Y \otimes R\| \\
&\quad \leq \|X - Y\| + \|R - R\| \\
&\quad = \|X - Y\|.
\end{aligned}$$

Taking the maximum over all sets S completes the proof. □

The next fact is useful when arguing that the statistical difference between two distributions is small.

FACT 2.5. *Suppose $X = (X_1, X_2)$ and $Y = (Y_1, Y_2)$ are probability distributions on a set $D \times E$ such that*

1. *X_1 and Y_1 are identically distributed, and*
2. *With probability $> 1 - \epsilon$ over $x \leftarrow X_1$ (equivalently, $x \leftarrow Y_1$),*

$$\|X_2|_{X_1=x} - Y_2|_{Y_1=x}\| < \delta \tag{2.1}$$

(where $B|_{A=a}$ denotes the conditional distribution of B given that $A = a$ for jointly distributed random variables A and B).

Then $\|X - Y\| < \epsilon + \delta$.

PROOF. Let $T \subset D$ be the set of x's for which Equation 2.1 holds. Now, let S be an arbitrary subset of $D \times E$ and, for every $x \in D$, define $S_x = \{y \in E\colon (x, y) \in S\}$. Then,

$$\begin{aligned}
\Pr[X \in S] &\leq \Pr[X_1 \notin T] + \sum_{x \in T} \Pr[X_2 \in S_x | X_1 = x] \Pr[X_1 = x] \\
&< \epsilon + \sum_{x \in T} (\Pr[Y_2 \in S_x | Y_1 = x] + \delta) \Pr[Y_1 = x] \\
&\leq \epsilon + \delta + \Pr[Y \in S].
\end{aligned}$$

By symmetry, we also have $\Pr[Y \in S] < \epsilon + \delta + \Pr[X \in S]$. Since S was arbitrary, $\|X - Y\| < \epsilon + \delta$. □

The next fact formalizes the intuition that if two distributions have small statistical difference, then their mass functions must be close at most points.

FACT 2.6. *If X and Y are any two distributions such that $\|X - Y\| < \epsilon$, then with probability $> 1 - 2\sqrt{\epsilon}$ over $x \leftarrow X$,*

$$(1 - \sqrt{\epsilon}) \Pr[X = x] < \Pr[Y = x] < (1 + \sqrt{\epsilon}) \Pr[X = x]. \tag{2.2}$$

PROOF. Let $S = \{x \colon (1 - \sqrt{\epsilon}) \Pr[X = x] \geq \Pr[Y = x]\}$, *i.e.* the set of x's for which the left-hand inequality in Equation 2.2 is violated. Then,

$$\begin{aligned} \Pr[Y \in S] &\leq (1 - \sqrt{\epsilon}) \Pr[X \in S] \\ &= \Pr[X \in S] - \sqrt{\epsilon} \Pr[X \in S]. \end{aligned}$$

Thus, $\|X - Y\| \geq \sqrt{\epsilon} \Pr[X \in S]$,so we must have $\Pr[X \in S] < \sqrt{\epsilon}$. A similar argument show that the right-hand inequality in Equation 2.2 is violated with probability less than $\sqrt{\epsilon}$. □

3. Direct Product and XOR Lemmas

In this section, we describe two simple constructions, and their effect on the statistical difference between a pair of distributions. These are essential building blocks for the more complex transformations in later sections. The first construction is known as the *direct product* construction. In this construction, one samples a distribution independently k times. When applied to a pair of distributions, this construction has the effect of increasing any noticeable statistical difference to one exponentially close to 1. This is formalized in the following lemma:

LEMMA 3.1 (Direct Product Lemma). *Let X and Y be distributions such that $\|X - Y\| = \epsilon$. Then for all k,*

$$k\epsilon \geq \| \otimes^k X - \otimes^k Y\| \geq 1 - 2e^{-k\epsilon^2/2}$$

PROOF. The upper bound of $k\epsilon$ follows immediately from Fact 2.3, so we proceed to the proof of the lower bound. Recall, from the definition of statistical difference, that there must exist a set S such that

$$\Pr[X \in S] - \Pr[Y \in S] = \epsilon.$$

Let $p = \Pr[Y \in S]$. Then, $\Pr[X \in S] = p + \epsilon$. Hence, in k independent samples of X, the expected number of samples that lie in S is $(p + \epsilon)k$, whereas in k independent samples of Y, the expected number of samples that lie in S is pk. The Chernoff bound[3] tells us that the probability that *at least* $(p + \frac{\epsilon}{2})k$ components of $\otimes^k Y$ lie in S is at most $\exp(-k\epsilon^2/2)$, whereas the probability that *at most* $(p + \frac{\epsilon}{2})k$ components of $\otimes^k X$ lie in S is at most $\exp(-k\epsilon^2/2)$. Let S' be the set of all k-tuples that contain more than $(p + \frac{\epsilon}{2})k$ components that lie in S. Then we have,

$$\| \otimes^k X - \otimes^k Y\| \geq \Pr[\otimes^k X \in S'] - \Pr[\otimes^k Y \in S'] \geq 1 - 2e^{-k\epsilon^2/2}.$$

□

[3]For the formulation of the Chernoff bound we use, see, for example, the formulation of Hoeffding's inequality in [**Hof95**, Sec. 7.2.1].

At first glance, it may seem that the proof above is unnecessarily loose, and that one might be able to prove that the statistical difference increases even for small values of k. However, Madhu Sudan [**Sud97**] has pointed out that for any $p \in [0,1]$, there exist distributions X and Y such that $\|X \otimes X - Y \otimes Y\| = \|X - Y\| = p$. Consider the following two distributions:

$$X = \begin{cases} 1 & \text{with probability } (1+p)/2 \\ 0 & \text{with probability } (1-p)/2 \end{cases}$$

$$Y = \begin{cases} 0 & \text{with probability } (1+p)/2 \\ 1 & \text{with probability } (1-p)/2 \end{cases}$$

Here, $\|X - Y\| = ((1+p) - (1-p))/2 = p$, and also

$$\|X \otimes X - Y \otimes Y\| = ((1+p)^2 - (1-p)^2)/4 = p.$$

Nevertheless, the direct product construction gives us an efficient and effective technique for increasing the statistical difference between two distributions. The two bounds in Lemma 3.1 show that large values go to 1 faster than small values. The following lemma provides a complementary technique which decreases the statistical difference to 0, with small values going to 0 faster than large values.

LEMMA 3.2 (XOR Lemma). *There is a polynomial-time computable function that maps a triple $(C_0, C_1, 1^k)$, where C_0 and C_1 are circuits, to a pair of circuits (D_0, D_1) such that $\|D_0 - D_1\| = \|C_0 - C_1\|^k$. Specifically, D_0 and D_1 are defined as follows:*

D_0: Uniformly select $(b_1, \ldots, b_k) \in \{0,1\}^k$ such that $b_1 \oplus \cdots \oplus b_k = 0$, and output a sample of $C_{b_1} \otimes \cdots \otimes C_{b_k}$.
D_1: Uniformly select $(b_1, \ldots, b_k) \in \{0,1\}^k$ such that $b_1 \oplus \cdots \oplus b_k = 1$, and output a sample of $C_{b_1} \otimes \cdots \otimes C_{b_k}$.

In order to prove this lemma, we employ a generalization of the technique used in [**DDPY94**] to represent the logical AND of statements about GRAPH NONISOMORPHISM. This tool is described in the following Proposition.

PROPOSITION 3.3. *Let X_0, X_1, Y_0, Y_1 be any random variables, and define the following pair of random variables:*

Z_0: Choose $a, b \in_R \{0,1\}$ such that $a \oplus b = 0$. Output a sample of $X_a \otimes Y_b$.
Z_1: Choose $a, b \in_R \{0,1\}$ such that $a \oplus b = 1$. Output a sample of $X_a \otimes Y_b$.

Then $\|Z_0 - Z_1\| = \|X_0 - X_1\| \cdot \|Y_0 - Y_1\|$.

The statistical difference between X_0 and X_1 (or Y_0 and Y_1) measures the advantage a computationally unbounded party has, over random guessing, of guessing b given a sample from X_b, where b is selected uniformly from $\{0,1\}$. Intuitively, the above Proposition says that the advantage one has in guessing the XOR of two independent bits is the product of the advantages one has for guessing each individual bit.

PROOF.

$$\begin{aligned}
\|Z_0 - Z_1\| &= \frac{1}{2}|Z_0 - Z_1|_1 \\
&= \frac{1}{2}\left|\left(\frac{1}{2}X_0 \otimes Y_0 + \frac{1}{2}X_1 \otimes Y_1\right) - \left(\frac{1}{2}X_1 \otimes Y_0 + \frac{1}{2}X_0 \otimes Y_1\right)\right|_1 \\
&= \frac{1}{4}|(X_0 - X_1) \otimes (Y_0 - Y_1)|_1 \\
&= \left(\frac{1}{2}|X_0 - X_1|_1\right) \cdot \left(\frac{1}{2}|Y_0 - Y_1|_1\right) \\
&= \|X_0 - X_1\| \cdot \|Y_0 - Y_1\|
\end{aligned}$$

□

Proposition 3.3 and an induction argument establish Lemma 3.2. Yao's XOR Lemma [**Yao82**] (see also [**GNW95**]) can be seen as an analogue of Lemma 3.2 in the computational setting, where the analysis is much more difficult.

4. Polarizing the Statistical Relationship

In this section, we combine the techniques from the previous section to yield the following lemma:

LEMMA 4.1 (Polarization Lemma).[4] *There is a polynomial-time computable function that takes a triple $(C_0, C_1, 1^k)$, where C_0 and C_1 are circuits, and outputs a pair of circuits (D_0, D_1) such that*

$$\begin{aligned}
\|C_0 - C_1\| < 1/3 &\Rightarrow \|D_0 - D_1\| < 2^{-k} \\
\|C_0 - C_1\| > 2/3 &\Rightarrow \|D_0 - D_1\| > 1 - 2^{-k}
\end{aligned}$$

The usefulness of the Polarization Lemma comes from the fact that the two distributions it produces can be treated almost as if they were identically distributed or disjoint (*i.e.* statistical difference 0 and 1, respectively). Indeed, it will be used in the constructions of the next two sections, and it was essential in proving that SD (with thresholds of 2/3 and 1/3, as we've defined it) is in SZK [**SV97**].

Recall that the Direct Product construction of Lemma 3.1 gives a way to increase statistical difference with large values going to 1 faster than small values. Similarly, the XOR Lemma (Lemma 3.2) shows how to decrease statistical difference with small values going to 0 faster than large values. Intuitively, alternating these procedures should "polarize" large and small values of statistical difference, pushing them closer to 1 and 0, respectively. A similar alternation between procedures with complementary effects was used by Ajtai and Ben-Or [**AB84**] to amplify the success probability of randomized constant-depth circuits.

PROOF. Let $\ell = \lceil \log_{4/3} 6k \rceil$. Apply the XOR Lemma (Lemma 3.2) to the triple $(C_0, C_1, 1^\ell)$ to produce (C_0', C_1') such that if

$$\begin{aligned}
\|C_0 - C_1\| < 1/3 &\Rightarrow \|C_0' - C_1'\| < (1/3)^\ell \\
\|C_0 - C_1\| > 2/3 &\Rightarrow \|C_0' - C_1'\| > (2/3)^\ell.
\end{aligned}$$

[4]The Polarization Lemma stated here is called the Amplification Lemma in [**SV97**]. We change the name here to stress that the Polarization Lemma does not merely increase statistical difference.

Let $m = 3^{\ell-1}$. Then apply the direct product construction, letting $C_0'' = \otimes^m C_0'$ and let $C_1'' = \otimes^m C_1'$. Then, by Lemma 3.1,

$$\|C_0 - C_1\| < 1/3 \Rightarrow \|C_0'' - C_1''\| < 1/3$$
$$\|C_0 - C_1\| > 2/3 \Rightarrow \|C_0'' - C_1''\| > 1 - 2e^{-3^{\ell-1}(2/3)^{2\ell}/2} > 1 - 2e^{-k}.$$

Finally, apply the XOR Lemma (Lemma 3.2) one more time to $(C_0'', C_1'', 1^k)$ to produce (D_0, D_1) such that

$$\|C_0 - C_1\| < 1/3 \Rightarrow \|D_0 - D_1\| < 3^{-k} < 2^{-k}$$
$$\|C_0 - C_1\| > 2/3 \Rightarrow \|D_0 - D_1\| > (1 - 2e^{-k})^k > 1 - 2ke^{-k} > 1 - 2^{-k}.$$

□

Notice that the above analysis relies on the fact that $(2/3)^2 > (1/3)$, so it will not work if $1/3$ and $2/3$ are replaced by, say, .49 and .51. We do not know how to prove such a polarization lemma for arbitrary constant thresholds. We can however extend it to thresholds α and β, where β^2/α is greater than 1. More precisely, we have the following lemma:

LEMMA 4.2 (General Polarization Lemma). *There is a function that takes as input a 5-tuple $(C_0, C_1, \alpha, \beta, 1^k)$, where $\beta^2 = \lambda \cdot \alpha$, with $\lambda > 1$, and C_0 and C_1 are circuits. The function is computable in time polynomial in the input size and $\alpha^{-1/\log(\lambda)}$, and outputs a pair of circuits (D_0, D_1) such that*

$$\|C_0 - C_1\| < \alpha \Rightarrow \|D_0 - D_1\| < 2^{-k}$$
$$\|C_0 - C_1\| > \beta \Rightarrow \|D_0 - D_1\| > 1 - 2^{-k}$$

PROOF. Let $\ell = \lceil \log_\lambda(6 \ln 6) \rceil$. Apply the XOR Lemma (Lemma 3.2) to the triple $(C_0, C_1, 1^\ell)$ to produce (C_0', C_1') such that if

$$\|C_0 - C_1\| < \alpha \Rightarrow \|C_0' - C_1'\| < \alpha^\ell$$
$$\|C_0 - C_1\| > \beta \Rightarrow \|C_0' - C_1'\| > \beta^\ell.$$

Let $m = 1/3\alpha^\ell = \lambda^\ell/(3\beta^{2\ell})$. Then apply the direct product construction, letting $C_0'' = \otimes^m C_0'$ and let $C_1'' = \otimes^m C_1'$. Then, by Lemma 3.1,

$$\|C_0 - C_1\| < \alpha \Rightarrow \|C_0'' - C_1''\| < 1/3$$
$$\|C_0 - C_1\| > \beta \Rightarrow \|C_0'' - C_1''\| > 1 - 2e^{-\frac{\lambda^\ell}{3\beta^{2\ell}} \cdot \frac{\beta^{2\ell}}{2}} \geq 2/3$$

An application of Lemma 4.1 finishes the proof. □

5. Reversing the Statistical Relationship

PROPOSITION 5.1 (Reversal Mapping). *There is a polynomial-time computable function that maps pairs of circuits (C_0, C_1) to pairs of circuits (D_0, D_1) such that*

$$\|C_0 - C_1\| < 1/3 \Rightarrow \|D_0 - D_1\| > 2/3$$
$$\|C_0 - C_1\| > 2/3 \Rightarrow \|D_0 - D_1\| < 1/3.$$

That is, SD *reduces to* $\overline{\text{SD}}$.

By Theorem 1.1 (and the closure of SZK under reductions [**SV97**]), Proposition 5.1 is equivalent to the closure of SZK under complement. In fact, the existence of such a transformation was originally deduced from the fact that SZK is closed under complement [**Oka96, SV97**]. This result motivated our search for a more explicit description of such a mapping. By extracting ideas used in the transformations of statistical zero-knowledge proofs given in [**Oka96**] and [**SV97**], we obtained the description of this transformation given below.

The Construction. Let (C_0, C_1) be any pair of circuits and let $n = |(C_0, C_1)|$. By the Polarization Lemma (Lemma 4.1), we can produce in polynomial time a pair of circuits (C'_0, C'_1) such that

$$\|C_0 - C_1\| < 1/3 \quad \Rightarrow \quad \|C'_0 - C'_1\| > 1 - 2^{-n}$$
$$\|C_0 - C_1\| > 2/3 \quad \Rightarrow \quad \|C'_0 - C'_1\| < 2^{-n}$$

Let $q = \text{poly}(n)$ be the number of input gates of C'_0 and C'_1 (w.l.o.g. we may assume they have the same number) and let $\ell = \text{poly}(n)$ be the number of output gates. For notational convenience, let $R = \{0,1\}^q$ and $L = \{0,1\}^\ell$. Let $m = n^3q^2$ and define a new distribution $\overline{C}\colon \{0,1\}^m \times R^m \to L^m$ as follows:

$$\overline{C}(\overline{b}, \overline{r}) = (C'_{b_1}(r_1), \ldots, C'_{b_m}(r_m)).$$

We use the notation $\overline{z} \leftarrow \overline{C}$ to denote $\overline{z}$ chosen according to $\overline{C}$, *i.e.* select $\overline{b}$ and $\overline{r}$ uniformly and let $\overline{z} = \overline{C}(\overline{b}, \overline{r})$.

Let $\mathcal{H}$ be a 2-universal family of hash functions from $\{0,1\}^m \times R^m \times L^m$ to $T = \{0,1\}^{(q+1)m - 2\Delta - n}$, where $\Delta = \sqrt{nmq^2} = m/n$. We can now describe the new distributions:

D_0: Let $(\overline{b}, \overline{r}) \in_R \{0,1\}^m \times R^m$, $\overline{y} \leftarrow \overline{C}$, and $h \in_R \mathcal{H}$. Output $(\overline{C}(\overline{b}, \overline{r}), \overline{b}, h, h(\overline{b}, \overline{r}, \overline{y}))$.

D_1: Let $(\overline{b}, \overline{r}) \in_R \{0,1\}^m \times R^m$, $h \in_R \mathcal{H}$, and $t \in_R T$. Output $(\overline{C}(\overline{b}, \overline{r}), \overline{b}, h, t)$.

The important things to note about these distributions are that $\overline{b}$ is part of the output, and that D_0 and D_1 only differ in the last component, where D_0 has the value of the hash function and D_1 has a truly random element of T. Also note that the size of T is chosen to be $|\{0,1\}^m \times R^m|/2^{2\Delta+n}$, which is essentially $|\{0,1\}^m \times R^m|$, scaled down by a factor of $2^{2\Delta+n}$, which can be thought of as a "fudge factor" needed to make the proof work. The introduction of the sample $\overline{y}$ in D_0 may at first seem superfluous; we explain below.

Intuition. For intuition, consider the case that $\overline{C}$ is a uniform distribution; that is, for every $\overline{z} \in \text{range}(\overline{C})$, the size of the preimage set $|\{(\overline{b}, \overline{r})\colon \overline{C}(\overline{b}, \overline{r}) = \overline{z}\}|$ is the same value N. (It turns out that $\overline{C}$ is actually "close enough" to uniform for these arguments to work.) Then the range of $\overline{C}$ has size $2^{(q+1)m}/N$. So, in D_0, conditioned on a value for $\overline{C}(\overline{b}, \overline{r})$, the triple $(\overline{b}, \overline{r}, \overline{y})$ is selected uniformly from a set of size $2^{(q+1)m}$. Since this is much greater than $|T|$, the Leftover Hash Lemma of [**ILL89**] implies that conditioned on any value for the first component of D_0, the last two components $(h, h(\overline{b}, \overline{r}, \overline{y}))$ are distributed close to the uniform distribution on $\mathcal{H} \times T$, which is the distribution that D_1 has in its last two components.[5] Thus, if their second components were missing, D_0 and D_1 would be statistically close.

[5]Here we see the importance of $\overline{y}$: Without $\overline{y}$, conditioned on some value of $\overline{C}(\overline{b}, \overline{r})$, the pair $(\overline{b}, \overline{r})$ would be selected uniformly from a space of size N. If we were only hashing this pair, for the distribution $h(\overline{b}, \overline{r})$ to be uniform by the Leftover Hash Lemma, T would have had to be chosen so that $|T| \ll N$. The value of N, however, depends on the inner workings of the circuit C, and is in general unknown. By including $\overline{y}$, which comes uniformly from a space of size $2^{(q+1)m}/N$, we balance the arguments to h so that they come from a space of size $2^{(q+1)m}$, a known quantity.

Now, consider the case that $\|C_0' - C_1'\| \approx 1$. Then $\bar{b}$ is essentially "determined" by $\overline{C}(\bar{b}, \bar{r})$. So the presence of $\bar{b}$ can be ignored, and the above argument says that D_0 and D_1 are statistically very close. Now, consider the case that $\|C_0' - C_1'\| \approx 0$. Then $\bar{b}$ is essentially "unrestricted" by $\overline{C}(\bar{b}, \bar{r})$. Since there are 2^m choices for $\bar{b}$, conditioning on $\bar{b}$ in addition to $\overline{C}(\bar{b}, \bar{r})$, cuts the number of triples $(\bar{b}, \bar{r}, \bar{y})$ down from $2^{m(q+1)}$ to roughly $2^{m(q+1)}/2^m$. Since $2^{m(q+1)}/2^m$ is much smaller than $|T|$, $h(\bar{b}, \bar{r}, \bar{y})$ will cover only a small fraction of $|T|$ and thus will be far from uniform (conditioned on values for $\overline{C}(\bar{b}, \bar{r})$, $\bar{b}$, and h).

Proof of Proposition 5.1. First we will argue that $\overline{C}$ is close to uniform, so that we can apply arguments like those given above. This is the case because $\overline{C}$ is composed of many independent, identically distributed random variables. For $\bar{z} \in L^m$, we say the weight of $\bar{z}$ is the logarithm of the size of the preimage set of $\bar{z}$. Formally, let $\mathrm{wt}(\bar{z}) = \log_2 |\{(\bar{b}, \bar{r}) \colon \overline{C}(\bar{b}, \bar{r}) = \bar{z}\}|$. Let w be the expected weight of an image, $w = \mathrm{E}_{\bar{z} \leftarrow \overline{C}}[\mathrm{wt}(\bar{z})]$. Then we can show the following:

LEMMA 5.2. $\Pr_{\bar{z} \leftarrow \overline{C}}[|\mathrm{wt}(\bar{z}) - w| > \Delta] < 2^{-\Omega(n)}$.

PROOF. For $z \in L$, let $\mathrm{wt}_0(z) = \log_2 |\{(b, r) \colon C_b(r) = z\}|$. Then, for $\bar{z} \in L^m$, $\mathrm{wt}(\bar{z}) = \mathrm{wt}_0(z_1) + \cdots + \mathrm{wt}_0(z_m)$. Observe that when $\bar{z}$ is selected according to $\overline{C}$, $z_1, \ldots, z_m$ are independent and identically distributed. Moreover, for any $z \in L$, $0 \leq \mathrm{wt}_0(z) \leq q$. So, by the Hoeffding inequality [**Hof95**, Sec. 7.2.1], we have

$$\Pr_{\bar{z} \leftarrow \overline{C}}[|\mathrm{wt}(\bar{z}) - w| > \Delta] < 2e^{-2\Delta^2/mq^2} = 2e^{-2n}.$$

□

It will be convenient to eliminate those $\bar{z} \in L^m$ that have weight far above or below the mean. Let $G = \{(\bar{b}, \bar{r}) \colon |\mathrm{wt}(\overline{C}(\bar{b}, \bar{r})) - w| \leq \Delta\}$ be the set of *good* pairs $(\bar{b}, \bar{r})$. The above Lemma says that $|G| \geq (1 - 2^{-\Omega(n)})|\{0,1\}^m \times R^m|$. Thus $\|G - \{0,1\}^m \times R^m\| \leq 2^{-\Omega(n)}$, where for simplicity of notation, we let the name of a set also refer to the uniform distribution on the same set. Define $\overline{C}'$ to be the distribution obtained by selecting $(\bar{b}, \bar{r}) \leftarrow G$ and outputting $\overline{C}(\bar{b}, \bar{r})$. Then, since $\overline{C}$ is a function, Fact 2.4 tells us that $\|\overline{C} - \overline{C}'\| = 2^{-\Omega(n)}$. Similarly, we define variants of D_0 and D_1 that sample from G instead of $\{0,1\}^m \times R^m$:

D_0': Let $(\bar{b}, \bar{r}) \in_R G$, $\bar{y} \leftarrow \overline{C}'$, and $h \in_R \mathcal{H}$. Output $(\overline{C}'(\bar{b}, \bar{r}), \bar{b}, h, h(\bar{b}, \bar{r}, \bar{y}))$.
D_1': Let $(\bar{b}, \bar{r}) \in_R G$, $h \in_R \mathcal{H}$, and $t \in_R T$. Output $(\overline{C}'(\bar{b}, \bar{r}), \bar{b}, h, t)$.

Since D_0' (or D_1') is a randomized procedure applied to two (or one) independent samplings from G, Fact 2.4 tells us that $\|D_0 - D_0'\| = 2^{-\Omega(n)}$ (and $\|D_1 - D_1'\| = 2^{-\Omega(n)}$). Hence, it suffices to prove that these modified distributions have the properties we want in each case. For the case when C_0 and C_1 are statistically far, we prove the following claim:

CLAIM 5.3. *If* $\|C_0' - C_1'\| > 1 - 2^{-n}$, *then* $\|D_0' - D_1'\| < 2^{-\Omega(n)}$.

This use of "dummy" samples to form a space whose size is known is the "complementary usage of messages" technique of Okamoto [**Oka96**].

PROOF. First we formalize the idea that $\overline{b}$ is "determined" by $\overline{C}$. Define $f\colon L \to \{0,1\}$ by

$$f(z) = \begin{cases} 0 & \text{if } \Pr[C'_0 = z] > \Pr[C'_1 = z] \\ 1 & \text{otherwise} \end{cases}$$

Then

$$\Pr_{b,r}[f(C'_b(r)) = b] = \frac{1}{2}\Pr_r[f(C'_0(r)) = 0] + \frac{1}{2}\Pr_r[f(C'_1(r)) = 1].$$

Now, by the definition of statistical difference, $\Pr_r[C'_0(r) \in f^{-1}(0)] \geq 1 - 2^{-n}$ and $\Pr_r[C'_1(r) \in f^{-1}(1)] \geq 1 - 2^{-n}$. Thus, $\Pr_{b,r}[f(C'_b(r)) = b] > 1 - 2^{-n}$. Now define $\overline{f}\colon L^m \to \{0,1\}^m$ by $\overline{f}(\overline{z}) = (f(z_1), \ldots, f(z_m))$. Then

$$\Pr_{\overline{b},\overline{r}}[\overline{f}(\overline{C}(\overline{b},\overline{r})) = \overline{b}] > (1 - 2^{-n})^m = 1 - 2^{-\Omega(n)}.$$

Since G is a $1 - 2^{-\Omega(n)}$ fraction of $\{0,1\}^m \times R^m$, the same is true when $(\overline{b},\overline{r})$ is selected uniformly from G. Thus, if we define:

D''_0: Let $(\overline{b},\overline{r}) \in_R G$, $\overline{y} \leftarrow \overline{C}'$, and $h \in_R \mathcal{H}$. Output $(\overline{C}'(\overline{b},\overline{r}), \overline{f}(\overline{C}'(\overline{b},\overline{r})), h, h(\overline{b},\overline{r},\overline{y}))$.
D''_1: Let $(\overline{b},\overline{r}) \in_R G$, $h \in_R \mathcal{H}$, and $t \in_R T$. Output $(\overline{C}'(\overline{b},\overline{r}), \overline{f}(\overline{C}'(\overline{b},\overline{r})), h, t)$.

Then, by Fact 2.5, $\|D'_0 - D''_0\| = 2^{-\Omega(n)}$ and $\|D'_1 - D''_1\| = 2^{-\Omega(n)}$. So it suffices to show that $\|D''_0 - D''_1\| = 2^{-\Omega(n)}$. Since the first components of D''_0 and D''_1 are identically distributed and the second components are determined by the first ones, it suffices to show (by Fact 2.5) that, conditioned on any value for the first coordinate, the third and fourth components have statistical difference $2^{-\Omega(n)}$. This will follow from the Leftover Hash Lemma [**ILL89**]:

LEMMA 5.4 (Leftover Hash Lemma [**ILL89**]). *Let $\mathcal{H}$ be a family of 2-universal hash functions from D to T. Let X by a probability distribution on D such that for all $x \in D$, $\Pr[X = x] \leq \epsilon/|T|$. Then the following two distributions have statistical difference at most $\epsilon^{1/3}$.*

1. *Choose $x \leftarrow X$, $h \in_R \mathcal{H}$. Output $(h, h(x))$.*
2. *Choose $h \in_R \mathcal{H}$, $t \in_R T$. Output (h, t).*

By the above argument and the Leftover Hash Lemma, it suffices to show that conditioned on any value $\overline{z}$ for $\overline{C}'(\overline{b},\overline{r})$, no triple $(\overline{b},\overline{r},\overline{y})$ has probability more than $2^{-\Omega(n)}/|T|$. The pair $(\overline{b},\overline{r})$ comes uniformly from a set of size $2^{\mathrm{wt}(\overline{z})} \geq 2^{w-\Delta}$, and $\overline{y}$ is selected independently according to $\overline{C}'$, so the probability of any triple $(\overline{b},\overline{r},\overline{y})$ is at most

$$\left(\frac{1}{2^{w-\Delta}}\right)\left(\frac{2^{w+\Delta}}{|G|}\right) \leq \frac{2^{2\Delta}}{(1 - 2^{-\Omega(n)})2^{(q+1)m}} = \frac{2^{-\Omega(n)}}{|T|}.$$

Thus, $\|D''_0 - D''_1\| \leq 2^{-\Omega(n)}$, and the claim is established. □

Now we treat the other case, when C_0 and C_1 are stastically close.

CLAIM 5.5. *If $\|C'_0 - C'_1\| < 2^{-n}$, then $\|D'_0 - D'_1\| > 1 - 2^{-\Omega(n)}$.*

PROOF. First, we formalize the idea that $\overline{b}$ is almost completely "undetermined" by $\overline{C}(\overline{b}, \overline{r})$. Since $\|C_0' - C_1'\| < 2^{-n}$, it follows from Fact 2.6 that with probability $1 - 2^{-\Omega(n)}$ over $z \leftarrow C_0'$,

$$(1 - 2^{-\Omega(n)}) \Pr[C_1' = z] \leq \Pr[C_0' = z] \leq (1 + 2^{-\Omega(n)}) \Pr[C_1' = z].$$

In other words,

$$1 - 2^{-\Omega(n)} \leq \frac{|\{r \colon C_0'(r) = z\}|}{|\{r \colon C_1'(r) = z\}|} \leq 1 + 2^{-\Omega(n)}.$$

The same is true with probability $1 - 2^{-\Omega(n)}$ when the roles of C_0' and C_1' are reversed. Thus, with probability $1 - m2^{-\Omega(n)} = 1 - 2^{-\Omega(n)}$ over $\overline{z} \leftarrow \overline{C}$, we have for *every* pair $\overline{b}, \overline{c} \in \{0,1\}^m$,

$$1 - 2^{-\Omega(n)} = (1 - 2^{\Omega(n)})^m \leq \frac{\left|\{\overline{r} \colon \overline{C}(\overline{b}, \overline{r}) = \overline{z}\}\right|}{\left|\{\overline{r} \colon \overline{C}(\overline{c}, \overline{r}) = \overline{z}\}\right|} \leq (1 + 2^{-\Omega(n)})^m = 1 + 2^{-\Omega(n)}.$$

Since there are 2^m choices for $\overline{c}$, this, combined with Lemma 5.2, implies that, with probability $1 - 2^{-\Omega(n)}$ over $\overline{z} \leftarrow \overline{C}$, the following holds for *every* $\overline{b} \in \{0,1\}^m$:

$$\left|\{\overline{r} \colon \overline{C}(\overline{b}, \overline{r}) = \overline{z}\}\right| \leq (1 + 2^{-\Omega(n)}) \cdot \frac{2^{\mathrm{wt}(\overline{z})}}{2^m} \leq (1 + 2^{-\Omega(n)}) \cdot 2^{w + \Delta - m}.$$

Since this is true with probability $1 - 2^{-\Omega(n)}$ for $\overline{z}$ selected according to $\overline{C}$, it is also true with probability $1 - 2^{-\Omega(n)}$ for $\overline{z}$ selected according to $\overline{C}'$. Fix any such $\overline{z}$ and fix any $\overline{b} \in \{0,1\}^m$ and $h \in \mathcal{H}$. Then, in D_0', conditioned on $\overline{C}'(\overline{b}, \overline{r}) = \overline{z}$, $\overline{b}$, and h, there are at most

$$\begin{aligned} (1 + 2^{-\Omega(n)}) \cdot 2^{w + \Delta - m} \binom{|G|}{2^{w - \Delta}} &\leq (1 + 2^{-\Omega(n)}) \cdot 2^{2\Delta - m}(2^{m + mq}) \\ &= (1 + 2^{-\Omega(n)}) \cdot 2^{4\Delta + n - m}|T| \\ &= 2^{-\Omega(m)}|T| \end{aligned}$$

possible values for $(\overline{r}, \overline{y})$. Thus, with probability $1 - 2^{-\Omega(n)}$, conditioned on values for the first three components of D_0', the fourth component $h(\overline{b}, \overline{r}, \overline{y})$ can cover at most a $2^{-\Omega(m)} \leq 2^{-\Omega(n)}$ fraction of T. In contrast, conditioned on values for the first three components of D_1', the fourth component is uniformly distributed on T. Therefore, $\|D_0' - D_1'\| \geq 1 - 2^{-\Omega(n)}$. □

6. An Application – Boolean Closure

In this section, we use the transformations developed in the prior sections to prove that any boolean formulae whose atoms are statements about membership in SD can be efficiently transformed into a single statement about SD. By Theorem 1.1 [**SV97**], this implies a very strong boolean closure property of SZK: given an arbitrary boolean formula whose atoms are statements about membership in *any* language in SZK, one can efficiently construct a statistical zero-knowledge interactive proof for its validity. Note that such a property does not follow immediately from the fact that a class is closed under intersection, union, and complementation, because applying these more than a constant number of times could incur a superpolynomial cost in efficiency, while we ask that the construction can be done efficiently with respect to the size of the formula. The procedure for doing this is based on work by De Santis, Di Crescenzo, Persiano, and Yung [**DDPY94**].

They show how to construct statistical zero-knowledge proofs for all monotone boolean formulae whose atoms are statements about a random self-reducible language. Their zero-knowledge proofs are constructed by producing two distributions which are either disjoint or identical, depending on whether or not the formula is true. Hence, their construction can be viewed as a reduction to an extreme case of SD, in which the thresholds are 1 and 0.

Using the direct product construction, the XOR Lemma, and the Polarization Lemma, we generalize their result to monotone formulae whose atoms are statements about membership in SD. Then, using our Reversal Mapping, we further generalize to non-monotone formulae.

DEFINITION 6.1. Let Π be any promise problem. Then we define a new promise problem $\Phi(\Pi)$ whose instances are $(\phi, x_1, \ldots, x_k)$ where $k \geq 0$, $x_1, \ldots, x_k \in \Pi_Y \cup \Pi_N$ (*i.e.* $x_1, \ldots, x_k$ satisfy the promise for Π), and $\phi(v_1, \ldots, v_k)$ is a k-ary propositional formula. The YES instances of $\Phi(\Pi)$ are those instances for which $\phi((x_1 \in \Pi_Y), \ldots, (x_k \in \Pi_Y))$ is true and the NO instances are those for which it is false.

Mon(Π) is defined analogously, except that only monotone ϕ are considered.[6]

The main result of this section follows:

THEOREM 6.2. Φ(SD) *reduces to* SD. *That is, there is a polynomial-time computable function that maps an instance* $x = (\phi, (C_0^1, C_1^1), \ldots, (C_0^k, C_1^k))$ *of* Φ(SD) *to an instance* $y = (D_0, D_1)$ *of* SD *such that*

$$x \in \Phi(\mathrm{SD})_Y \quad \Rightarrow \quad y \in \mathrm{SD}_Y$$
$$x \in \Phi(\mathrm{SD})_N \quad \Rightarrow \quad y \in \mathrm{SD}_N.$$

The main step in proving Theorem 6.2 is the following Lemma, which mimics the construction of [**DDPY94**] for Mon(GRAPH NONISOMORPHISM):

LEMMA 6.3. Mon(SD) *reduces to* SD.

PROOF. For intuition, consider two instances of statistical difference (C_0, C_1) and (D_0, D_1), both of which have statistical difference very close to 1 or very close to 0 (which can be achieved by the Polarization Lemma). Then $(C_0 \otimes D_0, C_1 \otimes D_1)$ will have statistical difference very close to 1 if either of the original statistical differences is very close to 1 and will have statistical difference very close to 0 otherwise. Thus, this operation represents OR. Similarly, the XOR operation in Proposition 3.3 represents AND. To obtain Lemma 6.3, we will recursively apply these constructions, taking care to keep the running time polynomial.

Let $w = (\phi, (C_0^1, C_1^1), \ldots, (C_0^k, C_1^k))$ be an instance of Mon(SD) and let $n = |w|$. By applying the Polarization Lemma (Lemma 4.1), we can constuct in polynomial time pairs of circuits $(D_0^1, D_1^1), \ldots, (D_0^k, D_1^k)$ such that the statistical difference between D_0^i and D_1^i is greater than $1 - 2^{-n}$ if $(C_0^i, C_1^i) \in \mathrm{SD}_Y$ and is less than 2^{-n} if $(C_0^i, C_1^i) \in \mathrm{SD}_N$.

Consider the randomized recursive procedure Sample(ψ, b) in Figure 6 which takes a subformula $\psi(v_{i_1}, \ldots, v_{i_k})$ of ϕ and a bit $b \in \{0, 1\}$ as input.

Executing Sample(ϕ, b) for $b \in \{0, 1\}$ takes time polynomial in n, because the number of recursive calls is equal to the number of subformulas of ϕ. For a

[6]In [**DDPY94**], only monotone formulae are treated. What they call $\Phi(L)$ is what we call Mon(L).

```
Sample(ψ, b)
    If ψ = v_i, sample z ← D_b^i.
    If ψ = τ ∨ μ,
        Sample z_1 ← Sample(τ, b);
        Sample z_2 ← Sample(μ, b);
        Let z = (z_1, z_2).
    If ψ = τ ∧ μ,
        Choose c, d ∈_R {0,1} subject to c ⊕ d = b;
        Sample z_1 ← Sample(τ, c);
        Sample z_2 ← Sample(τ, d);
        Let z = (z_1, z_2).
    Output z.
```

FIGURE 1

subformula τ of ϕ, let $\mathrm{Dif}(\tau) = \|\mathrm{Sample}(\tau, 0) - \mathrm{Sample}(\tau, 1)\|$. Then we can prove the following about Dif:

CLAIM 6.4. *For every subformula* $\tau = \tau(v_{i_1}, \ldots, v_{i_j})$ *of* ϕ, $\mathrm{Dif}(\tau) > 1 - m2^{-n}$ *if*

$$\tau((C_0^{i_1}, C_1^{i_1}) \in \mathrm{SD}_Y, \ldots, (C_0^{i_j}, C_1^{i_j}) \in \mathrm{SD}_Y)$$

is true and $\mathrm{Dif}(\tau) < m2^{-n}$ *if it is false, where* $m = |\tau|$.

PROOF OF CLAIM. By induction on subformulae of ψ. It holds for atomic subformulae (*i.e.* the variables v_i) by the properties of the D_b^i's.

Consider the case when $\psi = \tau \vee \mu$. If ψ is true (with the appropriate arguments), either τ or μ must be true. Without loss of generality, say τ is true. Then, by Fact 2.4 and induction,

$$\mathrm{Dif}(\psi) \geq \mathrm{Dif}(\tau) > 1 - |\tau|2^{-n} > 1 - |\psi|2^{-n}.$$

If ψ is false, then both τ and μ are false. By Fact 2.3 and induction,

$$\mathrm{Dif}(\psi) \leq \mathrm{Dif}(\tau) + \mathrm{Dif}(\mu) < |\tau|2^{-n} + |\mu|2^{-n} \leq |\psi|2^{-n}.$$

Now consider the case when $\psi = \tau \wedge \mu$. By Proposition 3.3, $\mathrm{Dif}(\psi) = \mathrm{Dif}(\tau) \cdot \mathrm{Dif}(\mu)$. If ψ is true, then, by induction,

$$\mathrm{Dif}(\psi) \geq (1 - |\tau|2^{-n})(1 - |\mu|2^{-n}) > 1 - (|\tau| + |\mu|)2^{-n} \geq 1 - |\psi|2^{-n}.$$

If ψ is false, then, without loss of generality, say τ is false. By induction,

$$\mathrm{Dif}(\psi) \leq \mathrm{Dif}(\tau) < |\tau|2^{-n} \leq |\psi|2^{-n}.$$

□

Now, let A and B be circuits describing the computations of $\mathrm{Sample}(\phi, 0)$ and $\mathrm{Sample}(\phi, 1)$, respectively, (which take the random bits each procedure uses as input). By the above claim, $\|A - B\| > 1 - n2^{-n} > 2/3$ if ϕ is true with the appropriate arguments, and $\|A - B\| < n2^{-n} < 1/3$ if ϕ is false. In other words, the construction of A and B from w describes a many-one reduction from Mon(SD) to SD. This reduction can be computed in polynomial time because Sample runs in polynomial time. □

Now it is straightforward to deduce Theorem 6.2

PROOF. Let $(\phi, x_1, \ldots, x_k)$ be any instance of $\Phi(\mathrm{SD})$, where $\phi = \phi(v_1, \ldots, v_k)$. Use DeMorgan's laws to propagate all negations of ϕ to its variables. Now replace all occurrences of the literal $\neg v_i$ with a new variable w_i. Let $\psi(v_1, \ldots, v_k, w_1, \ldots, w_k)$ be the resulting (monotone) formula. Then, letting f be the Reversal Mapping of Proposition 5.1 which reduces SD to $\overline{\mathrm{SD}}$,

$$(\phi, x_1, \ldots, x_k) \mapsto (\psi, x_1, \ldots, x_k, f(x_1), \ldots, f(x_k))$$

is a reduction from $\Phi(\mathrm{SD})$ to $\mathrm{Mon}(\mathrm{SD})$. Composing this with the reduction in Lemma 6.3, we obtain Theorem 6.2. □

By Theorems 1.1 and 6.2 (along with the fact that SZK is closed under many-one reductions [**SV97**]), we obtain the following:

COROLLARY 6.5. *For every language L, $L \in \mathsf{SZK} \Rightarrow \Phi(L) \in \mathsf{SZK}$.*

Corollary 6.5 is a strengthening of several previous results. In [**DDPY94**], it was shown that $\mathrm{Mon}(L) \in \mathsf{SZK}$ for any language L which is random self-reducible, whose complement is self-reducible, or whose complement has a noninteractive statistical zero-knowledge proof. They also gave statistical zero-knowledge proofs for some simple statements involving a random-self-reducible language and its complement. Damgård and Cramer [**DC96**] extended these results by showing that $\mathrm{Mon}(L) \in \mathsf{SZK}$ as long as L or its complement has a 3-round public coin statistical zero-knowledge proof, and also treat a larger class of monotone functions.

Corollary 6.5 can be generalized to work for all boolean formulae whose atoms are statements about membership in any finite set of languages in SZK, but we omit the notationally cumbersome formal statement. Corollary 6.5 can be viewed as demonstrating that SZK is closed under a weak form of (polynomial-time) Turing reducibility. In particular, if $\Phi(L)$ were defined in terms of circuits rather than formulae, then the analogue of Corollary 6.5, together with the closure of SZK under many-one reductions, would imply that SZK is closed under *nonadaptive* Turing reductions. We do not know whether this is true, but we can prove that SZK is closed under a weaker form of Turing reductions:

PROPOSITION 6.6. SZK *is closed under (adaptive) polynomial-time Turing reductions which make a $O(\log n)$ of oracle queries on inputs of length n. That is, if A is a language*[7] *in* SZK *and B reduces to A via such a reduction, then $B \in$* SZK.

PROOF. Let M be the polynomial-time oracle machine such that M decides B when given oracle access to A, and on every input x, M^A asks at most $m = c \log |x|$ oracle queries. We define a *transcript* to be a sequence $t = (y_1, \sigma_1, \ldots, y_m, \sigma_m)$ such that $M(x)$ would ask oracle queries $y_1, \ldots, y_m$ (in that order) when given oracle responses $\sigma_1, \ldots, \sigma_m \in \{0, 1\}$. (Note that we are not requiring $\sigma_1, \ldots, \sigma_m$ to be the *correct* responses which would indicate whether or not $x_i \in A$.) We call such a transcript *accepting* if it would make $M(x)$ accept. Notice that, given x, we can enumerate in polynomial time all accepting transcripts for x by simulating M on all 2^m possible sequences of oracle responses. Let $t_i = (y_1^i, \sigma_1^i, \ldots, y_m^i, \sigma_m^i)$ be the ith accepting transcript in this enumeration, and let s be the number of accepting

[7]This proposition extends to the case of promise problems if we assume the oracle queries in the reduction never violate the promise.

transcripts. Now consider the following formula ϕ on a total of $s \cdot m$ variables:

$$\phi(v_1^1, \ldots, v_m^1, \ldots, v_1^s, \ldots, v_m^s) = \bigvee_{i=1}^{s} \left(\ell_1^i \wedge \cdots \wedge \ell_m^i\right),$$

where the literal ℓ_j^i is v_j^i if $\sigma_j^i = 1$ and is $\neg v_j^i$ if $\sigma_j^i = 0$. Now we claim that

$$M^A(x) \text{ accepts} \Leftrightarrow \phi((y_1^1 \in A), \ldots, (y_m^1 \in A), \ldots, (y_1^s \in A), \ldots, (y_m^s \in A)) = 1.$$

If $M^A(x)$ accepts then the term corresponding to the transcript with the correct oracle responses will evaluate to true. However, if $M^A(x)$ does not accept, then any accepting transcript must have at least one σ_j which disagrees the response oracle A would give on input y_j, so all the terms evaluate to false. Thus,

$$x \mapsto (\phi, y_1^1, \ldots, y_m^1, \ldots, y_1^s, \ldots, y_m^s)$$

is a (many-one) reduction from B to $\Phi(A)$. By Corollary 6.5 and the fact that SZK is closed under many-one reductions [**SV97**], we conclude that $B \in$ SZK. □

It would be interesting to prove that SZK is closed under Turing reductions, adaptive or nonadaptive, that make polynomially many oracle calls, or give evidence that this is not the case.

Acknowledgments

We are grateful to Oded Goldreich for his guidance throughout this work, and for countless insightful comments and conversations. We also thank our advisor, Shafi Goldwasser, for encouragement throughout; Madhu Sudan, for helping us better understand the effect of parallel repetition on statistical difference; Luca Trevisan for a conversation which helped us see that Proposition 6.6 follows from Corollary 6.5; and Avi Wigderson for encouraging us to seek a more explicit description of the Reversal Mapping.

References

[AB84] Miklos Ajtai and Michael Ben-Or, *A theorem on probabilistic constant depth computations*, Proceedings of the Sixteenth Annual ACM Symposium on Theory of Computing (Washington, D.C.), 1984, pp. 471–474.

[DC96] Ivan Damgård and Ronald Cramer, *On monotone function closure of perfect and statistical zero-knowledge*, Theory of Cryptography Library: Record 96-03, 1996, `http://theory.lcs.mit.edu/~tcryptol`.

[DDPY94] Alfredo De Santis, Giovanni Di Crescenzo, Giuseppe Persiano, and Moti Yung, *On monotone formula closure of SZK*, Proceedings of the Thirty Fifth Annual Symposium on Foundations of Computer Science, 1994, pp. 454–465.

[ESY84] Shimon Even, Alan L. Selman, and Yacov Yacobi, *The complexity of promise problems with applications to public-key cryptography*, Information and Control **61** (1984), no. 2, 159–173.

[GG98] Oded Goldreich and Shafi Goldwasser, *On the limits of non-approximability of lattice problems*, Proceedings of the 30th Annual ACM Symposium on Theory of Computing (Dallas, TX), ACM, May 1998, To appear.

[GMR89] Shafi Goldwasser, Silvio Micali, and Charles Rackoff, *The knowledge complexity of interactive proof systems*, SIAM Journal on Computing **18** (1989), no. 1, 186–208.

[GMW91] Oded Goldreich, Silvio Micali, and Avi Wigderson, *Proofs that yield nothing but their validity or all languages in NP have zero-knowledge proof systems*, Journal of the Association for Computing Machinery **38** (1991), no. 1, 691–729.

[GNW95] Oded Goldreich, Noam Nisan, and Avi Wigderson, *On Yao's XOR-lemma*, ECCC Report TR95-050, March 1995, Available from `http://www.eccc.uni-trier.de/eccc/`.

[Hof95] Micha Hofri, *Analysis of algorithms: Computational methods and mathematical tools*, Oxford University Press, 1995.

[ILL89] Russell Impagliazzo, Leonid A. Levin, and Michael Luby, *Pseudo-random generation from one-way functions (extended abstract)*, Proceedings of the Twenty First Annual ACM Symposium on Theory of Computing (Seattle, Washington), 15–17 May 1989, pp. 12–24.

[Oka96] Tatsuaki Okamoto, *On relationships between statistical zero-knowledge proofs*, Proceedings of the Twenty Eighth Annual ACM Symposium on the Theory of Computing, 1996, See also preprint of full version, August 1997.

[Sud97] Madhu Sudan, September 1997, Personal Communication.

[SV97] Amit Sahai and Salil Vadhan, *A complete promise problem for statistical zero-knowledge*, Proceedings of the 38th Annual Symposium on the Foundations of Computer Science, IEEE, October 1997, pp. 448–457.

[Yao82] Andrew C. Yao, *Theory and application of trapdoor functions*, Proceedings of the Twenty Third Annual Symposium on Foundations of Computer Science, 1982, pp. 80–91.

MIT Laboratory for Computer Science, 545 Technology Sq., Cambridge, MA 02139
E-mail address: amits@theory.lcs.mit.edu, salil@math.mit.edu

DIMACS Series in Discrete Mathematics
and Theoretical Computer Science
Volume **43**, 1999

A Survey of the Role of Multicommodity Flow and Randomization in Network Design and Routing

Aravind Srinivasan

ABSTRACT. The combinatorial choice problem of connecting a node s to a node t by a suitable path in a network G has a natural linear relaxation: that of shipping one unit of flow from s to t. The more general problem of connecting s to at least one of the vertices in a target set T has a similar flow-based relaxation. When we have to make a multitude of such choices simultaneously on the same network G, we are naturally led to various types of *multicommodity flow* relaxations. It is often very useful to employ *randomization* to "round" such a relaxation to a feasible combinatorial (integral) solution. We survey this idea of combining multicommodity flow and randomization, in the setting of many network routing and design problems.

1. Introduction

A large number of routing and design problems in networks share a basic subproblem: deciding which path(s) to use (or maybe none at all, due to capacity or cost constraints) to connect up various pairs of nodes (or subsets of the nodes) in the network. Unfortunately, most such problems are computationally hard: NP-hard, typically. Thus, in situations where network size makes exhaustive enumeration, branch-and-bound etc. prohibitively time-consuming, one looks for approximation algorithms: *efficient* algorithms that produce feasible solutions with objective function value that is to within a *proven* factor of optimal. A rich variety of approximation techniques have been developed for network optimization and other problems. In this survey, we shall largely focus on the combination of multi-commodity flow and randomization in the design of such approximation algorithms for network design and routing.

The classical maxflow-mincut theorem that deals with sending flow from a node s to a node t through a network G, is basic to combinatorial optimization, and is associated with a fundamental minimax theorem (see the book by Ahuja, Magnanti & Orlin [**AMO**]). Such flow problems are often very helpful in path-selection and routing problems also. At the very basic level, connecting node s_i to t_i, $i = 1, 2, \ldots, k$ in a network G, has the natural "multi-commodity flow relaxation"

1991 *Mathematics Subject Classification.* Primary 05C85, 60C05, 68-02, 68M10, 90B10, 90B15, 90C08, 90C27; Secondary 68R05.

Supported in part by National University of Singapore Academic Research Fund Grant RP970607.

of sending one unit of flow from s_i to t_i for each i. Such a relaxation is typically a linear program (LP) and hence is efficiently solvable. Moreover, the importance of multi-commodity flow has led to the development of good algorithms and code for various versions of it: see, e.g., Leighton, Makedon, Plotkin, Stein, Tardos & Tragoudas [**LM+**] and Awerbuch & Leighton [**AL**]. In this survey, we study such multicommodity flow relaxations in various contexts, and show how randomization is often very useful in rounding multi-commodity flow solutions to feasible integral solutions: basically, the amounts of flow carried by the flow-paths often suggest appropriate probabilities with which these paths can be selected. For completeness, we also sketch a few situations where *deterministic* rounding schemes are useful.

We have not attempted to survey all the important advances in this area here, and sometimes do not discuss the best-known results; in most cases, we present the underlying intuition and basic proof ideas in as general a context as seemed appropriate. Our aim is to show-case the combined utility of multi-commodity flow and randomization in our context through a class of illustrative examples, with the hope that the interested reader will be spurred to delve into the details and references not included here; e.g., the book edited by Hochbaum [**Hoc**] and the articles of Goemans & Williamson [**GW**], Motwani, Naor & Raghavan [**MNR**] and Shmoys [**Sh1, Sh2**], should be helpful.

Two other major goals and recurring themes of this survey are as follows. The first is to provide a gentle introduction to some powerful tools for handling *correlations*, which have found much use in discrete probability. We show how many such methods are of use in computer science, and hope to make a case for the further development of correlation inequalities from the algorithmic viewpoint. Second, in various contexts, we show how "short" flow paths in multicommodity flow bring significant benefits to (approximation) algorithms.

We will consider an algorithm to be efficient if it runs in polynomial time, and will not explicitly present the running times of our algorithms. For maximization problems, we shall say that an efficient algorithm is a λ-approximation algorithm (for some $\lambda \geq 1$) if the algorithm always produces a feasible solution of value at least $1/\lambda$ times the optimal objective function value. The parameter λ is also called the "approximation guarantee" or "approximation bound" of the algorithm. For minimization problems, an efficient algorithm will be called a λ-approximation algorithm (again, for some $\lambda \geq 1$) if the algorithm always produces a feasible solution of value at most λ times the optimal objective function value. Our goal here will generally be to to develop polynomial-time algorithms with "good" approximation guarantees for the various problems considered; i.e., to push λ down as close to 1 as we can.

The rest of this survey is organized as follows. We start with some preliminaries in Section 2; one of the main concepts introduced here is that of various types of *dependency graphs*, which are some possible ways of quantifying dependencies among classes of events. In Section 3, we first define the NP-hard problem of routing to minimize congestion, and also introduce a family of integer linear programs that include this problem. We show how an analysis (through the Lovász Local Lemma and an extension of it) of the underlying *correlations* involved, facilitates improved approximation bounds for these problems. A few NP-hard relatives of the congestion-minimization problem are then introduced in Section 4, and it is shown how certain results from Section 3 are useful tools in developing good approximation algorithms for these problems. An approximation algorithm for a network design problem, the group Steiner tree problem, is presented in Section 5: correlation inequalities are shown to help in the process. Section 6 shows approximation algorithms for some routing problems that are related to the ones considered in Sections 3 and 4. Section 7 concludes.

2. Preliminaries

Given a non-negative integer k, $[k]$ will denote the set $\{1, 2, \ldots, k\}$; logarithms are to the base two unless specified otherwise. Let Z_+ denote the set of non-negative integers. The base of the natural logarithm will be denoted as usual by e, and $\exp(x)$ will denote e^x. We let "r.v." denote the phrase "random variable". $\mathbf{E}[X]$ will denote the expected value of an r.v. X, and $\Pr[A]$ the probability of event A. Event A is called *nonempty* if $\Pr[A] \neq 0$.

2.1. Large-deviation bounds. Let $X_1, X_2, \ldots X_n$ be *independent* r.v.s that take on values in $[0, 1]$. Let $X \doteq \sum_{i=1}^{n} X_i$, and $\mu \doteq \mathbf{E}[X]$. We recall the classical Chernoff-Hoeffding (CH) bounds [**Ch, Hoe**]:

$$\begin{aligned} \text{If } \delta \geq 0,\ \Pr[X \geq \mu(1+\delta)] &\leq \frac{\prod_{i\in[n]} \mathbf{E}[(1+\delta)^{X_i}]}{(1+\delta)^{\mu(1+\delta)}} \\ &\leq G(\mu, \delta) \doteq \left(\frac{\exp(\delta)}{(1+\delta)^{(1+\delta)}}\right)^{\mu}; \\ \text{If } \delta \in [0,1],\ \Pr[X \leq \mu(1-\delta)] &\leq \frac{\prod_{i\in[n]} \mathbf{E}[(1-\delta)^{X_i}]}{(1-\delta)^{\mu(1-\delta)}} \leq \exp(-\mu\delta^2/2). \end{aligned} \tag{2.1}$$

It can be checked that for $\delta \leq 1$, $G(\mu, \delta) \leq \exp(-\delta^2\mu/3)$; for $\delta > 1$, $G(\mu, \delta) \leq \exp(-(1+\delta)\ln(1+\delta)\mu/4)$.

2.2. Dependency graphs. In the design and analysis of randomized algorithms, we often require a good estimate of $\Pr[\bigvee_{i\in[m]} E_i]$ (and hence of $\Pr[\bigwedge_{i\in[m]} \overline{E_i}]$) for some given events E_i. If we wish an upper-bound, the most basic idea would be to apply the *union bound* $\Pr[\bigvee_{i\in[m]} E_i] \leq \sum_{i\in[m]} \Pr[E_i]$, which however can be quite weak in general. Often, especially in situations where we expect the E_i to be "not too dependent" on each other, it helps to quantify such dependence more precisely. One basic such quantification that will help us is the notion of a *dependency graph.* We construct a simple undirected graph with the E_i as vertices, where the (symmetric, irreflexive) edge relation $\sim$ intuitively spells out "dependencies". More precisely, $\sim$ should be such that for any $i \in [m]$ and any $S \subseteq \{E_j : j \neq i \text{ and } E_i \not\sim E_j\}$, E_i is independent of (any Boolean combination of) the events in S. (We often abuse notation by using "$i \sim j$" to mean "$E_i \sim E_j$".) Note that $\sim$ is not always unique.

Often, however, we will only be interested in certain *subsets* of the correlated events. Suppose we have two sets of events $S_1 = \{E_1, E_2, \ldots, E_m\}$ and $S_2 = \{F_1, F_2, \ldots, F_M\}$. We will say that a simple undirected graph with $S_1 \cup S_2$ as vertex-set is a *bipartite dependency graph w.r.t.* (S_1, S_2) iff:

(C1) the edge relation $\sim$ leads to a bipartite graph with S_1 and S_2 as the two sides of the graph; and

(C2) for any $i \in [M]$ and any $S \subseteq \{E_j : F_i \not\sim E_j\}$, F_i is independent of any Boolean combination of the events in S.

Finally, the following related notion of *refined bipartite dependency graph w.r.t.* (S_1, S_2) will be useful. Here, the events in S_2 are indexed as $F_{i,j}$, where i runs from 1 to $m = |S_1|$ (and each j runs over some finite range). The edge relation $\sim$ should still satisfy (C1); (C2) now gets modified to "(C2'): for any $F_{i,j}$ and any $S \subseteq \{E_k : k \neq i, F_{i,j} \not\sim E_k\}$, $F_{i,j}$ is independent of any Boolean combination of the events in S". In some situations, some of the $F_{i,j}$ can be, say, real-valued r.v.s rather than events, where the above definitions carry over in an obvious manner. (In (C2'), for any a, even conditional on any Boolean combination of the events in S, the probability of $F_{i,j}$ equaling a should remain unaltered.)

We shall see how such concepts of (specific types of) dependencies will be of quite some use.

2.3. Relaxations and (randomized) rounding. All the optimization problems we consider will have the property that every feasible solution has a non-negative objective function value. A *relaxation* of an optimization problem is, given an instance I of the problem, to enlarge the set of feasible solutions for I in a manner so that the objective function can be efficiently optimized over the enlarged set. In all our examples, the relaxation will be to replace "integrality" constraints such as "$x_i \in \{a_i, a_i + 1, \ldots, b_i\}$" (where a_i and b_i are given integers) by constraints that allow x_i to be an arbitrary *real* in the range $[a_i, b_i]$. (We will most often have $a_i = 0$ and $b_i = 1$.) The benefit of this in our context is that we will mostly study integer linear programs (ILPs); thus, such a relaxation leads to a *linear program* (LP), which can be solved efficiently. Hence, the type of relaxations we study are also called linear relaxations or LP relaxations. More specifically, most such relaxations of path-selection problems that we consider, can be viewed combinatorially as multi-commodity flow relaxations as explained in Section 1.

Returning to the general definition of relaxations, let x^* denote an optimal solution over the enlarged set. Randomized rounding is the process of using *randomization* in some suitable way to map x^* back to a feasible solution for I. Let y^* denote the objective function value at x^*. It is easy to check that $y^* \leq OPT(I)$ for minimization problems and that $y^* \geq OPT(I)$ for maximization problems. So, if we can prove that a given randomized rounding algorithm for a minimization problem delivers a solution of objective function value at most $y^*\lambda$ (in expectation or with high probability), then we have a λ-approximation algorithm (in expectation or with high probability). The idea is analogous for maximization problems (it suffices to show that the randomized algorithm yields a solution of value at least y^*/λ). In other words, rather than compare the final solution value to $OPT(I)$ (about which we may not have much information), we compare the final solution to an appropriate relaxation y^* of $OPT(I)$.

To specify more precisely the type of randomized rounding schemes that we will mostly employ, we next recall the classical notion of *flow-decomposition* [**AMO**].

2.4. Flow-decomposition and randomized rounding. Flow-decomposition is a simple and useful process of transforming "edge flows" into "path flows"; we give a brief description now. We start with the case where there is a flow f corresponding to one source-sink pair (s, t), and then sketch the obvious "multi-commodity extension" to the situation where there are k separate flows for k pairs (s_i, t_i), $i = 1, 2, \ldots, k$.

Suppose we are given a directed graph $G = (V, E)$ (if G is undirected, we can view it as directed by replacing each edge by two anti-parallel directed edges); let

$m = |E|$. A *non-negative flow* f in G is simply an assignment of a non-negative value $f_{u,v}$ to each edge $(u,v) \in E$. The "net outflow" $f_{out}(v)$ from any $v \in V$ is defined naturally to be $(\sum_{u:\ (v,u)\in E} f_{v,u}) - (\sum_{u:\ (u,v)\in E} f_{u,v})$. Suppose we have, for some given function c, some pair of vertices s and t and some $a \geq 0$, that:

(F1) $\forall(u,v) \in E,\ f_{u,v} \leq c_{u,v}$, and

(F2) $f_{out}(s) = -f_{out}(t) = a;\ \forall v \in (V - \{s,t\}),\ f_{out}(v) = 0$.

We now show how f can be transformed into a set of "path flows" which still basically satisfy (F1) and (F2). Find any (s,t)-path P_1 that is composed of only positive-flow edges; let $b_1 > 0$ be the minimum flow-value on the edges of P_1. Delete b_1 from the flow of all edges in P_1, declare P_1 to be a flow-path with flow-value b_1, remove all zero-flow edges, and repeat. Since at least one edge is removed in each iteration, we have at most m iterations. Thus, we finally have a set of flow-paths $\{P_i :\ 1 \leq i \leq \ell\}$, where each P_i is from s to t and has some flow value $b_i > 0$; we also have $\ell \leq m$. It is easy to check, e.g., by induction on the number of iterations of the above process, that

$$\sum_i b_i = a;\ \forall(u,v) \in E, \sum_{i:(u,v)\in P_i} b_i \leq c_{u,v}. \tag{2.2}$$

More commonly, we will require a generalization of the above situation to the case where we have k pairs (s_i, t_i) with one separate non-negative flow $f^{(i)}$ for each i. Here, for some $a_i \geq 0$, conditions (F1) and (F2) get replaced by

- $\forall(u,v) \in E,\ \sum_{i\in[k]} f^{(i)}_{u,v} \leq c_{u,v}$, and
- $\forall i \in [k]$: $f^{(i)}_{out}(s_i) = -f^{(i)}_{out}(t_i) = a_i;\ \forall v \in (V - \{s_i,t_i\}),\ f^{(i)}_{out}(v) = 0$.

By repeating the above-seen flow decomposition process separately for each i, we decompose each $f^{(i)}$ into a set of flow-paths $\{P_{i,j} :\ 1 \leq j \leq \ell_i\}$, where $\ell_i \leq m$ and each $P_{i,j}$ is an (s_i,t_i)-path carrying some non-negative flow of $b_{i,j}$. It is straightforward to check that the following analog of (2.2) holds:

$$\forall i, \sum_j b_{i,j} = a_i;\ \forall(u,v) \in E, \sum_{(i,j):(u,v)\in P_{i,j}} b_{i,j} \leq c_{u,v}. \tag{2.3}$$

Let $a_i, P_{i,j}, b_{i,j}$ etc. be as above. As sketched in Section 1, many problems we consider will require connecting each (or "many") of the vertices s_i to the corresponding t_i; informally speaking, one possible (multi-commodity flow) relaxation is to send some amount of flow from each s_i to t_i. We will retain this somewhat vague description for now, specializing the discussion when we come to specific problems; we now focus on how randomized rounding can be employed in such a context. We solve a suitable multi-commodity flow relaxation to (near-)optimality, and then do a flow-decomposition to the obtained solution. The problem remaining is to choose, for each i, one or at most one (s_i,t_i)-path $P_{i,j}$ appropriately – the notion of appropriateness, whether one (s_i,t_i)-path is really required, etc., will depend on the context. We now present two general randomized rounding schemes to do the path-selection.

In the first scheme, we will have $a_i = 1$ for each i, in the notation seen above. (These relaxations will usually correspond to problems where we **have to** connect s_i to t_i for each i.) Thus, we are sending a unit flow from s_i to t_i for each i, and so flow decomposition transforms this into a convex combination of (s_i,t_i)-paths. Such a convex combination can be viewed in a natural way as a probability distribution on the collection of (s_i,t_i)-paths. Motivated by this, our first randomized rounding scheme will be to, *independently for each* i, pick *exactly one* (s_i, t_i)-path $P_{i,j}$: the probability that $P_{i,j}$ is chosen equals $b_{i,j}$.

Our second randomized rounding scheme typically operates in situations where all the a_i are not necessarily one; in most such cases, while one would like to connect each s_i to t_i if possible, one may have to end up not connecting some of the s_i to the corresponding t_i due to cost, capacity, or other constraints. The second scheme turns this into an advantage: it just *independently* picks each $P_{i,j}$, with probability proportional to $b_{i,j}$. (The amount of proportionality here will vary depending upon the application.) Though we sacrifice the property that each s_i will get connected to the corresponding t_i with probability 1, the advantage here is that the independence can potentially bring us desirable correlations, as we will see in some situations. (If more than one (s_i, t_i)-path is chosen for some i, we can choose one of them using some rule; in all our applications, this "rule" will be to simply make an arbitrary choice.)

One common aspect of the two rounding schemes is that we *interpret flow values as probabilities*: path $P_{i,j}$ is chosen with probability equaling (or proportional to) its flow-value $b_{i,j}$. The intuitive appeal of this stems from the observation that the LP is probably implicitly suggesting "desirable" or "undesirable" flow-paths, based on their flow-values. Thus, it is promising that flow-paths with relatively large flow-values are preferred to those with lower flow-values. We will rigorously bound the behavior of such rounding schemes in various contexts.

3. Low-congestion routing, minimax integer programs, and the Lovász Local Lemma

We now present our first routing problem, the classical one of *routing to minimize congestion* (Raghavan & Thompson [**RT**], Kleinberg [**Kl1**]). Given a graph $G = (V, E)$ and a collection of pairs of vertices $T = \{(s_i, t_i) : 1 \leq i \leq k\}$, the NP-hard problem is to connect each s_i to t_i by some path through G, in such a way that the *congestion*—the maximum number of chosen paths using any edge of G—is minimized. This problem has also been studied for, e.g., global routing in VLSI gate arrays [**RT**]. If we view G as describing the set of channels through which we can route wires in a chip, the problem (approximately) models the issue of connecting up a given pairs of terminals in a chip, in a manner that minimizes the thickness of the chip.

A multicommodity flow relaxation for this problem is to route unit flow from each s_i to the corresponding t_i, so that the maximum total amount of flow using any edge is minimized. Since we have relaxed the given problem to a fractional counterpart, the maximum total amount of flow using any edge is called the *fractional congestion*. It is easy to see that the relaxation is a linear program, where the objective is to minimize a variable W that denotes the fractional congestion. Let y^* and OPT denote the minimum fractional congestion and minimum (integral) congestion respectively; clearly, $y^* \leq OPT$. We solve the LP, and then conduct flow-decomposition on the obtained sequence of k flows; in the notation of Section 2.4, this yields a set of paths $\{P_{i,j}\}$ with $a_i = 1$ for all i and with $\max_{(u,v) \in E} c_{u,v} = y^*$.

For the rest of this section, we will be concerned with analyzing the performance of the first randomized rounding scheme described in Section 2.4, for this problem. That is, we will see how good an upper-bound we can show on the final congestion produced by the randomized rounding scheme (with high probability or maybe just with positive probability). In Sections 3.1 and 3.3, we will in fact study this is the context of a more general class of problems involving *minimax integer programs* (MIPs). Sections 3.2 and 3.3 show how an analysis of the *correlations* involved leads to improved bounds on the performance of randomized rounding. A key notion here will be the *dilation* d, the maximum length of (i.e., number of edges in) any of the flow-paths $P_{i,j}$ output by flow-decomposition. In particular, we will show that for "small" d, randomized rounding performs well, at least with positive probability. The (fractional) congestion and dilation are classical parameters in routing, and will occur a few times in this survey.

3.1. Minimax integer programs and the first analysis of randomized rounding. The problem of choosing one $P_{i,j}$ for each i to minimize congestion, can be viewed more generally in the context of MIPs.

DEFINITION 3.1. An MIP has variables W and $\{x_{i,j} : i \in [n], j \in [\ell_i]\}$, for some integers $\{\ell_i\}$. Let $N = \sum_{i\in[n]} \ell_i$ and let x denote the N-dimensional vector of the variables $x_{i,j}$ (arranged in any fixed order). An MIP seeks to minimize a real W subject to:

(i) $\forall i \in [n]$ $\sum_{j\in[\ell_i]} x_{i,j} = 1$;

(ii) a system of linear inequalities $Ax \leq \vec{W}$, where $A \in [0,1]^{m\times N}$ and $\vec{W}$ is the m-dimensional vector with the variable W in each component, and

(iii) $\forall i, j,$ $x_{i,j} \in \{0,1\}$.

Constraints (i) and (iii) show that for each i, precisely one $x_{i,j}$ will be one, all the others being zero. Thus, MIPs model many situations where, for some "large" n, we have to choose one element each from n finite sets subject to some linear constraints, in order to optimize some linear function of the choices made. In particular, it is easy to check that the above-seen problem of picking one $P_{i,j}$ for each i to minimize congestion, is modeled by an MIP: $x_{i,j}$ is the indicator for choosing path $P_{i,j}$, and there is one (congestion) constraint for each edge of G. Other NP-hard problems such as some that involve hypergraph-partitioning, are also modeled by MIPs [**Sr1**]; so we now consider approximation algorithms for MIPs.

Let us generalize our discussion for low-congestion routing and the first randomized rounding approach, to the setting of MIPs. A natural LP relaxation for a given MIP is to allow the entries of x to be non-negative reals subject to the family of constraints (i) and (ii). Let x^* and y^* denote, resp., an optimal solution to, and the optimum value of, this LP relaxation. So if OPT denote the optimum objective function value for a given MIP, we have $y^* \leq OPT$. A natural randomized rounding approach to round such a solution, is suggested by our discussion of the first randomized rounding method for low-congestion routing. Independently for each i, randomly round *exactly one* $x_{i,j}$ to 1: the probability of choosing $x_{i,j}$ is $\{x^*_{i,j}\}$. We will now present an analysis of this scheme due to [**RT**]. An improved analysis via a more detailed study of the correlations involved, will be demonstrated in Section 3.3.

Henceforth, we will assume w.l.o.g. that $\{x^*_{i,j}\}$ is a basic feasible solution to the LP relaxation. So, at most m of the $x^*_{i,j}$ will be neither zero nor one, and clearly only these variables will participate in the rounding. Thus, we assume from now on that $\ell_i \le m$ for all $i \in [n]$. (In the context of the congestion minimization problem, this was already seen in the description of flow-decomposition.)

We now develop some common notation to help in the proofs of Theorems 3.2, 3.4, 3.9 and 3.10. Suppose we conduct randomized rounding on a generic given MIP. Let $z \in \{0,1\}^N$ denote the randomly rounded vector; by linearity of expectation,

$$\mu_i \doteq \mathbf{E}[(Az)_i] \le y^*, \ \forall i \in [m].$$

For some non-negative parameter ν, define "bad" events $E_1, E_2, \ldots, E_m$ by

$$E_i \equiv ((Az)_i \ge \mu_i + \nu);$$

ν will in general be a function of y^* and of certain parameters related to the coefficient matrix A. Our goal will be to show that $\bigwedge_i \overline{E_i}$ holds with reasonably high or at least nonzero probability, for as small a value of ν as we can come up with. If we are successful in this for some $\nu = \nu_0$, we will at least have that there exists an integral solution (to a generic MIP) of value at most ν_0 more than the fractional optimum y^*.

Some further notation to facilitate our analyses: for each $i \in [m]$, rewrite the ith constraint of the MIP as

$$\sum_{r\in[n]} X_{i,r} \le W, \text{ where } X_{i,r} = \sum_{s\in[\ell_r]} A_{i,(r,s)} x_{r,s}; \tag{3.1}$$

the notation $A_{i,(r,s)}$ assumes that the pairs $\{(r,s) : r \in [n], s \in [\ell_r]\}$ have been mapped bijectively to $[N]$, in some fixed way. In other words, for each $i \in [m]$ and $r \in [n]$, we collect together the contributions of $\{x_{r,s} : s \in [\ell_r]\}$ to the ith constraint of the MIP. Also define the r.v.

$$Z_{i,r} = \sum_{s\in[\ell_r]} A_{i,(r,s)} z_{r,s}. \tag{3.2}$$

It is evident from the description of randomized rounding here that for all $i \in [m]$,

(3.3) $Z_{i,1}, \ldots, Z_{i,n}$ lie in $[0,1]$ and are independent; $E_i \equiv (\sum_{r\in[n]} Z_{i,r} \ge \mu_i + \nu)$.

We start with Theorem 3.2 describing the analysis of [**RT**], which shows that $\bigwedge_i \overline{E_i}$ holds with a reasonably high probability for a certain choice of $\nu = \nu(y^*, m)$. Recall the CH bound function $G(\mu, \delta)$ of (2.1). The following proposition will be useful.

PROPOSITION 3.1. (a) $\forall \mu > 0\ \forall p \in (0,1)\ \exists \delta = \psi(\mu, p) > 0$ *such that: (i)* $G(\mu, \delta) \le p$, *and (ii)* $\psi(\mu, p) = \Theta(\frac{\log(p^{-1})}{\mu \log(2\log(p^{-1})/\mu)})$ *if* $\mu \le \log(p^{-1})$, *and is* $\Theta(\sqrt{\frac{\log(p^{-1})}{\mu}})$ *otherwise.*
(b) *If* $0 < \mu_1 \le \mu_2$, *then for any* $\delta > 0$, $G(\mu_1, \mu_2\delta/\mu_1) \le G(\mu_2, \delta)$.

THEOREM 3.2. ([**RT**]) *Given a generic MIP, randomized rounding produces a feasible solution of value at most* $y^*(1 + \psi(y^*, 1/(2m)))$ *with probability at least* $1/2$.

PROOF. Choose $\nu = y^*\psi(y^*, 1/(2m))$. We now show that for each i, $\Pr[E_i] \le 1/(2m)$; a union bound over all i will complete the proof. Recall, for each $i \in [m]$, that $\mu_i \doteq \mathbf{E}[(Az)_i] \le y^*$; define $\delta_i = \nu/\mu_i$. We deduce from (3.3) that

$$\Pr[E_i] \le G(\mu_i, \delta_i) \le G(y^*, \nu/y^*) = G(y^*, \psi(y^*, 1/(2m))) \le 1/(2m),$$

where the second inequality follows from part (b) of Proposition 3.1 and the last inequality holds by the definition of ψ (see part (a) of Proposition 3.1). As mentioned above, a union bound over all $i \in [m]$ finishes the proof. ■

The above algorithm can also be derandomized, i.e., turned into an efficient deterministic algorithm (Raghavan [**Rag**]). Specializing Theorem 3.2 to our low-congestion routing problem, we get

COROLLARY 3.1. *For a given congestion minimization problem, suppose y^* denotes the optimal fractional congestion. Then, we can efficiently do a path-selection so that with high probability, the integral congestion is at most:*

(i) $O(\frac{\log m}{y^* \log((2\log m)/y^*)})$ *if* $y^* \leq \log m$, *and*
(ii) $y^* + O(\sqrt{y^* \log m})$ *if* $y^* > \log m$.

In particular, part (i) shows that for families of instances where the fractional congestion y^* is at most 1 (or bounded by some absolute constant), the integral congestion is at most $O((\log m)/\log\log m)$. Part (ii), on the other hand, shows that if $y^* \geq \Omega(\log m)$, say, then the integral congestion is not much more than y^*. Some applications of this nice feature of part (ii) will be presented in Section 4.

If we just require $\Pr[\bigwedge_{i\in[m]} \overline{E_i}] > 0$, can we exploit the structure of MIPs to show that a value for ν that is smaller than given by Theorem 3.2 will suffice? (In applying the union bound to the E_i, the proof of Theorem 3.2 is potentially not exploiting possible *correlations* among them.) Such a result for low-congestion routing due to Leighton, Rao and the author [**LRS**], is now sketched in Section 3.2.

3.2. The Lovász Local Lemma. Recall the notion of dependency graphs from Section 2.2. We start our discussion here with the powerful Lovász Local Lemma (LLL) due to Erdős and Lovász ([**EL**]). The LLL, an important special case of which is shown below, helps prove that a set of ("bad") events U_i can be simultaneously avoided under some conditions:

LEMMA 3.3. ([**EL**]) *Let $U_1, U_2, \ldots, U_m$ be any events with $\Pr[U_i] \leq p\ \forall i$. Suppose a relation $\sim$ defines a dependency graph on the U_i, wherein each vertex has degree at most D. If $ep(D+1) \leq 1$, then $\Pr[\bigwedge_{i\in[m]} \overline{U_i}] \geq (1-ep)^m > 0$.*

PROOF. For $I \subseteq [m]$, let $Z(I)$ denote $\bigwedge_{j\in I} \overline{U_j}$. We will show by induction on $|I|$ that if $i \notin I$ then $\Pr[U_i \mid Z(I)] \leq e\Pr[i]$. This will suffice since

$$\Pr[\bigwedge_i \overline{U_i}] = \prod_{i\in[m]} (1 - \Pr[U_i \mid Z([i-1])]).$$

For the base case $I = \emptyset$, $\Pr[U_i \mid Z(I)] = \Pr[U_i]$. For the inductive step, let $S_{i,I} = \{j \in I : U_i \sim U_j\}$, and $S'_{i,I} = I - S_{i,I}$. (Note that $|S_{i,j,I}| \leq D$.) If $S_{i,I} = \emptyset$, then $\Pr[U_i \mid Z(I)] = \Pr[U_i]$ by the definitions of dependency graphs and of $S_{i,I}$.

Otherwise, suppose $S_{i,I} = \{j_1, j_2, \ldots, j_s\}$. Then,

$$\begin{aligned}
\Pr[U_i \mid Z(I)] &= \frac{\Pr[(U_i \wedge Z(S_{i,I})) \mid Z(S'_{i,I})]}{\Pr[Z(S_{i,I}) \mid Z(S'_{i,I})]} \\
&\leq \frac{\Pr[U_i \mid Z(S'_{i,I})]}{\Pr[Z(S_{i,I}) \mid Z(S'_{i,I})]} \\
&= \frac{\Pr[U_i]}{\Pr[Z(S_{i,I}) \mid Z(S'_{i,I})]} \quad (U_i \text{ is independent of } S'_{i,I}) \\
&= \frac{\Pr[U_i]}{\prod_{r\in[s]}(1 - \Pr[U_{j_r} \mid Z(\{j_1, j_2, \ldots, j_{r-1}\} \cup S'_{i,I})])} \\
&\leq \frac{\Pr[U_i]}{\prod_{r\in[s]}(1 - ep)} \quad \text{(induction hypothesis)} \\
&\leq \frac{\Pr[U_i]}{(1 - 1/(D+1))^D},
\end{aligned}$$

which is at most $e \Pr[U_i]$. ■

We now present an LLL-based improvement over Corollary 3.1 for low-congestion routing:

THEOREM 3.4. (**[LRS]**) *Suppose we are given any graph $G = (V, E)$, any collection $\{(s_i, t_i) : i \in [k]\}$ of pairs of vertices in it, and a unit flow from each s_i to t_i. Let $\{P_{i,j}\}$ denote the set of at most m flow-paths from s_i to t_i, and suppose $P_{i,j}$ carries a non-negative flow of $b_{i,j}$. Let y^* denote the* fractional congestion

$$\max_{(u,v)\in E} \sum_{(i,j):(u,v)\in P_{i,j}} b_{i,j};$$

let the dilation *(maximum length of any $P_{i,j}$) be d. Suppose B is any positive integer such that $3(B+1)d(y^*)^B \leq B!$. Then, there exists a way of choosing one flow-path $P_{i,j}$ for each (s_i, t_i), such that the maximum number of paths using any edge of G is at most B.*

PROOF SKETCH. We will start with an (almost) equivalent view of our randomized rounding process. To do so, we pre-process the flow-paths so that all the flow values become equal, as follows. Since $3(B+1)d(y^*)^B \leq B!$, there is some parameter $\epsilon > 0$ such that

$$e(B+1)d(y^*)^B \leq (1-\epsilon)^{B+1} B!. \tag{3.4}$$

For each (i, j), let $b'_{i,j}$ be the largest multiple of ϵ/m that is at most $b_{i,j}$; replace $P_{i,j}$ by $b'_{i,j}m/\epsilon$ copies of itself, each of the copies carrying a flow of precisely ϵ/m. It can be checked that after this, the fractional congestion is still at most y^*. Recall that $\sum_j b_{i,j} = 1$ for each i; also, $b'_{i,j} \geq b_{i,j} - \epsilon/m$. These observations, combined with the fact that there are at most m flow-paths $P_{i,j}$ for any given i, can be used to show that there are now at least $m(1-\epsilon)/\epsilon$ flow-paths $P'_{i,j}$ for each i. These new $P'_{i,j}$ will now be called the flow-paths.

Independently for each (s_i, t_i) pair, pick one of its flow paths uniformly at random. (Modulo the small error introduced by ϵ, the probability of picking $P_{i,j}$ is $b_{i,j}$; thus, we have an almost equivalent description of our original randomized rounding scheme. The correspondence will become exact if we let ϵ tend to 0 from above.) We now use the LLL to show that with positive probability, the integral congestion is at most B. For each edge $f \in E$ and every set of $B+1$ flow paths that use it, associate one bad event corresponding to all of these $B+1$ paths being chosen by the random process above. We now show that with positive probability, all of these bad events are avoided simultaneously.

Given a bad event, if two of the paths in it connect the same (s_i, t_i) pair, the probability of the event is 0. Otherwise, since there are at least $m(1-\epsilon)/\epsilon$ flow paths for each (s_i, t_i), the probability of the event is at most

$$(\epsilon/(m(1-\epsilon)))^{B+1}. \tag{3.5}$$

Next, let us construct a dependency graph on the set of bad events. Let U and U' be generic distinct bad events that corresponding respectively to the sets of flow-paths

$$\{P'_{i_1,j_1}, P'_{i_2,j_2}, \ldots, P'_{i_{B+1},j_{B+1}}\} \text{ and } \{P'_{r_1,s_1}, P'_{r_2,s_2}, \ldots, P'_{r_{B+1},s_{B+1}}\}.$$

We will define $U \sim U'$ iff

$$\{i_1, i_2, \ldots, i_{B+1}\} \cap \{r_1, r_2, \ldots, r_{B+1}\} \neq \emptyset.$$

Since each (s_i, t_i)-pair *independently* chooses its flow-path, it is not hard to check that we have a valid dependency graph. In order to apply the LLL, we now need to bound the degree of, say, U in this dependency graph. Using the facts that: (a) each (s_i, t_i)-pair has at most m/ϵ flow-paths, (b) at most my^*/ϵ flow-paths use any given edge, and (c) each flow-path passes through at most d edges, a simple calculation shows that the degree of U in the dependency graph is at most

$$(B+1)(m/\epsilon)^{B+1} d(y^*)^B/B! - 1. \tag{3.6}$$

So, (3.4), (3.5) and (3.6), in conjunction with the LLL, show that all the bad events can be avoided with positive probability. ■

The above improves on similar results of Leighton & Rao [**LR2**]. One noteworthy feature of this improvement is that in typical applications of the LLL, the definition of "bad events" will be fairly obvious. However, the improvement above comes from a careful choice of the bad events; the LLL had also been applied in [**LR2**], but with a different choice of bad events. Furthermore, with a constant-factor loss in B, the result above can also be turned into a polynomial-time algorithm via the work of Beck [**Be**] and Alon [**Al**].

To get a more convenient idea of the parameter B of Theorem 3.4, note that we can take B to be: (i) $O((\log d)/\log\log d)$ if $y^* \leq O(1)$, and (ii) $y^* + O(\log d)$ if $y^* \geq \Omega(\log d)$. Thus, for situations where d is "small", say, $O(\text{polylog}(n))$, we get significant improvements over Corollary 3.1. This is our first illustration of the utility of multicommodity flow along "short" paths.

Are there natural settings where d is "small"? An important family of such settings due to Leighton & Rao [**LR1**], will be shown in Section 4; we now present another family of examples shown by Kleinberg & Rubinfeld [**KR**]. If $G = (V, E)$ is an undirected graph, define $\delta(S)$ to be the set of edges of G that have precisely one end-point in S, for any $S \subseteq V$. Let $n = |V|$. The edge-expansion of G, $\beta = \beta(G)$, is defined as

$$\min_{S \subseteq V : |S| \leq n/2} |\delta(S)|/|S|.$$

Expansion is an important parameter in many routing situations. (Informally, if a graph G has "low" edge-expansion, then there is a "bottleneck" set S such that the number of edges leaving S is "small" in comparison with the number of vertices in S. This can often be used to show that if G is an interconnection network, say, then some basic problems will require much time to be solved using G; such lower-bound proofs are typically accomplished by constructing input instances where a fair amount of information needs to be communicated between a bottleneck set S and its complement. See Leighton's book [**Lei**] for precise versions of this informal statement. Conversely, networks such as the butterfly that have "reasonable" (say, $\Omega(1/\text{polylog } n)$) values of β are good choices for use as interconnection networks [**Lei**]. Furthermore, if β is "high", then G has good fault-tolerance properties in the face of a certain number of vertices and/or edges being removed; see Alon & Chung [**AC**].) Let $\Delta = \Delta(G)$ denote the maximum degree of the vertices of G.

Let $\epsilon > 0$ be any parameter. Consider the LP relaxation of our low-congestion routing problem. A special case of a result of [**KR**] shows that there exists a feasible solution to this relaxation with fractional congestion at most $y^*(1+\epsilon)$, and with dilation at most $d_0 = O(\epsilon^{-1}\Delta^2\beta^{-2}\log^3 n)$.

Fix ϵ to be, say, 1. It is easy to add the *linear* constraints that "all flow paths are of length at most d_0" to our LP relaxation; thus, we can effectively replace d by d_0 in Theorem 3.4. So, graphs with "low" maximum degree and "high" expansion have a relatively low ratio of integral to fractional congestion. Thus we see our first example of the benefits of multicommodity flow using short paths. Networks with "low" maximum degree and "high" expansion are often good choices for, e.g., interconnection networks; for instance, we note that the n-node butterfly has constant degree and expansion $\Theta(1/\log n)$.

Our result here has been accomplished by a study of some of the *correlations* involved, and usage of the LLL. We now present a more general version of this in Section 3.3.

3.3. An extension of the Lovász Local Lemma. The work described here is from [**Sr1**]. The condition that "all flow-paths $P_{i,j}$ have length at most d", when interpreted in the more general context of MIPs, translates to "all columns of the coefficient matrix A have at most d nonzero entries". Motivated by this, we can ask: if the coefficient matrix of a given MIP is *column-sparse*, can it be approximated well? To make this more concrete, let us parametrize the matrix A of an MIP as follows:

DEFINITION 3.5. For a given MIP, let d denote the maximum number of non-zero entries in any column of its coefficient matrix A. Also let $g \le d$ be $\max_j \sum_i A_{i,j}$.

So, a natural question is to derive approximations (or at least integrality gaps) for MIPs parametrized by d or g. One such example is Theorem 3.6, which can be shown using a key rounding theorem of [**KL+**]:

THEOREM 3.6. *Given an MIP, we can efficiently construct an integral solution for it with value less than $y^* + g$.*

This is clearly very useful if g is "small" or "comparable" to y^*; we just mention here that Theorem 3.6 can be used to give an alternative proof of the main rounding theorem of Lenstra, Shmoys & Tardos [**LST**], since $g = y^*$ there. However, Theorem 3.6 is not so useful if g is somewhat large compared to y^*. (For instance, $g = d$ for the low-congestion routing problem. However, if $y^* \leq O(1)$, say, then direct use of Theorem 3.6 only guarantees an integral congestion of $d + O(1)$, whereas Theorem 3.4 yields a bound of $O((\log d)/\log\log d)$.) We present an extension of the LLL, Theorem 3.8; Theorem 3.9 shows how Theorem 3.8 can help for certain types of column-sparse MIPs. We will then bootstrap Theorem 3.9 to get the further improved Theorem 3.10.

For real x and any positive integer r, let $\binom{x}{r}$ denote $x(x-1)\cdots(x-r+1)/r!$ as usual, with $\binom{x}{0} \doteq 1$. Define, for $z = (z_1, z_2, \ldots, z_n) \in \Re^n$, a family of symmetric multi-linear polynomials $\phi_j(z), j = 0, 1, \ldots, n$, where $\phi_0(z) \equiv 1$, and for $1 \leq j \leq n$,

$$\phi_j(z) \doteq \sum_{1 \leq i_1 < i_2 \cdots < i_j \leq n} z_{i_1} z_{i_2} \cdots z_{i_j}.$$

A small extension of a result of [**SSS**], presented in [**Sr1**], is as follows:

THEOREM 3.7. *Let $X_1, \ldots, X_n \in [0,1]$ be r.v.s with $X = \sum_i X_i$ and $\mu = E[X]$.*

(a) *For any $\delta > 0$, any nonempty event Z and any $k \leq \mu(1+\delta)$,*

$$\Pr[X \geq \mu(1+\delta) \mid Z] \leq \mathbf{E}[Y_k \mid Z], \text{ where } Y_k = \phi_k(X_1, \ldots, X_n) / \binom{\mu(1+\delta)}{k}.$$

(b) *If the X_i are all independent and $k = \lceil \mu\delta \rceil$, then* $\Pr[X \geq \mu(1+\delta)] \leq \mathbf{E}[Y_k] \leq G(\mu, \delta)$.

Suppose we have "bad" events $U_1, U_2, \ldots, U_m$ such that any dependency graph for the events has a fairly "high" degree D. This will preclude direct application of the LLL. However, as mentioned in Section 2.2, one often does not have to consider all possible dependencies; this is one of the reasons why the bipartite and refined bipartite versions of dependency graphs were introduced. Theorem 3.8 exploits this, by placing some sufficient conditions on a refined bipartite dependency graph (RBDG) related to the events U_i. It is easily seen to generalize Lemma 3.3, and its proof is similar to the one seen for Lemma 3.3. We will use Theorem 3.8 to present an improved analysis of randomized rounding for column-sparse MIPs via Theorems 3.9 and 3.10. Theorem 3.7 will be seen to provide some of the requirements of Theorem 3.8, in our applications.

THEOREM 3.8. *Suppose a set of events $S_1 = \{U_1, U_2, \ldots, U_m\}$ and a set of non-negative valued r.v.s $S_2 = \{F_{i,j} : i \in [m], j \in \eta_i\}$ (each η_i is some given finite number) satisfy*

$$\forall i \in [m]\ \forall \text{ nonempty events } \mathbf{A},\ \Pr[U_i \mid \mathbf{A}] \leq \sum_j \mathbf{E}[F_{i,j} \mid \mathbf{A}]. \tag{3.7}$$

Let $\sim$ denote the edge relation of any particular RBDG w.r.t. (S_1, S_2). Then if $p = \max_i \sum_j \mathbf{E}[F_{i,j}]$ and $D = \max_{i,j} |\{k \neq i : U_k \sim F_{i,j}\}|$ satisfy $ep(D+1) \leq 1$, we have $\Pr[\bigwedge_i \overline{U_i}] \geq (D/(D+1))^m > 0$. (Note that $\max_i \Pr[U_i] \leq p$: set $\mathbf{A}$ to be the tautology in (3.7).)

PROOF SKETCH. As in the proof of Lemma 3.3, we let $Z(I) = \bigwedge_{j \in I} \overline{U_j}$ and show by induction on $|I|$ that if $i \notin I$ then $\Pr[U_i \mid Z(I)] \leq e \Pr[U_i]$. The base case $I = \emptyset$ is straightforward as before. For the inductive step, in a manner similar to the proof of Lemma 3.3, we show, for all j, that $\mathbf{E}[F_{i,j} \mid Z(I)] \leq e\mathbf{E}[F_{i,j}]$. The induction step is completed by invoking (3.7). ■

The tools behind the MIP application are Theorems 3.8 and 3.7. Fix a generic MIP. Recalling the notation of (3.1), define, for each $r \in [n]$,

$$Q_r = \{i \in [m] : \text{ row } i \text{ has a nonzero contribution from variable } X_{i,r}\}. \tag{3.8}$$

Next define

$$t \doteq \max_{r\in[n]} |Q_r|.$$

Note from Definition 3.5 that

$$g \le d \le t \le \min\{m, d \cdot \max_{i\in[n]} \ell_i\}. \tag{3.9}$$

A simple calculation using the CH bound function G yields

PROPOSITION 3.2. *$\forall \mu > 0$ $\forall p \in (0,1)$, $\exists \delta = L(\mu, p) > 0$ such that: (i) $\lceil \mu\delta \rceil \cdot G(\mu, \delta) \le p$, and (ii) $L(\mu, p) = \Theta(\frac{\log(p^{-1})}{\mu \log(\log(p^{-1})/\mu)})$ if $\mu \le \log(p^{-1})/2$, and is $\Theta(\sqrt{\frac{\log(\mu+p^{-1})}{\mu}})$ otherwise.*

Our first cut at an approximation for MIPs is parametrized by t:

THEOREM 3.9. *Given a generic MIP, randomized rounding produces a feasible solution of value at most $y^*(1 + L(y^*, 1/(et)))$, with non-zero probability.*

PROOF. We choose $\nu = y^* L(y^*, 1/(et))$, and show that $\Pr[\bigwedge_{i\in[m]} \overline{E_i}] > 0$ using Theorem 3.8.

For the rest of this proof, fix any $i \in [m]$. Define R_i to be the set of all $r \in [n]$ such that row i gets a nonzero contribution from the variable $X_{i,r}$, i.e., $R_i = \{r \in [n] : i \in Q_r\}$. Recall from (3.3) that the event E_i is a large-deviation event for $\sum_{r\in R_i} Z_{i,r}$; the r.v.s $Z_{i,r}$ lie in $[0, 1]$ and are independent. Since Theorem 3.7 handles precisely such situations, we will now see that Theorem 3.7 suggests a suitable choice for the crucial r.v.s $F_{i,j}$ in order to apply Theorem 3.8.

Let $\nu' = \lceil \nu \rceil$ and $u_i = \binom{|R_i|}{\nu'}$. We now define the r.v.s $\{F_{i,j} : i \in [m], j \in [u_i]\}$ as follows, motivated by Theorem 3.7. Identify each $j \in [u_i]$ with some distinct ν'-element subset $T(j)$ of R_i, and let

$$F_{i,j} \doteq \frac{\prod_{v\in T(j)} Z_{i,v}}{\binom{\mu_i+\nu}{\nu'}}. \tag{3.10}$$

Let $S_1 = \{E_1, E_2, \dots, E_m\}$, and S_2 be the set of all the $F_{i,j}$. There is a natural candidate for an RBDG H w.r.t. (S_1, S_2). In the notation of (3.10), define the set of neighbors of $F_{i,j}$ in H to be $\bigcup_{v\in T(j)} Q_v$. Using the definition of our randomized rounding process, a moment's reflection shows that H is indeed a valid RBDG w.r.t. (S_1, S_2). We will make two observations that will facilitate application of Theorem 3.8.

(a) Let $\delta_i = \nu/\mu_i$. We have

$$G(\mu_i, \delta_i) \le G(y^*, \nu/y^*) = G(y^*, L(y^*, 1/(et))) \le 1/(e\nu' t),$$

where the first inequality follows from part (b) of Proposition 3.1 (since $\mu_i \le y^*$), and the last inequality follows from Proposition 3.2. Combining this with Theorem 3.7, we get

PROPOSITION 3.3. *For all $i \in [m]$ and for all nonempty events Z, $\Pr[E_i \mid Z] \le \sum_{j\in[u_i]} \mathbf{E}[F_{i,j} \mid Z]$. Also, $p_i \doteq \sum_{j\in[u_i]} \mathbf{E}[F_{i,j}] \le G(\mu_i, \delta_i) \le 1/(e\nu' t)$.*

(b) In order to use Theorem 3.8, we now bound the degree of $F_{i,j}$ in H. Since $F_{i,j}$ involves ν' terms, the definition of t shows that the degree of $F_{i,j}$ in H is at most $\nu' t$. Let $\sim$ denote the edge relation of H. Since E_i is a neighbor of $F_{i,j}$ in H, we see that $|\{\ell \neq i : E_\ell \sim F_{i,j}\}| \leq \nu' t - 1$. Thus we have

PROPOSITION 3.4. *$D \doteq \max_{i,j}\ |\{\ell \neq i : E_\ell \sim F_{i,j}\}|$ is at most $\nu' t - 1$. Also, for all i and j, $F_{i,j}$ takes on values in $[0,1]$.*

From Propositions 3.3 and 3.4 and by noting that $ep_i(D+1) \leq e(\nu' t - 1 + 1)/(e\nu' t) \leq 1$, we invoke Theorem 3.8 to conclude that $\Pr[\bigwedge_{i\in[m]} \overline{E_i}] > 0$. ■

As shown in [**Sr1**], an idea suggested by É. Tardos can help bootstrap Theorem 3.9 to give

THEOREM 3.10. ([**Sr1**]) *For MIPs, there is an integral solution of value at most $y^* + \mathrm{O}(y^* L(y^*, 1/d)) + \mathrm{O}(1)$.*

PROOF SKETCH. Recall how our observation on basic feasible solutions helped us assume w.l.o.g. that $\max_i \ell_i \leq m$: variables $x_{i,j}$ that are already 0 or 1 do not participate in the rounding. Thus, rounding a large number of variables $x_{i,j}$ appropriately to 0 and 1 effectively cuts down the value of $\max_i \ell_i$. Our idea now is to do such a rounding that eventually effectively ensures that $\max_i \ell_i \leq \mathrm{O}(\mathrm{poly}(d))$. By (3.9), this would imply that $t \leq \mathrm{O}(\mathrm{poly}(d))$, at which point we can employ Theorem 3.9.

Our rounding process that cuts down the value of t will involve a series of usages of Theorem 3.9; that is, we will bootstrap Theorem 3.9. We first show how we may assume certain bounds on y^*, t and d w.l.o.g. If

$$(y^* \geq t^{1/7}) \text{ or } (t \leq d^4) \tag{3.11}$$

is true, then we will be done, by Theorem 3.9. So suppose (3.11) is false. Next if $y^* \leq t^{-1/7}$, one can check that $y^* L(y^*, 1/(et)) = \mathrm{O}(1)$; thus, if $y^* \leq t^{-1/7}$, Theorem 3.9 guarantees a solution of value $\mathrm{O}(1)$. So, we can also assume that $y^* > t^{-1/7}$.

The idea, as mentioned above, is to round many $x^*_{i,j}$ to zero for each i without blowing up y^* significantly, so that $\max_i\ \ell_i$ will essentially get reduced. This is iterated until (3.11) holds or until t becomes smaller than a certain absolute constant, whichever happens first.

We consider two cases, depending on the value of y^*. Suppose $y^* \geq 1$. For each i,j, we set $x'_{i,j} := (y^*)^2(\log^5 t)x^*_{i,j}$. We now conduct randomized rounding on the $x'_{i,j}$ in a style similar to our *second* randomized rounding approach of Section 2.4. *Independently* for each (i,j), we will round $x'_{i,j}$ to a random $z_{i,j} \in \{\lfloor x'_{i,j}\rfloor, \lceil x'_{i,j}\rceil\}$ as follows: $\Pr[z_{i,j} = \lceil x'_{i,j}\rceil] = x'_{i,j} - \lfloor x'_{i,j}\rfloor$, and $\Pr[z_{i,j} = \lfloor x'_{i,j}\rfloor] = 1 - \Pr[z_{i,j} = \lceil x'_{i,j}\rceil]$. (If $x'_{i,j}$ is an integer, then $z_{i,j} = x'_{i,j}$ deterministically.) That is, we interpret the fractional part of $x'_{i,j}$ as a probability.

Note that $\mathbf{E}[z_{i,j}] = x'_{i,j}$. Appropriate use of the proof technique of Theorem 3.2 show that with nonzero probability, we have:

(i) $\forall i \in [m]$, $(Az)_i \leq (y^*)^3 \log^5 t(1 + \mathrm{O}(1/((y^*)^{1.5}\log^2 t)))$, and
(ii) $\forall i \in [n]$, $|(\sum_j z_{i,j}) - (y^*)^2 \log^5 t| \leq \mathrm{O}(y^* \log^3 t)$.

Suppose we have such a rounding z. For each $i \in [n]$ and $j \in [\ell_i]$, set $x''_{i,j} := z_{i,j}/\sum_u z_{i,u}$. We can deduce the following two crucial properties from (i) and (ii).

(a). For all $i \in [m]$,

$$(Ax'')_i \leq \frac{y^*(1 + O(1/((y^*)^{1.5} \log^2 t)))}{1 - O(1/(y^* \log^2 t))} = y^*(1 + O(1/(y^* \log^2 t))).$$

In other words, we have lost just a little in the value of y^*.

(b). For any fixed $i \in [n]$, since the $z_{i,j}$ are non-negative integers summing to at most $(y^*)^2 \log^5 t(1 + O(1/(y^* \log^2 t)))$, at most $O((y^*)^2 \log^5 t)$ of the $x''_{i,j}$ are nonzero. Thus, by losing a little in y^*, we now have a vector x'' with a much-reduced value $O((y^*)^2 \log^5 t)$ for $\max_{i \in [n]} \ell_i$. So (3.9) shows that t is now at most $O(d(y^*)^2 \log^5 t) = O(t^{1/4+2/7} \log^5 t)$, since (3.11) was assumed false.

Repeating this $O(\log \log t)$ times makes t small enough to satisfy (3.11), or for t to become smaller than a certain absolute constant.

Finally, the ideas are similar for the case where $t^{-1/7} < y^* < 1$. Recall that in the above-seen case where $y^* \geq 1$, we started by scaling up all the $x^*_{i,j}$ by $(y^*)^2(\log^5 t)$. Now, we scale up all the $x^*_{i,j}$ by $(\log^5 t)/y^*$ and then do a "randomized rounding followed by scaling down" appropriately as above. The proof proceeds similarly as for the case where $y^* \geq 1$: we can show here that with a slight blow-up in the value of y^*, the value of $\max_i \ell_i$ can be brought down to $O((\log^5 t)/y^*) = O(t^{1/7} \log^5 t)$. ■

Thus, column-sparsity in MIPs, a generalization of the notion of multicommodity flow along short paths, is seen to be fruitful from the viewpoint of approximation.

4. Applications of high-bandwidth routing

Part (ii) of Corollary 3.1 shows that the optimal integral and fractional congestions are not far away when the fractional congestion y^* is "high" (at least logarithmic in the size of the network G). The parameter y^* being high can be viewed as the network offering reasonably large *bandwidth*—allowing a fairly large number of paths to use any edge. In this short section, we will sketch a few applications of part (ii) of Corollary 3.1.

We start by presenting the classical edge-disjoint paths problem that is closely related to low-congestion routing. One major routing approach is *circuit switching*, where a connection path from source to destination is established for an appropriate duration of time, for each routing request that is admitted. Given a graph $G = (V, E)$, let $n = |V|$ and $m = |E|$. Suppose we are given G and a (multi-)set $T = \{(s_i, t_i) : 1 \leq i \leq k\}$ of pairs of vertices of G. $T' \subseteq T$ is said to be *realizable* iff the pairs of vertices in T' can be connected in G by mutually edge-disjoint paths. The classical maximum disjoint paths problem (henceforth MDP) is to find a realizable T' of maximum cardinality. This is probably the most basic circuit-switching problem, and is one of the early combinatorial optimization problems to be shown NP-hard.

Another problem that is closely related to low-congestion routing and the MDP that is motivated, e.g., by minimizing the number of wavelengths in wavelength-division multiplexing, is the following problem of *routing in rounds* or *path coloring*. In the definition of the MDP, what if we do need to connect up each s_i to t_i? Since the "disjointness" constraint may preclude this, a natural question is: what is the minimum number of *rounds* in which T can be routed, if the routing should be accomplished in an edge-disjoint manner in each round? In other words, we want the minimum number of realizable sets into which T can be partitioned. This NP-hard problem is that of routing in rounds. We will now briefly see some good approximation algorithms for the MDP and routing in rounds, if G has good edge-expansion, or if G belongs to a certain family of "densely embedded" networks that includes the two-dimensional mesh.

4.1. Expander graphs. For the purposes of Section 4.1, we will refer to any family of graphs that have maximum degree Δ and edge-expansion β upper- and lower-bounded respectively by some absolute constants, as *expanders*. For instance, it is known that for any fixed $\Delta \geq 3$, a natural model of random Δ-regular multigraphs leads to expanders with high probability. As mentioned in Section 3.2, expanding graphs are of basic importance to routing.

Seminal work of Leighton & Rao initiated work on approximate max-flow min-cut theorems for multicommodity flow [**LR1**]; see [**Sh2**] for a survey of extensions and applications of this. Let G be any n-node expander. We will call a collection $T = \{(s_i, t_i) : i \in [k]\}$ of routing requests in G a *partial permutation*, if all the s_i are distinct and all the t_i are distinct. One of the major consequences of the work of [**LR1**] is that for any partial permutation T, there is a unit flow from each s_i to t_i that has fractional congestion and dilation bounded by $O(\log n)$. (The two constants hidden in the $O(\cdot)$ notation for the congestion and dilation depend on $\Delta(G)$ and $\beta(G)$, but not on n.) Given the existence of such a flow, it is not hard to see that it can be computed in polynomial time. By part (ii) of Corollary 3.1, we also see that we can efficiently connect each s_i to t_i by one path in such a way that the (integral) congestion and dilation are both bounded by $O(\log n)$. This has seen many applications.

Here is an application, due to Aumann & Rabani [**AR**]. Suppose T is a partial permutation, and that we want to do a "routing in rounds" efficiently for it on G. First construct the above-seen set of (s_i, t_i)-paths that have congestion and dilation $O(\log n)$. Define an undirected graph with these paths as vertices, with two vertices adjacent iff the corresponding paths intersect in at least one edge. It is easy to check that this graph has maximum degree $\Delta' = O((\log n) \cdot (\log n)) = O(\log^2 n)$; hence it can be colored using $\Delta' + 1 = O(\log^2 n)$ colors. That is, we can do a feasible "routing in rounds" of T in at most $O(\log^2 n)$ rounds.

Next, a useful notion of "strong expansion" has been defined by Broder, Frieze & Upfal [**BFU1**], generalizing that of expansion. Informally, if $S \subseteq V$ is "small", we want it to "expand" very well; see [**BFU1**] for details. We just note here the fact from [**BFU1**] that random regular graphs (with high probability) as well as explicit families of expanders such as those of Lubotzky, Phillips & Sarnak [**LST**] are strong expanders. By using the rapid-mixing property of random walks on expanders, it has been shown by Broder, Frieze & Upfal that among other things, any constant-degree strong expander G can route any partial permutation T in an edge-disjoint manner if $|T| \leq c_0 n / \log^2 n$, for some constant $c_0 > 0$ [**BFU1, BFU2**]. By using the above-seen work of [**LR1**] in conjunction with a more general version of Theorem 3.4, this fact has been generalized in [**LRS**]: if $|T| \leq c_0' n / \log^{1+1/B} n$ for some positive integer B, then T can be routed with integral congestion at most B.

The above-seen path-selection result of [**LR1**] has also led to an *on-line* MDP algorithm that has a polylogarithmic competitive ratio on expanders [**KR**].

4.2. A class of networks generalizing the mesh. Kleinberg & Tardos have presented constant-factor approximation algorithms for the MDP, for a family of *densely embedded, nearly-Eulerian graphs* that generalize the two-dimensional mesh [**KT**]. We give a very brief sketch of their approach now. Suppose the network G is the $N \times N$ mesh. For some constant $c_1 > 1$, a maximal set of $c_1 \log N \times c_1 \log N$–sized subsquares of G is chosen, such that each chosen pair is separated by at least $2c_1 \log N$. Construct a certain graph with these subsquares as the vertices, with each edge replicated $\Omega(\log N)$ times (i.e., the "capacity" of the edges is $\Omega(\log N)$). By the same argument as for Corollary 3.1, one can show that the integral and fractional optima for a certain routing problem on this high-capacity network are close to each other. Starting with the path-selection output by an integral routing in this network, a good path-selection for the original problem is accomplished via a number of additional ideas in [**KT**].

Next, it is not hard to check that a λ-approximation for the MDP translates to an $O(\lambda \log n)$-approximation for routing in rounds. Thus, in particular, the above result immediately implies an $O(\log n)$-approximation for routing in rounds on, e.g., the mesh. Using some ideas from above (such as working with an appropriate high-capacity simulated network) and also making involved use of the LLL, better bounds have been achieved, e.g., for the mesh, by Rabani [**Rab**]. In [**Rab**], it is shown that on the mesh, the optimal number of rounds can be efficiently estimated to within a constant factor, and that the actual routing can be constructed efficiently in a certain number of rounds that is to within a poly$(\log \log n)$ factor of optimal.

5. Correlation inequalities and the group Steiner tree problem

The main problem considered here will be the *Group Steiner Tree* problem, an NP-hard problem in network design. We will sketch a recent approximation algorithm for this problem, due to Garg, Konjevod & Ravi [**GKR**]. To analyze the algorithm, certain correlation inequalities will be useful, which we describe in Section 5.1. These inequalities are powerful and are of independent interest, and will also be of use in Section 6.3. The group Steiner tree problem and the algorithm of [**GKR**] are discussed in Section 5.2.

5.1. A family of correlation inequalities. Our study of the LLL should be convincing evidence that it is helpful to study correlations. However, even in describing the LLL, much of our focus was on a coarse notion of correlation: classifying families of events as independent or dependent. It is often of much use to make this notion finer, when the events are dependent: to classify them, say, as "positively" or "negatively" correlated (we shall make this formal below). Even finer study is certainly useful, which we shall not undertake much here. We now start with some useful correlation inequalities, working up to a family of inequalities that are presented in Section 8.1 of Alon & Spencer [**AS**].

To motivate some of the main correlation issues that will arise presently, suppose we wish to estimate a probability such as

$$\Pr[A \mid (\bigwedge_{i \in [t]} \overline{B_i}) \wedge C], \tag{5.1}$$

where A and C are *independent.* If we desire an *upper-bound* on this probability, recall that the approach of the LLL is as follows:

$$\begin{aligned}
\Pr[A \mid (\bigwedge_{i\in[t]} \overline{B_i}) \wedge C] &= \frac{\Pr[A \wedge (\bigwedge_{i\in[t]} \overline{B_i}) \mid C]}{\Pr[(\bigwedge_{i\in[t]} \overline{B_i}) \mid C]} \\
&\leq \frac{\Pr[A \mid C]}{\Pr[(\bigwedge_{i\in[t]} \overline{B_i}) \mid C]} \\
&= \frac{\Pr[A]}{\Pr[(\bigwedge_{i\in[t]} \overline{B_i}) \mid C]} \quad (A \text{ and } C \text{ are independent}) \\
&= \frac{\Pr[A]}{\prod_{i\in[t]} \Pr[\overline{B_i} \mid (\bigwedge_{j\in[i-1]} \overline{B_j}) \wedge C]}.
\end{aligned}$$

Thus, in particular, if we have a good lower-bound on $\Pr[\overline{B_i} \mid (\bigwedge_{j\in[i-1]} \overline{B_j}) \wedge C]$ for each i, we can get an upper-bound for (5.1).

However, what if we want a *lower-bound* on (5.1)? We present here an idea of Boppana & Spencer [**BoS**] for this, whose work presents a simple proof of a very useful result from the powerful work of Janson, Łuczak & Ruciński [**JLR**]. One of the basic inequalities used here will be $\Pr[X \mid (Y \wedge Z)] \geq \Pr[(X \wedge Y) \mid Z]$, which can be easily checked using Bayes' theorem. Thus we have

$$\begin{aligned}
\Pr[A \mid (\bigwedge_{j\in[t]} \overline{B_j}) \wedge C] &\geq \Pr[(A \wedge (\bigwedge_{j\in[t]} \overline{B_j})) \mid C] \\
&\geq \Pr[A \mid C] - \sum_{j\in[t]} \Pr[(A \wedge B_j) \mid C] \quad \text{(inclusion-exclusion)} \\
&= \Pr[A] - \sum_{j\in[t]} \Pr[(A \wedge B_j) \mid C]. \qquad (5.2)
\end{aligned}$$

Let us now present a general context in which probabilities such as (5.1) need to be lower-bounded. To do so, we need a few definitions. Suppose $X_1, X_2, \ldots, X_\ell$ are *independent* r.v.s, each taking values in $\{0,1\}$. Let $\vec{X}$ denote $(X_1, X_2, \ldots, X_\ell)$. A special case of the powerful FKG inequality [**FKG**] can be summarized as follows. Suppose U is an event that depends only on $\vec{X}$. Define U to be *increasing* iff for all $\vec{a} = (a_1, a_2, \ldots, a_\ell) \in \{0,1\}^\ell$ such that U holds when $\vec{X} = \vec{a}$, U also holds when $\vec{X} = \vec{b}$ for any $\vec{b} = (b_1, b_2, \ldots, b_\ell) \in \{0,1\}^\ell$ such that $a_i \leq b_i$ for all i. (For instance, "$X_2 + 0.7X_3 + 0.4X_6 + X_9 \geq 2$" is an increasing event.) Define U to be *decreasing* iff $\overline{U}$ is increasing. The FKG inequality proves a class of intuitively appealing results, such as that any set of increasing events are *positively correlated* with each other.

THEOREM 5.1. **(FKG inequality)** *Suppose $\vec{X}$ is a vector of finitely many* independent *binary r.v.s. Let $I_1, I_2, \ldots, I_t$ be any sequence of* increasing *events and $D_1, D_2, \ldots, D_t$ be any sequence of* decreasing *events (each I_i and D_i completely determined by $\vec{X}$). Then for any $i \in [t]$ and any $S \subseteq [t]$,*

(i) $\Pr[I_i \mid \bigwedge_{j\in S} I_j] \geq \Pr[I_i]$ *and* $\Pr[D_i \mid \bigwedge_{j\in S} D_j] \geq \Pr[D_i]$*;*
(ii) $\Pr[I_i \mid \bigwedge_{j\in S} D_j] \leq \Pr[I_i]$ *and* $\Pr[D_i \mid \bigwedge_{j\in S} I_j] \leq \Pr[D_i]$.

Letting $\vec{X}$ be a vector of r.v.s as above, let us say that a collection of events (each being completely determined by $\vec{X}$) are of the *same type* if they are all increasing or if they are all decreasing. Suppose $E_1, E_2, \ldots, E_m$ is any collection of events of the same type. Then, part (i) of the FKG inequality immediately yields

$$\Pr[\bigwedge_{i\in[m]} \overline{E_i}] \geq \prod_{i\in[m]} \Pr[\overline{E_i}].$$

But what if we want an *upper bound* on the left-hand-side of this expression? (For instance, suppose each E_i is a "good" event, and that we wish to lower-bound the probability of at least one good event occurring.) We now show an approach of [**BoS**] and Spencer (see the Bonus Lecture in the second edition of [**Sp**]) which handles this issue, based on (5.2).

THEOREM 5.2. *Suppose $\vec{X}$ is a vector of finitely many independent binary r.v.s. Let $E_1, E_2, \ldots, E_m$ be any events of the* same type, *each E_i fully determined by $\vec{X}$. Suppose $\sim$ is any relation defining a* dependency graph *on the E_i (see Section 2). Then:*

(i) $\forall S \subseteq [m]\ \forall i \notin S$, $\Pr[\overline{E_i} \mid (\bigwedge_{j\in S} \overline{E_j})] \leq \Pr[\overline{E_i}] + \sum_{j\in S:\ i\sim j} \Pr[E_i \wedge E_j]$.

(ii) *Let $\mu = \sum_i \Pr[E_i]$ and $\Delta = \sum_{(i,j):\ i\sim j} \Pr[E_i \wedge E_j]$. (Each pair i, j such that $i \sim j$ is counted twice in this sum.) Then,* $\Pr[\bigwedge_{i\in[m]} \overline{E_i}] \leq \exp(-\mu + \Delta/2)$. *Also, if $\mu \leq \Delta$,* $\Pr[\bigwedge_{i\in[m]} \overline{E_i}] \leq \exp(-\mu^2/(2\Delta))$.

PROOF. (i) Let $S_0 = \{j \in S : i \sim j\}$, and $S_1 = S - S_0$. Let $t = |S_0|$. In the notation of (5.1), let A denote E_i, $\overline{B_1 B_2} \cdots \overline{B_t}$ denote the event $\wedge_{j\in S_0} \overline{E_j}$, and C denote $\wedge_{j\in S_1} \overline{E_j}$. By the definition of $\sim$ and S_1, A is independent of C. Thus, by (5.2),

$$\begin{aligned}\Pr[A \mid (B \wedge C)] &\geq \Pr[E_i] - \sum_{j\in S:\ i\sim j} \Pr[(E_i \wedge E_j) \mid \bigwedge_{j\in S_1} \overline{E_j}] \\ &\geq \Pr[E_i] - \sum_{j\in S:\ i\sim j} \Pr[E_i \wedge E_j];\end{aligned}$$

the last inequality follows from part (ii) of the FKG inequality, since the event "$E_i \wedge E_j$" is of the type opposite to that of C.

(ii) Direct application of Bayes' theorem to part (i) yields

$$\begin{aligned}\Pr[\bigwedge_{i\in[m]} \overline{E_i}] &\leq \prod_{i\in[m]} (1 - \Pr[E_i] + \sum_{j<i:i\sim j} \Pr[E_i \wedge E_j]) \\ &\leq \exp((-\sum_i \Pr[E_i]) + \sum_{(i,j):\ j<i, i\sim j} \Pr[E_i \wedge E_j]) \\ &= \exp(-\mu + \Delta/2).\end{aligned}$$

However, this is not a good bound if Δ is "large" in comparison with μ. If $\Delta \geq \mu$, a cute application of the probabilistic method (to choose a *random* set S in order to use the result of (i)) shown in the second edition of [**Sp**], yields the upper bound $\exp(-\mu^2/(2\Delta))$ on $\Pr[\bigwedge_{i\in[m]} \overline{E_i}]$. ■

We are now ready to consider the application of such ideas to the group Steiner tree problem [**GKR**].

5.2. The group Steiner tree problem. This network design problem is a generalization of the well-known NP-hard Steiner tree problem, motivated by a family of VLSI design problems (Reich and Widmayer [**RW**]). Given: (i) an undirected graph $G = (V, E)$, (ii) a family of subsets $S_1, S_2, \ldots, S_k$ of V, and (iii) a non-negative cost c_f for each edge $f \in E$, the problem is to construct a minimum-cost subtree of G that spans at least one vertex from each S_i. This problem also generalizes the classical set-cover problem, since the latter easily reduces to instances of the group Steiner tree problem where, in fact, G is a star-graph. See [**RW, GKR**] for these and related points. We now sketch the approximation algorithm due to [**GKR**] for this problem. Once again, the key idea will be to start with an appropriate multicommodity flow relaxation and use randomization to round it well; Theorem 5.2 will be crucial for analyzing the algorithm.

Let $n = |V|$ and $N \doteq \max_i |S_i|$. Also recall that there are k "groups" S_i. The approximation bound will be $O(\log^2 n \log N \log k) = O(\log^3 n \log k)$. First of all, the breakthrough work of Bartal [**Ba1**] can be used appropriately to reduce the problem to the case where G is a tree, with an $O(\log^2 n)$ factor loss in the approximation bound. Thus, we will now focus on the case where G is a tree, and present an $O(\log N \log k)$-approximation algorithm. In fact, we can make a further specialization: considering the *rooted* version where we are required to include a particular vertex r in the constructed tree. Then, by searching over all possible choices for r, we can choose the best solution.

Consider the rooted version with root r. We need to construct a min-cost tree T contained in the rooted tree G, such that r is included and at least one vertex of each S_i is spanned. We can assume w.l.o.g. that all nodes in $\bigcup_i S_i$ are leaves in G: if some $u \in S_i$ is an internal node, we construct a dummy node u' and:

- make u' a child of the parent (if any) of u in G;
- make all children of u the children of u', and
- make u a leaf whose parent is u', and connect u and u' with a zero-cost edge.

We can check that this results in an equivalent problem.

It is not hard to write the given problem as an integer linear program, with an indicator random variable $x(f)$ for including edge f in the tree T; the objective function is to minimize $\sum_f c_f x(f)$. There is now a natural multicommodity flow relaxation for this problem: sending flow from the root r, such that at least one unit of flow reaches each S_i. More precisely, we allow each $x(f)$ to lie in $[0, 1]$; interpreting $x(f)$ as the capacity of edge f, the relaxation is to choose these capacities such that for each individual S_i, the capacities can support one unit of flow from r to S_i. (It is **not** required that the capacities be able to support a unit flow *simultaneously* for all the S_i; for any *given* S_i, it suffices for the capacities to support unit flow from r to S_i.) We can check that this is a valid (flow-based) relaxation of the problem: indeed, if all the $x(f)$ lay in $\{0, 1\}$, we would get a tree in G that includes r and includes at least one vertex from each S_i.

Let $\{x^*(f) : f \in E\}$ denote an optimal solution to the above multicommodity flow relaxation, with $y^* = \sum_f c_f x^*(f)$. We now exhibit a random process to round the $x^*(f)$ to $z(f) \in \{0, 1\}$, and show that with reasonable probability, it leads to a feasible solution of objective function value $\sum_f c_f z(f) = O(y^* \log N \log k)$. The random process is as follows: each edge f incident with the root r is chosen with probability $x^*(f)$. Every other edge f is chosen with probability $x^*(f)/x^*(g)$, where g is the "parent edge" of f. (It is easy to check that $x^*(f) \leq x^*(g)$ holds w.l.o.g.) All these choices are made independently. Finally, we choose the connected component of the chosen edges that includes r, as our tree T. What is the intuition for such a randomized rounding scheme? It can be verified via a telescoping product that the probability of edge f appearing in T is $x^*(f)$. This is in keeping with the general randomized rounding notion of "interpreting values in $[0,1]$ as probabilities", and hence the scheme seems promising. By linearity of expectation, the expected cost of T is y^*. We now move on to the more interesting problem of lower-bounding the probability that T has good "spanning" properties.

Fix a group $S_i = \{u_1, u_2, \ldots, u_t\}$; recall that each u_j is a leaf of G and that $t \leq N$. Let f_j denote the unique edge of G incident with u_j, and E_j be the "good" event that the path from r to u_j is chosen by our random process. Note that each E_j is an *increasing* event as a function of the underlying random choices; we now proceed to use Theorem 5.2 in order to upper-bound $\Pr[\bigwedge_{j\in[t]} \overline{E_j}]$.

Let $p_j = \Pr[E_j] = x^*(f_j)$. Since the vector x^* can support unit flow from r to S_i, we have

$$\mu \doteq \sum_{j\in[t]} p_j \geq 1. \tag{5.3}$$

In the notation of Theorem 5.2, we now need to estimate (upper-bound) Δ. First define $E_j \sim E_\ell$ (and, by abuse of notation, $j \sim \ell$) for $j \neq \ell$ iff the least common ancestor of u_j and u_ℓ in G is different from r. It is readily checked that this leads to a valid dependency graph. In order to estimate

$$\Delta = \sum_{(j,\ell):\ j\neq\ell, j\sim\ell} \Pr[E_j \wedge E_\ell],$$

fix any distinct j, ℓ such that $j \sim \ell$. Letting $\mathrm{lca}(j,\ell)$ denote the least-common ancestor of u_j and u_ℓ in G and $\mathrm{lce}(j,\ell)$ be the edge in G connecting $\mathrm{lca}(j,\ell)$ to its parent, we can verify that

$$\Pr[E_j \wedge E_\ell] = x^*(f_j)x^*(f_\ell)/x^*(\mathrm{lce}(j,\ell)).$$

Thus,

$$\Delta = \sum_{j\in t} x^*(f_j) \sum_{\ell: j\sim\ell} \frac{x^*(f_\ell)}{x^*(\mathrm{lce}(j,\ell))}. \tag{5.4}$$

Just for the purposes of analysis, it is shown in [**GKR**] that by a small constant-factor loss in various parameters, we can contract certain edges of G and make small changes to the vector x^* in such a way that:

(P1) the modified tree has depth $d = \mathrm{O}(\log N)$, and
(P2) the right-hand-side of (5.4) is $\mathrm{O}(\mu d)$.

Thus, $\Delta = \mathrm{O}(\mu \log N)$. This bound, combined with (5.3) and part (ii) of Theorem 5.2, shows that

$$\Pr[\bigwedge_{j\in[t]} \overline{E_j}] \leq \exp(-\Omega(1/\log N)) = 1 - \Omega(1/\log N).$$

So, suppose we repeat our basic random process independently $c_2(\log N)\ln(4k)$ times (for a suitably large absolute constant c_2), the probability of never spanning any vertex of S_i is at most $\exp(-\ln(4k)) = 1/(4k)$. Hence, the probability of there existing an uncovered group after this many iterations is at most $k/(4k) = 1/4$. Thus, if we overlay all these trees and remove any cycles, we will span at least one vertex of each group with probability at least $3/4$. Furthermore, as observed above, the expected cost of each tree is y^*. So the expected cost of the final "overlaid" tree is at most $O(y^*(\log N)\log(4k))$; the probability of this cost being more than twice its expectation is at most $1/2$ by Markov's inequality. Thus, with probability at least $1-(1/4+1/2)=1/4$, we finally have a feasible solution (tree) of cost $O(y^*(\log N)\log(4k))$; as usual, the success probability can be boosted by repetition.

Thus we see how the multicommodity flow relaxation, along with correlation inequalities that tackle "negative" correlations, can be a useful tool for network design problems such as the group Steiner tree problem. Furthermore, (P1) and (P2) demonstrate yet another situation where keeping all flow-paths "short" is possible and beneficial.

Bartal's recent work improves the above approximation ratio to $O((\log n)\cdot(\log N)\cdot(\log k)\cdot\log\log n)$ [**Ba2**]. Also, recent work of Charikar, Chekuri, Goel & Guha leads to good *deterministic* approximation algorithms for the group Steiner tree and other problems [**CC+**]. The reader is referred to [**GW**] for a survey of some powerful approximation methods in network design.

6. Unsplittable and splittable flow

We consider a class of problems that generalize the disjoint-paths problem, in Sections 6.1 and 6.2. In Section 6.3, we study a different class of routing problems motivated by routing power in electric power networks. All the problems considered are NP-hard, and the combination of multicommodity flow and randomization will be seen to be quite fruitful once again, in Sections 6.1 and 6.3. For completeness, Section 6.2 describes an interesting *deterministic* method of rounding multi-commodity flow, for a class of low-congestion routing problems.

6.1. Approximation algorithms for unsplittable flow and related problems. Recall the MDP and related notation introduced in Section 4. A natural generalization of the MDP is considered in [**Kl1**]. We allow each request pair (s_i,t_i) to have a *demand* $\rho_i>0$, and each edge f of G to have a *capacity* $c_f>0$; we also assume that

$$\forall i\ \forall f,\ \rho_i\le c_f. \tag{6.1}$$

$T'\subseteq T$ is now called realizable iff each pair of vertices in T' can be connected by **one path** in G, such that the total demand using any edge f does not exceed c_f. The *unsplittable flow problem* (UFP) is to find a realizable $T'\subseteq T$ that maximizes $\sum_{i:(s_i,t_i)\in T'}\rho_i$ [**Kl1**]. Thus, the MDP is a special case of the UFP.

A fair amount of progress has been made recently on approximating the MDP, UFP, and related problems: see, e.g., [**KT, Kl1, KS1, Sr2, KS2, BaS**] and the references therein. We will in fact work with the general *weighted* case of the UFP, where each (s_i,t_i) has a weight $w_i>0$, with the objective being to find a realizable $T'\subseteq T$ that maximizes $\sum_{i:(s_i,t_i)\in T'}w_i$. (We assume w.l.o.g. that $\max_i w_i=1$.)

Let OPT denote the optimal value for a given UFP instance. An efficient algorithm to construct a feasible path selection of value $\Omega(\max\{OPT^2/m, OPT/\sqrt{m}\})$ is shown in the work of Baveja and this author [**BaS**]. We now sketch another approximation result, Theorem 6.2, which has been developed independently using different methods by Kolliopoulos & Stein [**KS2**] and by [**BaS**]. We will specialize some of the discussion to the MDP following [**Sr2**]; we will then sketch how this is generalized to the UFP in [**BaS**].

There is a natural ILP formulation for a given UFP instance. Define variables $\{x_i \in \{0,1\} : i \in [k]\}$, where x_i will be 1 iff we connect s_i to t_i by a path. Also define variables $\{y_{i,f} \in \{0,1\} : i \in [k],\ f \in E\}$, where $y_{i,f}$ is 1 iff the path chosen (if any) for (s_i, t_i), passes through f. For each $f \in E$, we have a capacity constraint $\sum_i \rho_i y_{i,f} \leq c_f$. Linear equalities relating the x_i's and $y_{i,f}$'s model, in an obvious way, the understanding that x_i is the indicator for choosing an (s_i, t_i)-path. Subject to the above constraints, the objective is to maximize $\sum_i w_i x_i$. Relaxing each x_i and $y_{i,f}$ to be a real lying in $[0,1]$, we get a natural LP (multicommodity flow) relaxation.

Let $\{x_i^*, y_{i,f}^* \in [0,1] : i \in [k], f \in E\}$ denote some feasible (not necessarily optimal) solution to the above relaxation, with objective function value $y^* = \sum_i w_i x_i^*$. As described in Section 2.4, use flow-decomposition to transform x^* into a set $\{P_{i,1}, P_{i,2}, \ldots, P_{i,\ell_i}\}$ of (s_i, t_i)-paths for each i, where $\ell_i \leq m$. Each path $P_{i,j}$ carries a non-negative amount $\rho_i z_{i,j}^*$ of flow from s_i to t_i, where $x_i^* = \sum_j z_{i,j}^*$. We have by the capacity constraints that

$$\forall f \in E, \quad \sum_{(i,j):\ f \in P_{i,j}} \rho_i z_{i,j}^* \leq c_f. \tag{6.2}$$

The rounding process employed is the second randomized rounding scheme shown in Section 2.4. For a parameter $\gamma > 1$, construct, independently for all i, j, a $\{0,1\}$ r.v. $z_{i,j}$ with $\Pr[z_{i,j} = 1] = z_{i,j}^*/\gamma$. Then, for each i: (i) if at least one of the $z_{i,j}$'s is 1, we shall arbitrarily choose one of them, say $z_{i,k}$, and connect s_i to t_i by path $P_{i,k}$; (ii) if all the $z_{i,j}$'s are 0, we do not connect s_i to t_i. Let d denote the dilation of the paths $P_{i,j}$. Our main approximation algorithm here, Theorem 6.2, is parametrized by d; some reasons for d being an interesting parameter are sketched following the proof of Theorem 6.2. The work described from now on is basically from [**Sr2**]; we will finally very briefly sketch how it is generalized in [**BaS**]. Also, as mentioned above, a different proof of Theorem 6.2 has been obtained in [**KS2**].

For any $f \in E$, the total demand requested of it is at most

$$D_f \doteq \sum_{(i,j):\ f \in P_{i,j}} \rho_i z_{i,j}. \tag{6.3}$$

We have by (6.2) that

$$\mathbf{E}[D_f] = \sum_{(i,j):\ f \in P_{i,j}} \rho_i z_{i,j}^*/\gamma \leq c_f/\gamma. \tag{6.4}$$

For any $f \in E$, let E_f denote the bad event that $D_f > c_f$. We will show that for a certain $\gamma = \Theta(d)$, with positive probability we can avoid all the bad events **and** have objective function value $\Omega(y^*/d)$. Constructive versions of this are shown in [**Sr2, BaS**].

We now work toward a proof of Theorem 6.2; one of the key points will be the handling of "negative" correlations. In this sense, the setting is similar to that of Section 5.2. However, the type of negative correlations involved will be different; a useful lemma to handle these correlations now is Lemma 6.1. This lemma is implicit in the work of [**Sr2**] and is related to (5.2).

LEMMA 6.1. *Given an underlying collection of independent* $\{0,1\}$ *r.v.s, let* $S_1 = \{U_1, \ldots, U_m\}$ *and* $S_2 = \{y_1, y_2, \ldots, y_M\}$ *be sets of events,* all U_i and y_j

of the same type. *Let* $\sim$ *be a relation defining any particular bipartite dependency graph w.r.t.* (S_1, S_2). *Then,* $\Pr[(\exists j : y_j) \mid \bigwedge_{r\in[m]} \overline{U_r}]$ *is at least*

$$\Big(\sum_{j\in[M]} (\Pr[y_j] - \sum_{s:U_s\sim y_j} \Pr[(y_j = 1) \wedge U_s])\Big) - \sum_{1\le j<j'\le M} \Pr[y_j \wedge y_{j'}].$$

PROOF. By inclusion-exclusion,

(6.5)
$$\Pr[\exists j : y_j \mid \bigwedge_{r\in[m]} \overline{U_r}] \ge \Big(\sum_{j\in[M]} \Pr[y_j \mid \bigwedge_{r\in[m]} \overline{U_r}]\Big) - \sum_{1\le j<j'\le M} \Pr[y_j \wedge y_{j'} \mid \bigwedge_{r\in[m]} \overline{U_r}].$$

Fix j. By (5.2) and by the definition of a bipartite dependency graph,

$$\begin{aligned}\Pr[y_j \mid \bigwedge_{r\in[m]} \overline{U_r}] &\ge \Pr[y_j] - \sum_{s:U_s\sim y_j} \Pr[((y_j = 1) \wedge U_s) \mid \bigwedge_{r:U_r\not\sim y_j} \overline{U_r}] \\ &\ge \Pr[y_j] - \sum_{s:U_s\sim y_j} \Pr[(y_j = 1) \wedge U_s] \qquad (6.6)\end{aligned}$$

by part (ii) of the FKG inequality, since the event "$(y_j = 1) \wedge U_s$" is of a type opposite to that of "$\bigwedge_{r:U_r\not\sim y_j} \overline{U_r}$".

Once again by part (ii) of the FKG inequality, we have for any (j, j') that

$$\Pr[y_j \wedge y_{j'} \mid \bigwedge_{r\in[m]} \overline{U_r}] \le \Pr[y_j \wedge y_{j'}]. \qquad (6.7)$$

Substituting (6.6) and (6.7) into (6.5) completes the proof. ■

We return to the UFP. Recall that $z_{i,j}$ is the indicator for path $P_{i,j}$ being chosen. Define, for all (f, i, j) such that $f \in P_{i,j}$,

$$D_f(i,j) \doteq D_f - \rho_i z_{i,j} = \sum_{(r,s)\ne(i,j):\ f\in P_{r,s}} \rho_r z_{r,s}. \qquad (6.8)$$

Since $\gamma > 1$, there is a positive probability that in fact no path is chosen by the randomized rounding. Thus,

$$\Pr[\bigwedge_{f\in E} \overline{E_f}] > 0. \qquad (6.9)$$

For each $i \in [k]$, let Z_i be the indicator r.v. for choosing at least one (s_i, t_i)-path. Note next that all the events E_f and all the events $z_{i,j} = 1$, are *increasing* as a function of the underlying independent binary r.v.s. Thus, a consequence of Lemma 6.1 is

COROLLARY 6.1. *For any* $i \in [k]$, $\mathbf{E}[Z_i \mid \bigwedge_{f\in E} \overline{E_f}]$ *is lower-bounded by*

$$\Big(\sum_j \Pr[z_{i,j} = 1] \cdot (1 - \sum_{f\in P_{i,j}} \Pr[D_f(i,j) > (c_f - \rho_i)])\Big) - \sum_{j<j'} \Pr[z_{i,j} = z_{i,j'} = 1].$$

Suppose for now that we have an MDP instance, i.e., that all the ρ_i and c_f are one. Then, for any (f, i, j),

$$\Pr[D_f(i,j) > (c_f - \rho_i)] = \Pr[D_f(i,j) > 0] = \Pr[D_f(i,j) \ge 1] \le 1/\gamma, \qquad (6.10)$$

the inequality following by an application of Markov's inequality to (6.4). Since $\sum_j z^*_{i,j} = x^*_i/\gamma$ for each i, we have

$$\sum_{j<j'} \Pr[z_{i,j} = z_{i,j'} = 1] \leq (x^*_i)^2/(2\gamma^2) \leq x^*_i/(2\gamma^2). \tag{6.11}$$

Recall that any path $P_{i,j}$ uses at most d edges. Thus, choosing $\gamma = 2d$, substituting (6.10) and (6.11) into Corollary 6.1 and applying linearity of expectation, we get

$$\begin{aligned} \mathbf{E}[(\sum_{i\in[k]} w_i Z_i) \mid \bigwedge_{f\in E} \overline{E_f}] &\geq \sum_{i\in[k]} w_i(x^*_i/\gamma)(1 - d/\gamma - 1/(2\gamma)) \\ &\geq \sum_{i\in[k]} w_i(x^*_i/\gamma)(1 - 1/2 - 1/8) \\ &= \sum_{i\in[k]} 3w_i x^*_i/(16d) \\ &= 3y^*/(16d). \end{aligned}$$

In conjunction with (6.9), this shows that there exists a feasible solution of value $\Omega(y^*/d)$.

For the more general UFP, the idea of [**BaS**] is as follows. For any $S \subseteq [k]$ and any (f, i, j), define

$$D_f(S, i, j) \doteq \sum_{r\in S,\ (r,s)\neq(i,j):\ f\in P_{r,s}} \rho_r z_{r,s}.$$

In [**BaS**], a suitable set $S \subseteq [k]$ is chosen so that:

(a) $\sum_{i\in S} w_i x^*_i \geq y^*/2$, and such that
(b) for all $i \in S$ and for all (f, j), $\Pr[D_f(S, i, j) > (c_f - \rho_i)] \leq O(1/\gamma)$.

All pairs (s_i, t_i) such that $i \notin S$ are removed from consideration, i.e., we choose to not connect them. Part (a) shows that this does not bring us much loss. Choosing $\gamma = \Theta(d)$, using part (b) and arguing as we did for the MDP above, we can show once again that there is a feasible solution of value $\Omega(y^*/d)$. Furthermore, this can be made constructive, i.e., turned into a (deterministic) polynomial-time algorithm [**BaS**].

THEOREM 6.2. ([**KS2, BaS**]) *For any* UFP *instance, a feasible solution of value $\Omega(y^*/d)$ can be computed in deterministic polynomial time.*

In addition to others, one reason for parametrization via d is as follows. Recall the parameters Δ and β of G, introduced in Section 3.2. Suppose all capacities are the same in our UFP instance. Then, a result of [**KR**] implies that we can take d to be $O(\Delta^2\beta^{-2}\log^3 n)$ with just a small constant-factor loss in the LP optimum. Thus, for graphs of low degree and good edge-expansion, good approximations are possible for the uniform-capacity version of the UFP: we see yet another utility of multicommodity flow using short paths.

Also, a class of integer programs (*column-restricted packing integer programs*) related to packing problems and to the UFP have been proposed and analyzed in [**KS2**]. See [**KS2, BaS**] for approximation algorithms for this class, and for applications to certain families of packing integer programs.

6.2. Minimizing congestion in single-source unsplittable flow. Suppose, as in Section 6.1, that we are given a network G with capacities and a collection T of pairs with demands, where each demand needs to be routed through a single path. Recalling the basic problem considered in Section 3, a natural problem that is related to the UFP when we do need to route a path for each (s_i, t_i), is to minimize the *relative congestion*: the maximum value, over all edges f, of the total demand using f divided by c_f. In the NP-hard *single-source* case where all the s_i are the same, *constant-factor* approximation algorithms for this and related problems were developed by Kleinberg [**Kl2**]. Many of these algorithms were simplified and the (constant) approximation guarantees improved, in subsequent work of Kolliopoulos & Stein [**KS1**]. Let s be the common source; we can assume w.l.o.g. that all sinks t_i are different. Consider the problem of minimizing relative congestion. A natural multi-commodity flow relaxation is to send ρ_i flow from s to t_i for each i, in such a way that the fractional relative congestion (the maximum value, over all edges f, of the total flow on f divided by c_f) is minimized. We now describe a special case of a deterministic rounding scheme employed in [**KS1**] to round this.

One of the key ideas will be the following well-known fact.

PROPOSITION 6.1. ([**Kl2, KS1**]) *Fix a single-source* UFP *instance. Suppose all demands were the same, and that we have a feasible solution to the multicommodity flow relaxation in which for all edges f, the total demand using f is upper-bounded by some D_f that is an integral multiple of the common demand. Then, we can efficiently find an* integral *solution (i.e., an unsplittable routing) in which the total demand using any edge f is at most D_f.*

This suggests grouping the sinks t_i to their according demands ρ_i, and handling each group separately. The grouping is done in a suitable way so that there is only a constant-factor loss in the objective function value.

Given the underlying network $G = (V, E)$, we let $n = |V|$ and $m = |E|$. We start with a preliminary lemma, the first part of which handles the case where all demands are "very small"; the second part formalizes the "grouping by demands" idea.

LEMMA 6.3. ([**KS1**]) *Suppose we are given a single-source* UFP *instance with arbitrary capacities.*
(i) If all demands lie in $[0, \rho)$, then we can efficiently find an unsplittable flow in which the demand using any edge is at most $n\rho$.
(ii) Suppose all demands lie in $(\rho, 2\rho]$, and that we have a feasible solution **F** *to the multicommodity flow relaxation in which the total flow using any edge $f \in E$ is D_f. Then, we can efficiently find an unsplittable flow in which the demand using any $f \in E$ is at most $2D_f + 2\rho$.*

PROOF. For (i), we route the demand along an arbitrary (s, t_i)-path for each i.

For (ii), let D'_f be the smallest multiple of 2ρ that is no smaller than $2D_f$, for each $f \in E$; note that $D'_f \leq 2D_f + 2\rho$. Round up all demands to 2ρ. Multiplying all flow values in **F** by at most 2, we can ensure that there is a flow of 2ρ from s to each sink. Furthermore, the new flow value on any $f \in E$ is at most D'_f. Now apply Proposition 6.1. ∎

Using (6.1), we assume w.l.o.g. that $\max_i \rho_i \leq 1$ and $\min_f c_f \geq 1$. The algorithm starts by solving the multicommodity flow relaxation to optimality; let $\mathbf{F}^*$ be the optimal solution obtained, in which the total flow using any $f \in E$ is D_f^*. Do a flow decomposition on $\mathbf{F}^*$, and let $\{P_{i,j} : j \in [\ell_i]\}$ be the set of (s, t_i)-paths obtained, with path $P_{i,j}$ carrying a flow of $b_{i,j}$. We will just consider a special case here where the optimal fractional relative congestion (i.e., the relative congestion in the flow $\mathbf{F}^*$) is at least 1.

THEOREM 6.4. (**[KS1]**) *Suppose the fractional relative congestion y^* of $\mathbf{F}^*$ is at least* 1. *Then for any desired fixed $a > 0$, we can efficiently round $\mathbf{F}^*$ in such a way that the relative congestion is at most $(4 + n^{-a})y^*$.*

PROOF. Let $\lambda = 1 + \lceil (a+1) \log n \rceil$. Partition the sinks t_i into λ groups, depending on which of the intervals $(0, 1/2^{\lambda-1}]$, $(1/2^{\lambda-1}, 1/2^{\lambda-2}], \ldots, (1/4, 1/2]$, $(1/2, 1]$ their demands lie in; the sets of sinks with demands in the above sequence of intervals will be denoted $S_1, S_2, \ldots, S_\lambda$ respectively. We create one distinct copy G_i of the graph G for each S_i; in G_i, only the sinks that lie in S_i will be considered. Crucially, we will let the capacity of each edge f in G_i to be

$$c_f^i \doteq \sum_{(\ell,j):\ t_\ell \in S_i,\ f \in P_{\ell,j}} b_{\ell,j}.$$

That is, we set the capacity of f in G_i to be the total amount of demand imposed on f by the sinks in S_i, in the fractional optimum $\mathbf{F}^*$. It is in this sense that the relaxation provides us crucial information here.

In G_1, we run the algorithm implied by part (i) of Lemma 6.3; for each of $G_2, G_3, \ldots, G_\lambda$, we separately run the method of part (ii) of Lemma 6.3. The final routing is given by the union of these routings. What is the relative congestion that we end up with? Fix $f \in E$. By Lemma 6.3, we see that the total demand using it is at most

$$\begin{aligned} n/2^{\lambda-1} + \sum_{i=2}^{\lambda} 2c_f^i + \sum_{i=2}^{\lambda} 2^{-(\lambda-i)} &\leq 1/n^a + 2\sum_{i=2}^{\lambda} c_f^i + 2 \\ &= 1/n^a + 2D_f + 2. \end{aligned}$$

Combined with the facts that $y^* \geq 1$ and that $c_f \geq 1$ for all f, this can be used to show that the obtained (integral) relative congestion is at most $(4 + n^{-a})y^*$. ■

See [**Kl2, KS1**] for further improvements and related results. In particular, constant-factor approximation algorithms have been presented in these papers for the NP-hard problem of finding a maximum-cardinality realizable $T' \subseteq T$ in the single-source case.

We briefly remark that many of the routing problems considered so far have been related to circuit switching. Another basic mode of message transmission is *packet switching*, where the messages are divided into atomic packets that move through the network, getting queued when necessary. Via a deterministic approach (related to Theorem 3.6) for rounding a certain multicommodity flow relaxation and by using some seminal results of, e.g., Leighton, Maggs & Rao and Leighton, Maggs & Richa [**LMR1, LMR2**], a constant-factor approximation algorithm for a basic problem here has been presented by this author and Teo [**ST**].

6.3. Electric power networks and splittable flow. Via Menger's theorem, it is easy to decide if a given *single-source* version of the MDP is feasible. One generalization of the MDP, as seen above, is to the *unsplittable* case, where the reasonable assumption of (relatively) high capacities is usually made: no single demand exceeds any capacity (see (6.1)). A related natural generalization of the MDP is in the other direction, to the *low bandwidth* case: we have a network with fixed capacities that may have to route large demands when necessary. Individual demands are allowed to exceed capacities, and thus flows for some requests need to be *split* into different flow-paths. This is the framework for computational problems arising from, e.g., electrical power networks. We refer the reader to [**CF+**] for a discussion of this framework, and just present a sample result of [**CF+**] here to show how splittable flow problems exhibit some sharp differences from the "high bandwidth" case (the UFP).

Suppose we have a capacitated network $G = (V, E)$ with m capacitated edges, and a collection $T = \{(s_1, t_1), \ldots, (s_k, t_k)\}$ of request pairs with pair (s_i, t_i) having a demand ρ_i and weight $w_i \in [0, 1]$. The crucial difference from the UFP is that it is possible for some ρ_i to exceed all individual capacities c_f, thus necessitating splittable flows. We now say that $T' \subseteq T$ is realizable iff we can send an (s_i, t_i)-flow of value ρ_i for each $(s_i, t_i) \in T'$ in such a way that no edge's capacity is exceeded.

Consider the problem of choosing a maximum weight realizable $T' \subseteq T$. It is shown in [**CF+**] that even for the *single-source* version of this problem, there is no polynomial-time approximation algorithm with an approximation bound of $m^{1/2-\epsilon}$ for any fixed $\epsilon > 0$, unless $NP = ZPP$. This holds even if all capacities c_f and weights w_i are one, and all demands ρ_i are integral. Thus we see a sharp contrast with the corresponding single-source versions of the MDP and the UFP.

On the positive side, let OPT denote the optimum value of a given instance of this problem. The following bi-criteria result is shown in [**CF+**]. For any given parameter $\delta \in (0, 1)$, we can efficiently construct a flow in which the total weight of all the (s_i, t_i) for which at least $(1 - \delta)$ of their demand is satisfied, is at least: (a) $\Omega(\delta \cdot OPT^2/m)$ if $\delta \leq 1/2$, and (b) $\Omega(OPT \cdot (OPT/m)^{(1-\delta)/\delta})$ if $\delta > 1/2$. Note that if δ is "small", say 0.1, then we almost satisfy the demands of a fair number of the request pairs, while remaining close to the $m^{1/2-\epsilon}$-hardness-of-approximation result.

This result is accomplished once again via an appropriate multicommodity flow relaxation followed by randomized rounding; a key difference from before is the addition of some extra *valid constraints* to the relaxation. Correlation inequalities (the FKG inequality) are once again useful in the analysis.

7. Conclusion

We have seen a family of approximation results for various network optimization problems, where the combination of multicommodity flow relaxation followed by randomized rounding has been fruitful. Two main themes here have been the use of *correlation inequalities* in the analysis and design of such algorithms, and the utility of multi-commodity flow on *short* paths for various algorithmic problems.

The development of further tools for the design and analysis of various relevant random processes will be a welcome addition to the active field surveyed here. We have analyzed rounding methods for various *linear* relaxations; general approaches to design and analyze various (randomized) rounding methods for important non-linear relaxations such as semidefinite programming relaxations, would also be of much value.

References

[AB+] A. Aggarwal, A. Bar-Noy, D. Coppersmith, R. Ramaswami, B. Schieber, and M. Sudan, *Efficient routing in optical networks*, Journal of the ACM **46** (1996), 973–1001.

[AMO] R. K. Ahuja, T. L. Magnanti, and J. B. Orlin, *Network flows: theory, algorithms, and applications*, Prentice Hall, Englewood Cliffs, New Jersey, 1993.

[Al] N. Alon, *A parallel algorithmic version of the Local Lemma*, Random Structures & Algorithms **2** (1991), 367–378.

[AC] N. Alon and F. R. K. Chung, *Explicit construction of linear sized tolerant networks*, Discrete Mathematics **72** (1989), 15–19.

[AS] N. Alon and J. H. Spencer, *The Probabilistic Method*, Wiley–Interscience Series, John Wiley & Sons, Inc., New York, 1992.

[AR] Y. Aumann and Y. Rabani, *Improved bounds for all-optical routing*, Proc. ACM-SIAM Symposium on Discrete Algorithms, 1995, pp. 567–576.

[AL] B. Awerbuch and F. T. Leighton, *Improved approximations for the multi-commodity flow problem and local competitive routing in networks*, Proc. ACM Symposium on Theory of Computing, 1994, pp. 489–496.

[Ba1] Y. Bartal, *Probabilistic approximation of metric spaces and its algorithmic applications*, Proc. IEEE Symposium on Foundations of Computer Science, 1996, pp. 184–193.

[Ba2] Y. Bartal, *On approximating arbitrary metrics by tree metrics*, Proc. ACM Symposium on Theory of Computing, 1998.

[BaS] A. Baveja and A. Srinivasan, *Approximation algorithms for disjoint paths and related routing and packing problems*, Manuscript, January 1998 (submitted for publication).

[Be] J. Beck, *An algorithmic approach to the Lovász Local Lemma*, Random Structures & Algorithms **2** (1991), 343–365.

[BV] D. Bertsimas and R. Vohra, *Linear programming relaxations, approximation algorithms and randomization; a unified view of covering problems*, Technical Report OR 285–94, MIT, 1994.

[BoS] R. B. Boppana and J. H. Spencer, *A useful elementary correlation inequality*, Journal of Combinatorial Theory, Series A **50** (1989), 305–307.

[BFU1] A. Broder, A. Frieze and E. Upfal, *Existence and construction of edge-disjoint paths on expander graphs*, SIAM J. Comput. **23** (1994), 976–989.

[BFU2] A. Broder, A. Frieze and E. Upfal, *Static and dynamic path selection on expander graphs: a random walk approach*, Proc. ACM Symposium on Theory of Computing, 1997, pp. 531–539.

[CC+] M. Charikar, A. Chekuri, A. Goel, and S. Guha, *Rounding via trees: deterministic approximation algorithms for group Steiner trees and k-median*, Proc. ACM Symposium on Theory of Computing, 1998.

[Ch] H. Chernoff, *A measure of asymptotic efficiency for tests of a hypothesis based on the sum of observations*, Annals of Mathematical Statistics **23** (1952), 493–509.

[CF+] D. Cook, V. Faber, M. Marathe, A. Srinivasan and Y. J. Sussmann, *Low-bandwidth routing and electrical power networks*, Proc. International Colloquium on Automata, Languages, and Programming, 1998.

[EL] P. Erdős and L. Lovász, *Problems and results on 3-chromatic hypergraphs and some related questions*, in Infinite and Finite Sets, A. Hajnal et. al., editors, Colloq. Math. Soc. J. Bolyai 11, North Holland, Amsterdam, 1975, pp. 609–627.

[FKG] C. M. Fortuin, J. Ginibre, and P. N. Kasteleyn, *Correlational inequalities for partially ordered sets*, Communications of Mathematical Physics **22** (1971), 89–103.

[GKR] N. Garg, G. Konjevod and R. Ravi, *A polylogarithmic approximation algorithm for the Group Steiner Tree problem*, Proc. ACM-SIAM Symposium on Discrete Algorithms, 1998, pp. 253–259.

[GW] M. X. Goemans and D. P. Williamson, *The primal-dual method for approximation algorithms and its applications to network design problems*, in Approximation Algorithms for NP-Hard Problems, D. S. Hochbaum, editor, PWS Press, 1997.

[Hoc] D. S. Hochbaum, Editor, *Approximation algorithms for NP-hard problems*, PWS Press, 1997.

[Hoe] W. Hoeffding, *Probability inequalities for sums of bounded random variables*, American Statistical Association Journal **58** (1963), 13–30.

[JLR] S. Janson, T. Łuczak, and A. Ruciński, *An exponential bound for the probability of nonexistence of a specified subgraph in a random graph*, in Random Graphs '87 (M. Karoński, J. Jaworski and A. Ruciński, eds.), John Wiley & Sons, Chichester, 1990, pp. 73–87.

[KL+] R. M. Karp, F. T. Leighton, R. L. Rivest, C. D. Thompson, U. V. Vazirani, and V. V. Vazirani, *Global wire routing in two-dimensional arrays*, Algorithmica **2** (1987), 113–129.

[Kl1] J. Kleinberg, *Approximation algorithms for disjoint paths problems*, PhD Thesis, Department of EECS, MIT, 1996.

[Kl2] J. Kleinberg, *Single-source unsplittable flow*, in Proc. IEEE Symposium on Foundations of Computer Science, 1996, pp. 68–77.

[KR] J. Kleinberg and R. Rubinfeld, *Short paths in expander graphs*, in Proc. IEEE Symposium on Foundations of Computer Science, 1996, pp. 86–95.

[KT] J. Kleinberg and É. Tardos, *Disjoint paths in densely embedded networks*, in Proc. IEEE Symposium on Foundations of Computer Science, 1995, pp. 52–61.

[KS1] S. G. Kolliopoulos and C. Stein, *Improved approximation algorithms for unsplittable flow problems*, in Proc. IEEE Symposium on Foundations of Computer Science, 1997, pp. 426–435.

[KS2] S. G. Kolliopoulos and C. Stein, *Approximating disjoint-path problems using greedy algorithms and packing integer programs*, Technical Report PCS TR97-325, Department of Computer Science, Dartmouth College, 1997.

[Lei] F. T. Leighton, *Introduction to Parallel Algorithms and Architectures: Arrays • Trees • Hypercubes*, Morgan Kaufmann, San Mateo, California, 1992.

[LMR1] F. T. Leighton, B. Maggs, and S. Rao, *Packet routing and jobshop scheduling in* O(*congestion + dilation*) *steps*, Combinatorica **14** (1994), 167–186.

[LMR2] F. T. Leighton, B. Maggs, and A. Richa, *Fast algorithms for finding* O(*congestion + dilation*) *packet routing schedules*, Technical Report CMU-CS-96-152, School of Computer Science, Carnegie-Mellon University, 1996. To appear in Combinatorica.

[LM+] F. T. Leighton, F. Makedon, S. Plotkin, C. Stein, É. Tardos, and S. Tragoudas, *Fast approximation algorithms for multicommodity flow problems*, Journal of Computer and System Sciences **50** (1995), 228–243.

[LR1] F. T. Leighton and S. B. Rao, *An approximate max-flow min-cut theorem for uniform multicommodity flow problems with applications to approximation algorithms*, in Proc. IEEE Symposium on Foundations of Computer Science, 1988, pp. 422–431.

[LR2] F. T. Leighton and S. B. Rao, *Circuit switching: a multi-commodity flow based approach*, in Proc. Workshop on Randomized Parallel Computing, 1996.

[LRS] F. T. Leighton, S. B. Rao, and A. Srinivasan, *Multicommodity flow and circuit switching*, in Proc. Hawaii International Conference on System Sciences, 1998.

[LST] J. K. Lenstra, D. B. Shmoys, and É. Tardos, *Approximation algorithms for scheduling unrelated parallel machines*, Mathematical Programming **46** (1990), 259–271.

[LST] A. Lubotzky, R. Phillips and P. Sarnak, *Ramanujan graphs*, Combinatorica **8** (1988), 261–277.

[MNR] R. Motwani, J. (Seffi) Naor, and P. Raghavan, *Randomized approximation algorithms in combinatorial optimization*, in Approximation Algorithms for NP-Hard Problems, D. S. Hochbaum, editor, PWS Press, 1997.

[MR] R. Motwani and P. Raghavan, *Randomized Algorithms*, Cambridge University Press, 1995.

[Rab] Y. Rabani, *Path coloring on the mesh*, in Proc. IEEE Symposium on Foundations of Computer Science, 1996, pp. 400–409.

[Rag] P. Raghavan, *Probabilistic construction of deterministic algorithms: approximating packing integer programs*, Journal of Computer and System Sciences **37** (1988), 130–143.

[RT] P. Raghavan and C. D. Thompson, *Randomized rounding: a technique for provably good algorithms and algorithmic proofs*, Combinatorica **7** (1987), 365–374.

[RW] G. Reich and P. Widmayer, *Beyond Steiner's problem: a VLSI oriented generalization*, in Graph Theoretic Concepts of Computer Science (WG89), Lecture Notes in Computer Science **411**, Springer-Verlag, 1990, pp. 196–210.

[SSS] J. P. Schmidt, A. Siegel, and A. Srinivasan, *Chernoff-Hoeffding bounds for applications with limited independence*, SIAM J. Discrete Math. **8** (1995), 223–250.

[Sh1] D. B. Shmoys, *Computing near-optimal solutions to combinatorial optimization problems*, in *Combinatorial Optimization* (W. Cook, L. Lovász and P. D. Seymour, eds.), American Mathematical Society, 1995, pp. 355–397.

[Sh2] D. B. Shmoys, *Cut problems and their application to divide-and-conquer*, in Approximation Algorithms for NP-Hard Problems, D. S. Hochbaum, editor, PWS Press, 1997, pp. 192–235.

[Sp] J. H. Spencer, *Ten Lectures on the Probabilistic Method*, SIAM, Philadelphia, 1987. (Second edition published in 1994.)

[Sr1] A. Srinivasan, *An extension of the Lovász Local Lemma, and its applications to integer programming*, in Proc. ACM-SIAM Symposium on Discrete Algorithms, 1996, pp. 6–15.

[Sr2] A. Srinivasan, *Improved approximations for edge-disjoint paths, unsplittable flow, and related routing problems*, in Proc. IEEE Symposium on Foundations of Computer Science, 1997, pp. 416–425.

[ST] A. Srinivasan and C.-P. Teo, *A constant-factor approximation algorithm for packet routing, and balancing local vs. global criteria*, in Proc. ACM Symposium on Theory of Computing, 1997, pp. 636–643.

DEPT. OF INFORMATION SYSTEMS & COMPUTER SCIENCE, NATIONAL UNIVERSITY OF SINGAPORE, SINGAPORE 119260.

E-mail address: aravind@iscs.nus.edu.sg

DIMACS Series in Discrete Mathematics
and Theoretical Computer Science
Volume **43**, 1999

Some Remarks on the Optimal Level of Randomization in Global Optimization

Theodore V. Theodosopoulos

ABSTRACT. For a class of stochastic restart algorithms we address the effect of a nonzero level of randomization in maximizing the convergence rate for general energy landscapes. The resulting characterization of the optimal level of randomization is investigated computationally for random as well as parametric families of *rugged* energy landscapes.

1. Introduction

The question at the center of this short paper arose as a byproduct of the author's doctoral dissertation. In [**The95**] the author studied a class of stochastic restart algorithms for global optimization and developed upper and lower bounds to their asymptotic convergence. These algorithms were then tested against selected variants of the Simulated Annealing (SA) algorithm on a gamut of global optimization problems.

The first step in the analysis performed in [**The95**] was the martingale representation of the moment generating function of certain exit times of the Markov process describing the algorithm. Subsequently, this representation was used to establish asymptotic estimates of the Legendre transform of the moment generating functions in question which finally led to the large deviations bounds on the convergence rate.

The present paper deals with the study of the representation of the moment generating function. Aided by the specific representation, we investigate the dependence of the asymptotic convergence rate on the level of global mixing. This level of randomization is an explicit design parameter for the class of algorithms we describe.

The fundamental message in this paper is that a nonzero level of randomness often improves performance robustness in an unknown *rugged* landscape. The qualitative behavior of the convergence rate with varying levels of randomness is largely insensitive to detailed characteristics of the energy landscape. Thus, when faced with a global optimization problem for which we have limited knowledge of the

1991 *Mathematics Subject Classification.* Primary 49J55, 60G40; Secondary 90C30, 93E23.

energy landscape, a critical amount of *randomization by dcsign* is likely to maximize the expected convergence rate while maintaining a consistently competitive performance over a wide range of varied energy landscapes.

The first section of the paper reviews the problem setup and the main results from [**The95**]. An appendix includes all the relevent nomenclature from [**The95**] which is used throughout the paper. The second section develops the criterion for the existence of a nonzero level of randomness which maximizes the convergence rate. This optimal level of randomness is represented as the solution to a pair of polynomial equations whose order is an increasing function of the relative depth and steepness of the global minimum well and the deepest strictly local minimum well. The next section exhibits the dependence of the convergence rate on the level of randomness for a set of random energy landscapes as well as three parametric families of energy landscapes. Finally, we briefly contrast the findings of our study to results regarding parallel implementations of simulated annealing by Azencott et.al. in [**Aze92**].

2. Class of Algorithms and Convergence Rate Estimates

Let $f : \mathcal{X} \to \mathcal{R}$ be a bounded, real-valued function on a discrete set $\mathcal{X}$ (the analysis in the paper applies irrespective of the finiteness of $\mathcal{X}$ but we choose to concentrate the discussion in this paper to the finite case which offers an ample set of applications). Let's assume that $\mathcal{X}$ is equipped with a probability measure $\mu \in \mathcal{M}_1(\mathcal{X})$ and a neighborhood structure $\{\mathcal{N}(x) \subseteq \mathcal{X},\, x \in \mathcal{X}\}$. The problem is to locate the set

$$\arg\min_{x\in\mathcal{X}} f(x) \triangleq \{y \in \mathcal{X} :\ f(y) \le f(x),\, \forall x \in \mathcal{X}\}\,.$$

Unless otherwise noted, we will assume from now on that $\min_{x\in\mathcal{X}} f(x) = 0$. With this in mind, our problem can be rephrased as searching for $f^{-1}(0)$ or its ϵ-approximation $\mathcal{L}(\epsilon)$.

The family of stochastic restart algorithms we study is denoted by $\mathcal{A}$ and comprises of Markov processes on $\mathcal{X}$ with generators of the general form

$$[\mathcal{G}\phi](x) \triangleq p\mathbf{1}_A(x)\phi(\mathcal{D}f(x)) + (1 - p\mathbf{1}_A(x))\,\mathbf{E}^{\mu}[\phi] - \phi(x), \tag{2.1}$$

for every $\phi : \mathcal{X} \to \mathcal{R}$, where $p \in [0,1]$, $A \subseteq \mathcal{X}$ and $\mathcal{D}f(x) = \arg\min_{y\in\mathcal{N}(x)} f(y)$. We will say that the algorithm is generated by $\mathcal{G}$. From now on we will assume that $\mathcal{D}f(\cdot)$ is a well-defined mapping of $\mathcal{X}$ to itself; when

$$\text{card}\left(\arg\min_{y\in\mathcal{N}(x)} f(y)\right) > 1$$

we will implicitly assume that a deterministic choice is made.

Let $\{X_k,\, k \ge 0\}$ be the stochastic process generated by $\mathcal{G}$. For any $\epsilon \in f(\mathcal{X})$, let

$$\tau(\epsilon) = \inf\{k \ge 0 : X_k \in \mathcal{L}(\epsilon)\}\,.$$

We are interested in evaluating the following performance measure for the class $\mathcal{A}$ of algorithms defined above:

$$\mathcal{C}(\mathcal{G}, \epsilon, x) \triangleq \lim_{N\to\infty} \frac{1}{N} \log \mathrm{P}_x(\tau(\epsilon) > N),$$

where $P_x \in \mathcal{M}_1(\mathcal{X})$ is the measure induced by the process when $X_0 = x$. We see immediately that when $x \in \mathcal{X}_2$, $p = 1$ and $N \geq d(x)$ we clearly have $P_x(\tau(\epsilon) > N) = 0$ and so $\mathcal{C}(\mathcal{G}, \epsilon, x)$ is not defined in that case.

The main results in [**The95**] are summarized in the following theorem:

THEOREM 2.1. *Fix $\epsilon \in f(\mathcal{X})$. Let*

$$\begin{aligned} Q(\xi,p) \quad &\triangleq \quad 1 - \frac{(1-p)e^{\xi}}{1-pe^{\xi}} \sum_{j=0}^{a} p_1(j) - e^{\xi} \sum_{j=0}^{a} p_2(j) p^j e^{j\xi} \\ &\quad -(1-p)e^{\xi} \sum_{j=0}^{a \vee b} (q(j) + p_2(j)) \left(\sum_{i=0}^{j-1} p^i e^{i\xi} \right), \end{aligned} \tag{2.2}$$

where we use the convention that $\sum_{i=0}^{c} a_i = 0$ when $c < 0$. Then the following statements hold:

1. *When $\mathcal{X}_3 = \emptyset$, $Q(\cdot, p)$ has a unique positive root for all $p \in [0,1]$. When $\mathcal{X}_3 \neq \emptyset$, $Q(\cdot, p)$ has a unique root in $(0, -\log p)$ for all $p \in [0,1)$. Let the unique root defined above be denoted by $\xi_{\mathrm{crit}}(p)$.*
2. *Let $f(y) = ye^{1-y}$. The following set of equations*

$$\left\{ \begin{array}{l} \gamma f(\gamma) + \frac{1}{\alpha} f\left(\frac{1}{\alpha}\right) = 1 \\ f(\gamma) + f\left(\frac{1}{\alpha}\right) = 1 \end{array} \right. . \tag{2.3}$$

has a unique solution in $(1, \infty)^2$. Let (α^*, γ^*) be this unique solution of (2.3). Then

$$-\alpha^* \gamma^* \xi_{\mathrm{crit}} \leq \mathbf{E}^{\mu} \left[\mathcal{C}(\mathcal{G}, \epsilon, x) \right] \leq -\xi_{\mathrm{crit}} \tag{2.4}$$

A numerical evaluation of (2.3) leads to the approximation $\alpha^* \gamma^* \approx 8$.

The proof of Theorem 2.1 consists of four steps as presented in [**The95**]:

(i) We formulate a Dirichlet problem for $\mathcal{G}$ on $\overline{B(\epsilon)}$ whose solution will provide a martingale representation of the moment-generating function for the distribution of $\tau(\epsilon)$, i.e. $\mathbf{E}^x \left[e^{\xi \tau(\epsilon)} \right]$, where $\mathbf{E}^x$ is shorthand for $\mathbf{E}^{P_x}$.

(ii) We solve the above Dirichlet problem. The martingale representation we obtain is valid for $\xi < \xi_{\mathrm{crit}}$ where ξ_{crit} is the unique root defined in Theorem 2.1 above. Specifically, we have $\mathbf{E}^x \left[e^{\xi \tau(\epsilon)} \right] = P(\xi, p) / Q(\xi, p)$, where $P(\cdot, p)$ is another polynomial in e^{ξ}.

(iii) The idea is to use Cramér's large deviations theorem (as described on pp. 22–31 of [**Str93**]) in order to estimate the tails of the distribution of $\tau(\epsilon)$. As a first step in that direction we have to estimate the Legendre transform $I_x(y)$ of $\log \left(\mathbf{E}^x \left[e^{\xi \tau(\epsilon)} \right] \right)$. In [**The95**] we prove that

$$I_x(y) = y \Xi_x(y) - \log \left(\mathbf{E}^x \left[e^{\Xi_x(y) \tau(\epsilon)} \right] \right),$$

where

$$\lim_{y \to \infty} \Xi_x(y) = \xi_{\mathrm{crit}}, \ \forall x \in \mathcal{X}.$$

Actually, the following rate estimate holds:

LEMMA 2.2. *For every $x \in \mathcal{X}$, there exists a positive constant $c(x)$ such that*

$$\Xi_x(y) = \xi_{\mathrm{crit}} - \frac{c(x)}{y} + \mathcal{O}\left(\frac{1}{y^2} \right).$$

Using the upper bound in Cramér's theorem we obtain the easy direction of Theorem 2.1; namely, for each $x \in \mathcal{X}$,

$$\mathcal{C}(\mathcal{G}, \epsilon, x) \leq -\xi_{\text{crit}}.$$

(iv) In order to obtain the lower bound in Theorem 2.1 we need to strengthen Cramér's lower bound. In particular, instead of using a variance estimate in conjunction with Chebyshev's inequality, we apply Cramér's upper bound to the appropriately normalized random variable. Even though Cramér's upper bound uses Chebyshev's inequality, it proves to be stronger because of the maximization involved. This difference is sufficient to provide us with the desired result.

3. Dependence of $\xi_{\text{crit}}(\mathcal{G})$ on p

The equation defining the critical exponent $\xi_{\text{crit}}(\mathcal{G})$ can be rewritten as:

$$\sum_{j=0}^{(a\vee b)+2} e^{j\xi_{\text{crit}}} \sum_{i=0}^{j} c_{ji} p^i = 0 \tag{3.1}$$

where c_{ji} are constants which depend only on $p_1(\cdot)$, $p_2(\cdot)$ and $q(\cdot)$. Differentiating (3.1) with respect to p we obtain

$$\sum_{j=0}^{(a\vee b)+2} e^{j\xi_{\text{crit}}} \sum_{i=0}^{j} c_{ji} i \,(p^*)^i = 0. \tag{3.2}$$

Changing the order of summation makes (3.2) into a polynomial in p^*:

$$\sum_{i=0}^{(a\vee b)+2} (p^*)^i \, i \sum_{j=i}^{(a\vee b)+2} c_{ji} e^{j\xi_{\text{crit}}} = 0. \tag{3.3}$$

Thus we can solve (3.1) and (3.3) simultaneously for $\xi_{\text{crit}}(\mathcal{G})$ and p^*.

Let $\mathcal{G}_1$ denote the algorithm obtained by (2.1) when $p = 1$ and $A = \mathcal{X} \backslash \mathcal{F}(\mathcal{D}f)$. Using (2.2) we see that in this case (3.1) becomes (since $\xi_{\text{crit}} > 0$)

$$Q_1(\xi) \triangleq 1 - e^{\xi} \sum_{j=0}^{a} p_2(j) e^{j\xi} = 0. \tag{3.4}$$

This case corresponds to the minimum randomization that still guarantees asymptotic convergence to $\mathcal{L}(\epsilon)$.

On the other hand let $\mathcal{A}_2$ denote the subclass of algorithms obtained by (2.1) when $A = \mathcal{X}$. This condition implies that $p_2(\cdot) \equiv 0$. Using (2.2) in this case we see that (3.1) simplifies to

$$Q_2(\xi, p) \triangleq \frac{1 - e^{\xi} + (1-p)e^{\xi} \sum_{j=0}^{b} q(j) p^j e^{j\xi}}{1 - pe^{\xi}} = 0. \tag{3.5}$$

The algorithms in $\mathcal{A}_2$ choose between a steepest descent step and a global jump with fixed probability p irrespective of current location. In order to guarrantee asymptotic convergence to $\mathcal{L}(\epsilon)$ we must restrict $\mathcal{A}_2$ to have $p < 1$.

We will show when $\mathcal{A}_2$ is preferred to $\mathcal{G}_1$ and for which p.

LEMMA 3.1. *Let*

$$p^* = \arg \sup_{p \in [0,1), \mathcal{G} \in \mathcal{A}_2} \xi_{\text{crit}}(\mathcal{G}).$$

If $q(1) - q(0)(1 - q(0)) > 0$*, then,* $p^* \in (0,1)$ *and it solves equation (3.3) with* c_{ji} *corresponding to* $A = \mathcal{X}$*.*

PROOF. Let $\tilde{Q}_2 = \left(1 - pe^{\xi}\right) Q_2$. Differentiating $Q_2(\xi_{\text{crit}}) = 0$ with respect to p we obtain

$$\frac{d\xi_{\text{crit}}}{dp} = -\frac{\frac{\partial Q_2}{\partial p}}{\frac{\partial Q_2}{\partial \xi}}(\xi_{\text{crit}}(p), p) = -\frac{\frac{\partial \tilde{Q}_2}{\partial p}}{\frac{\partial \tilde{Q}_2}{\partial \xi}}(\xi_{\text{crit}}(p), p). \tag{3.6}$$

Using (3.5) we see that

$$\frac{\partial \tilde{Q}_2}{\partial p}(\xi, p) = \sum_{j=0}^{b} q(j) p^{j-1} e^{(j+1)\xi} \left[j - (1+j)p\right] \tag{3.7}$$

which implies that

$$\frac{\partial \tilde{Q}_2}{\partial p}(\xi, 0) = e^{\xi}\left(e^{\xi} q(1) - q(0)\right).$$

Evaluating the above equation at $\xi = \xi_{\text{crit}}(0)$ and noticing that

$$\xi_{\text{crit}}(0) = -\log\left(1 - q(0)\right)$$

we conclude that

$$\frac{\partial \tilde{Q}_2}{\partial p}(\xi_{\text{crit}}(0), 0) = \frac{q(1) - q(0)(1 - q(0))}{(1 - q(0))^2} > 0 \tag{3.8}$$

where we have used the assumption in the statement of the lemma. Furthermore, (3.7) implies that

$$\lim_{p \to 1} \frac{\partial \tilde{Q}_2}{\partial p}(\xi, p) = -\sum_{j=0}^{b} q(j) e^{(j+1)\xi} < 0. \tag{3.9}$$

At the same time differentiating (3.5) with respect to ξ we obtain

$$\frac{\partial \tilde{Q}_2}{\partial \xi}(\xi, p) = -e^{\xi} + (1-p)\sum_{j=0}^{b} q(j)(j+1) p^j e^{(j+1)\xi}$$

which implies that

$$\frac{\partial \tilde{Q}_2}{\partial \xi}(\xi, 0) = -e^{\xi}\left(1 - q(0)\right) < 0 \tag{3.10}$$

and

$$\lim_{p \to 1} \frac{\partial \tilde{Q}_2}{\partial \xi}(\xi, p) = -e^{\xi} < 0. \tag{3.11}$$

Using (3.8), (3.9), (3.10) and (3.11) we see that (3.6) implies

$$\frac{d\xi_{\text{crit}}}{dp}(0) > 0 \text{ and } \lim_{p \to 1} \frac{d\xi_{\text{crit}}}{dp}(p) < 0. \tag{3.12}$$

Lemma 3 on p.25 of [**The95**] tells us more generally that

LEMMA 3.2. $\frac{\partial Q}{\partial \xi} < 0$ *for all* $\xi \in [0, -\log p)$ *and* $p \in [0, 1)$*.*

From (3.6) we observe that $\left|\frac{d\xi_{\text{crit}}}{dp}\right| < \infty \Longleftrightarrow \frac{\partial Q_2}{\partial \xi} = 0$. As a consequence of Lemma 3.2 we conclude that $\xi_{\text{crit}}(p)$ is differentiable for all $p \in [0, 1)$ and therefore continuous. Then, (3.12) leads to the conclusion that there exists a $p^* \in (0, 1)$ which solves the equation $\frac{d\xi_{\text{crit}}}{dp} = 0$. In particular we can conclude that there is a $p^* \in (0, 1)$ which maximizes ξ_{crit}. Specializing (3.1) and (3.3) to this case we can represent

$$\left(\sup_{\mathcal{G}\in\mathcal{A}_2} \xi_{\text{crit}}(\mathcal{G}), p^*\right)$$

as a solution of the following set of equations:

$$\begin{cases} \sum_{j=0}^{b} q(j)p^{j-1}e^{(j+1)\xi}\left[j-(1+j)p\right] = 0 \\ 1 - e^{\xi} + (1-p)e^{\xi}\sum_{j=0}^{b} q(j)p^j e^{j\xi} = 0 \end{cases}.$$

□

The following theorem contains the main result of this paper:

THEOREM 3.3. *For any energy landscape which satisfies*

$$\sum_{j=0}^{a} \frac{p_2(j)}{(1-q(0))^{j+1}} \geq 1$$

there exists a nonzero level of randomness which optimizes the asymptotic convergence rate of algorithms in class $\mathcal{A}_2 \cup \{\mathcal{G}_1\}$, that is

$$\sup_{\mathcal{G}\in\mathcal{A}_2} \xi_{\text{crit}}(\mathcal{G}) \geq \xi_{\text{crit}}(\mathcal{G}_1).$$

PROOF. Let's consider the joint solvability of (3.4) and (3.5). Specifically, we obtain $\xi_{\text{crit}}(\mathcal{G}_1)$ by solving (3.4) and then we solve

$$Q_2\left(\xi_{\text{crit}}(\mathcal{G}_1), \cdot\right) = 0. \tag{3.13}$$

Let's assume that $\hat{p} \in (0, 1)$ exists which solves (3.13). In this case, let

$$\check{p} = \begin{cases} \hat{p}+\delta & \text{if } \frac{\partial Q_2}{\partial p}\left(\xi_{\text{crit}}(\mathcal{G}_1), \hat{p}\right) > 0 \\ \hat{p}-\delta & \text{if } \frac{\partial Q_2}{\partial p}\left(\xi_{\text{crit}}(\mathcal{G}_1), \hat{p}\right) < 0 \\ \hat{p} & \text{otherwise} \end{cases} \tag{3.14}$$

for some sufficiently small $\delta > 0$. Then, by construction, $Q_2\left(\xi_{\text{crit}}(\mathcal{G}_1), \check{p}\right) \geq 0$. Theorem 2.1 guarrantees the existence of $\xi_{\text{crit}}(\check{p})$ which solves $Q_2(\cdot, \check{p}) = 0$. Lemma 3.2 then implies that

$$\sup_{\mathcal{G}\in\mathcal{A}_2} \xi_{\text{crit}}(\mathcal{G}) \geq \xi_{\text{crit}}(\check{p}) > \xi_{\text{crit}}(\mathcal{G}_1).$$

Conversely, let's assume that there exists a $\check{p} \in (0, 1)$ such that $\xi_{\text{crit}}(\check{p}) > \xi_{\text{crit}}(\mathcal{G}_1)$. Then, Lemma 3.2 implies that $Q_2\left(\xi_{\text{crit}}(\mathcal{G}_1), \check{p}\right) > 0$. The question then becomes one of existence of a solution to

$$Q_2(\xi, \cdot) = 0 \tag{3.15}$$

in $\left[0, e^{-\xi}\right)$. From (3.5) we see that $Q_2(\xi, 0) = 1 - e^{\xi}(1 - q(0))$ and

$$\lim_{p \nearrow e^{-\xi}} Q_2(\xi, p) = -\infty.$$

So, when $\xi < -\log(1 - q(0))$, there is always a solution to (3.15).

There are three corner cases. First, if $\xi_{\text{crit}}(\mathcal{G}_1) = -\log(1-q(0))$, then, as we saw in the proof of Lemma 3.1, $\xi_{\text{crit}}(\mathcal{G}_1) = \xi_{\text{crit}}(0) \leq \sup_{\mathcal{G}\in\mathcal{A}_2} \xi_{\text{crit}}(\mathcal{G})$.

The second corner case is when $\hat{p} = 1$. This means that $\lim_{p\to 1} \xi_{\text{crit}}(p) = \xi_{\text{crit}}(\mathcal{G}_1)$. But Lemma 3.1 tells us that there exists a $p^* < 1$ such that, $\xi_{\text{crit}}(p^*) > \lim_{p\to 1} \xi_{\text{crit}}(p)$ and thus, $\sup_{\mathcal{G}\in\mathcal{A}_2} \xi_{\text{crit}}(\mathcal{G}) > \xi_{\text{crit}}(\mathcal{G}_1)$.

Finally, the case $\frac{\partial Q_2}{\partial p}(\xi_{\text{crit}}(\mathcal{G}_1), \hat{p}) = 0$ implies that $\xi_{\text{crit}}(\mathcal{G}_1) = \xi_{\text{crit}}(\hat{p}) \leq \sup_{\mathcal{G}\in\mathcal{A}_2} \xi_{\text{crit}}(\mathcal{G})$.

Therefore we have proved that

$$\xi_{\text{crit}}(\mathcal{G}_1) \leq -\log(1-q(0)) \tag{3.16}$$

is a sufficient condition for $\sup_{\mathcal{G}\in\mathcal{A}_2} \xi_{\text{crit}}(\mathcal{G}) \geq \xi_{\text{crit}}(\mathcal{G}_1)$. Plugging (3.16) into (3.4) we see that

$$\sum_{j=0}^{a} \frac{p_2(j)}{(1-q(0))^{j+1}} \geq 1 \Longrightarrow \sup_{\mathcal{G}\in\mathcal{A}_2} \xi_{\text{crit}}(\mathcal{G}) > \xi_{\text{crit}}(\mathcal{G}_1)$$

which completes the proof of the theorem. □

Let $p_2^{\min} \triangleq \min\{p_2(j) | j \in [0,a]\}$. Notice that $p_2^{\min} > 0$, and therefore

$$\sum_{j=0}^{a} \frac{p_2(j)}{(1-q(0))^{j+1}} \geq \frac{p_2^{\min}}{q(0)} \left(\frac{1}{(1-q(0))^{a+1}} - 1 \right).$$

Also, observe that the general case ($\mathcal{G} \in \mathcal{A}$) is always preferable to $\mathcal{A}_2$ and that

$$\lim_{p\searrow 0} (\mathcal{A} \setminus \mathcal{A}_2) = \lim_{p\nearrow 1} (\mathcal{A} \setminus \{\mathcal{G}_1\}) = \emptyset.$$

Finally, let

$$p_{\text{best}} \triangleq \arg \sup_{p\in[0,1],\mathcal{G}\in\mathcal{A}_2\cup\{\mathcal{G}_1\}} \xi_{\text{crit}}(\mathcal{G}).$$

The above discussion leads to the following corollary.

COROLLARY 3.4. *Any energy landscape with a sufficiently deep strictly local minimum has a nonzero level of randomness which optimizes the asymptotic convergence rate of any algorithm in $\mathcal{A}$. Specifically, for any fixed $q(0)$ and $p_2^{\min}$,*

$$a \geq -\frac{\log\left(1 + \frac{q(0)}{p_2^{\min}}\right)}{\log(1-q(0))} - 1$$

is sufficient to guarantee that $p_{\text{best}} \in (0,1)$. This further implies that

$$\arg \sup_{p\in[0,1],\mathcal{G}\in\mathcal{A}} \xi_{\text{crit}}(p) \in (0,1).$$

In practice, it is useful to note that the measure of depth used in Corollary 3.4 is a function of the discretezation level. Specifically, given any energy function f, increasing the desired level of accuracy for the determination of $\mathcal{L}(\epsilon)$ leads to an effective increase in the energy landscape parameters a and b. Thus, Corollary 3.4 assures us that a nonzero optimal level of randomization is ubiquitous in global optimization.

4. Dependence of p^* on the Energy Landscape

So far we have exhibited the dependence of the convergence rates in $\mathcal{A}$ on the level of *randomization by design* in the algorithm. We have seen that under a rather general condition on the energy landscape, there is a nonzero level of imposed randomization which maximizes the convergence rate, irrespective of the detailed structure of the energy landscape. In this section we focus on the dependence of the optimal level of randomization on global characteristics of the energy landscape. Specifically we are interested in the question of optimizing the design of the algorithm for particular energy landscapes.

The outcome of this investigation is a surprising lack of sensitivity in the qualitative characteristics of the way in which the optimal level of randomization varies across wide ranges of energy landscapes. Specifically we study four families of energy landscapes: random, exponential, polynomial, and logarithmic, where the latter three refer to the steepness of the energy wells in the landscape. Random landscapes were constructed by identifying $q(\cdot)$, $p_1(\cdot)$ and $p_2(\cdot)$ with appropriately normalized uniform random variables. On the other hand, the exponential, polynomial and logarithmic landscapes were constructing by assigning (after appropriate normalization) $q(\cdot), p_1(\cdot), p_2(\cdot) \sim \beta^j$ for $\beta \geq 1$, $q(\cdot), p_1(\cdot), p_2(\cdot) \sim j^\alpha$ for $\alpha > 0$ and $q(\cdot), p_1(\cdot), p_2(\cdot) \sim (\log j)^\gamma$ for $\gamma > 0$ respectively. The parameters β, α and γ quantify the *steepness* of the basins of attraction in the respective parametric family. Specifically, in all three cases, the basins of attraction become steeper as the relevent parameter *decreases* in value.

Another parameter that controls the geometry of the energy landscape is

$$c \triangleq \frac{\mu\left(\Gamma(\epsilon)\right)}{\mu\left(\mathcal{W}\left(\mathcal{L}(\epsilon)\right)\right)}.$$

All computational experiments reported below have been performed using the values $a = 20$, $b = 10$ and $c = 1000$.

Figure 1 illustrates the optimal level of randomization for a randomly generated energy landscape. We see that the optimal convergence rate can be significantly faster than the convergence rate corresponding to the minimum amount of randomization or to a randomly chosen level of randomization. We can also observe that there is a range of p around p^* for which $\mathcal{A}_2$ is preferable to $\mathcal{G}_1$. Outside this range, $\mathcal{G}_1$ is preferable to any member of $\mathcal{A}_2$.

Figure 2 illustrates the observed tradeoff when facing an unknown energy landscape. Specifically, what is required is a high expected convergence rate and at the same time a low variance for the convergence rate. This is the performance robustness problem. A Monte Carlo simulation was performed with 100 independent randomly generated landscapes. In Figure 2 we have suppressed the third dimension which represents the variation of p. As p increases from 0 to 1, we move along the curve in Figure 2 from the top right-hand corner to the bottom left-hand corner. Two *regimes* appear prominent:

- ▷ a *liquid* regime in which the variance of the convergence rate decreases steadily with decreasing levels of randomization while the expected convergence rate remains relatively constant, and
- ▷ a *solid* regime in which the expected convergence rate decreases rapidly over a very short range of randomization levels while the variance of the convergence rate remains largely unchanged.

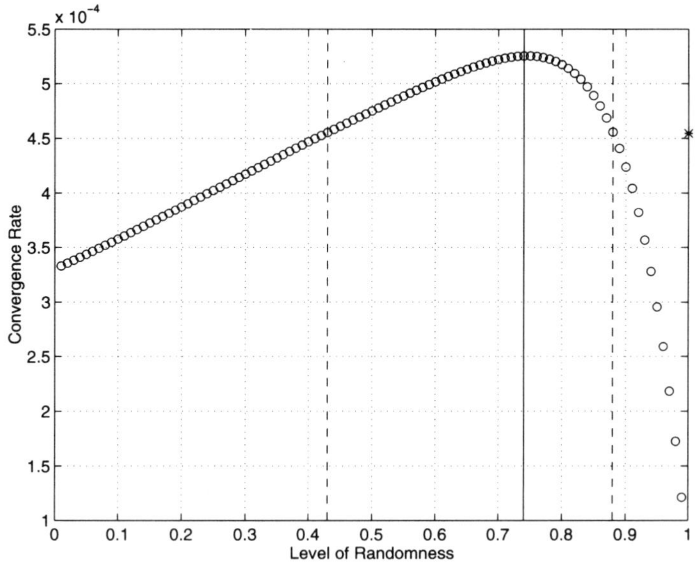

FIGURE 1. Convergence rate as a function of p for a randomly generated energy landscape

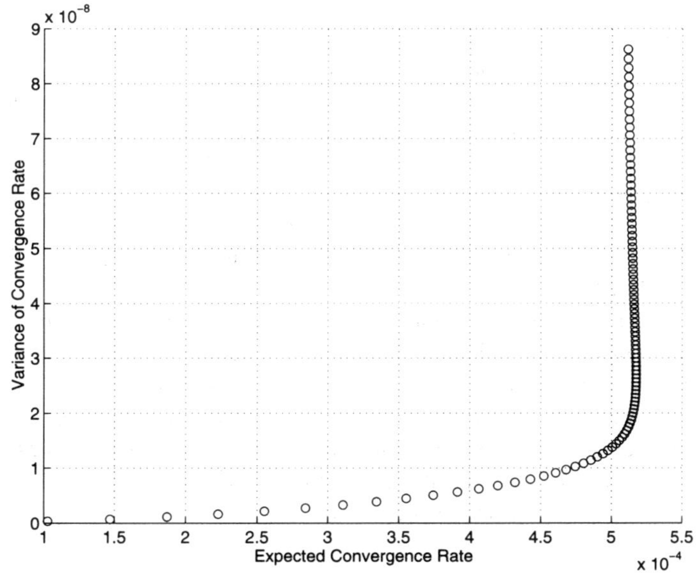

FIGURE 2. *Phase Transition* for the Algorithm Class $\mathcal{A}_2$

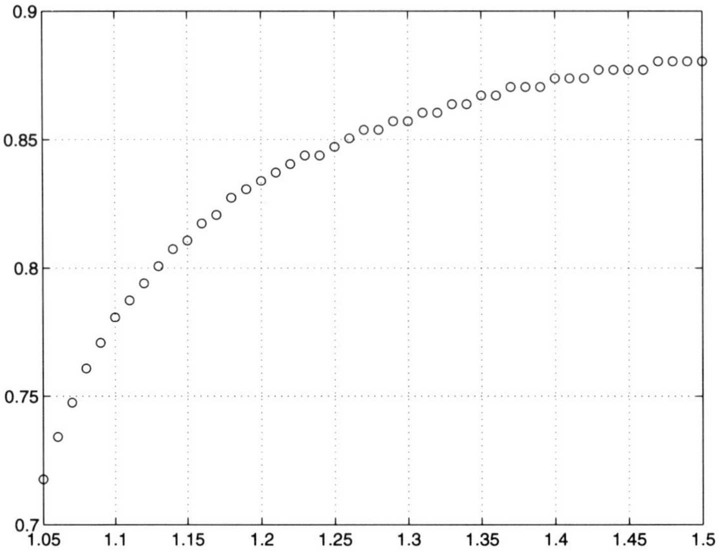

FIGURE 3. Dependence of p_{best} on β for exponential energy landscapes

We refer to this empirically observed phenomenon as a *Phase Transition.* There appears to exist a deeper relationship between the values of p that are poised between the two regimes and p^*. At this juncture there is limited computational evidence in support of this conjecture. This conjecture appears to be related to the identification of the *edge of chaos* in [**Kau95**] as well as to the critical level of parallelism investigated in [**MSK96**]. Our approach of examining the performance robustness tradeoff in a population of randomly generated energy landscapes is related to the *efficiency frontiers* described in [**HLH97**] and the discussion of *ensemble of landscapes* in [**Dit96**].

Figures 3, 4 and 5 show the way in which p_{best} varies with the parameters of exponential, polynomial and logarithmic energy landscapes respectively. We see that the qualitative characteristics in all three cases are indistinguishable.

Similarly, Figures 6, 7 and 8 illustrate the way in which $\xi_{\text{crit}}\,(p_{\text{best}})$ varies with the parameters of exponential, polynomial and logarithmic energy landscapes respectively. Once more, the qualitative behavior of the optimal convergence rate is indistinguishable between the three cases.

5. Conclusions

Using the methodology developed in [**The95**] we have studied the desirability of *randomization by design* to improve the convergence rate of global optimization algorithms. Theorem 3.3 describes a sufficient condition for the usefulness of such imposed randomness. This condition has been shown to hold generically for energy landscapes with sufficiently deep strictly local minima. We have also shown how to represent the optimal level of randomization as the solution to a pair of polynomial

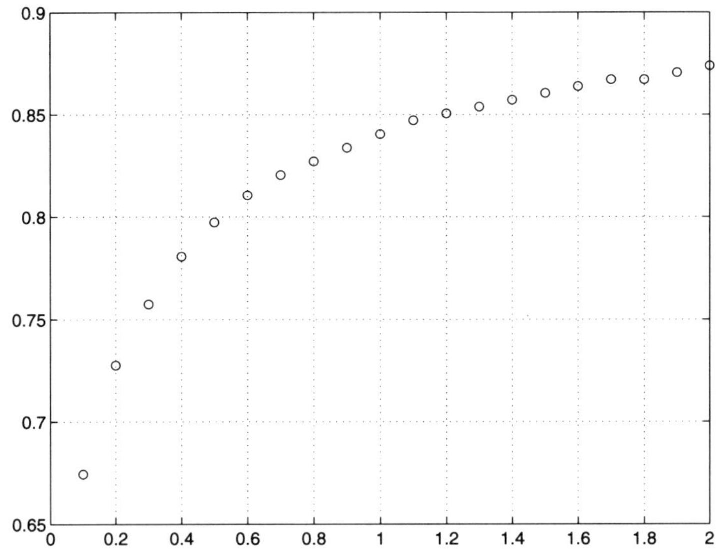

FIGURE 4. Dependence of p_{best} on α for polynomial energy landscapes

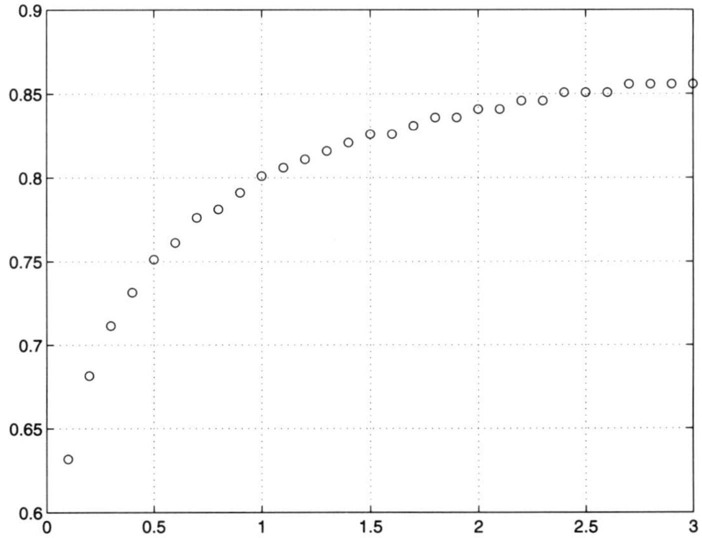

FIGURE 5. Dependence of p_{best} on γ for logarithmic energy landscapes

equations whose orders are related to the depths of the basins of attraction in the energy landscape in question.

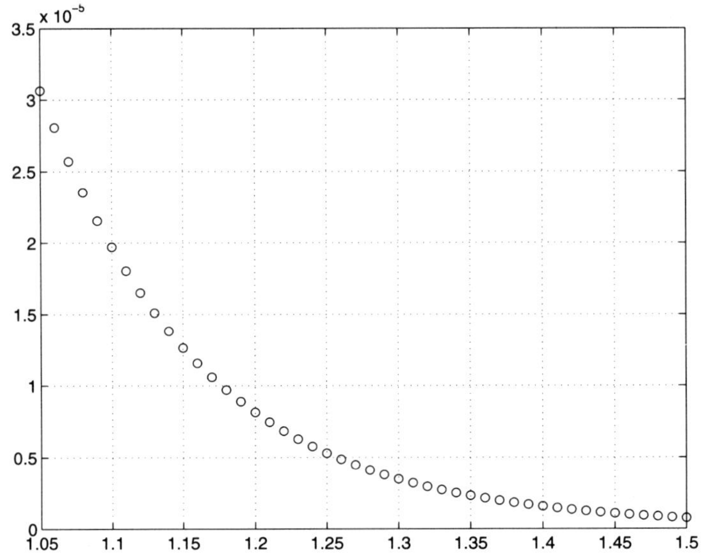

FIGURE 6. Dependence of $\xi_{\text{crit}}\,(p_{\text{best}})$ on β for exponential energy landscapes

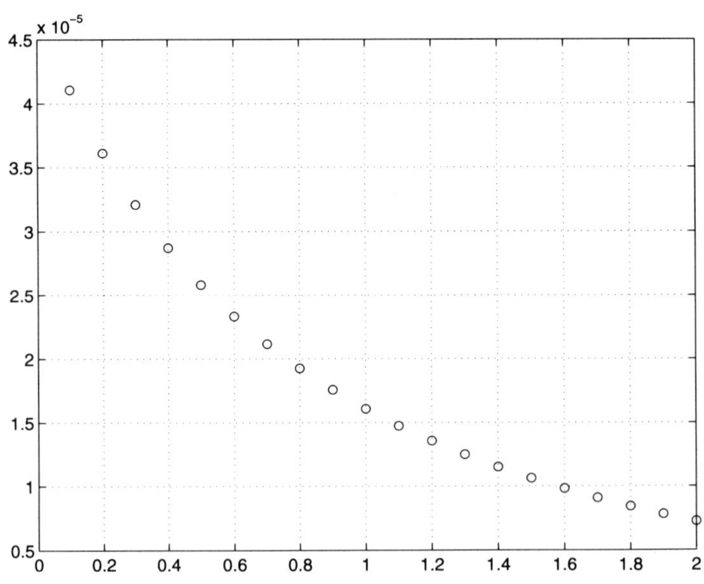

FIGURE 7. Dependence of $\xi_{\text{crit}}\,(p_{\text{best}})$ on α for polynomial energy landscapes

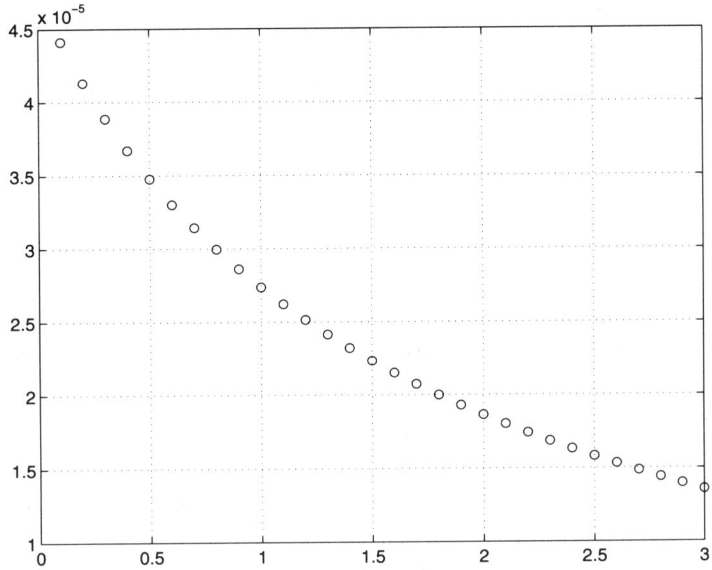

FIGURE 8. Dependence of $\xi_{\text{crit}}\,(p_{\text{best}})$ on γ for logarithmic energy landscapes

The study of randomly generated energy landscapes has led to the characterization of a *Phase Transition* associated with the performance robustness problem. Specifically, there is a narrow range of randomization levels which combine competitive expected convergence rates with minimal variance of that convergence rate. If we increase the level of randomization, we fall into a *liquid* phase which increases the variance of the convergence rate. If on the other hand we decrease the level of randomization, we fall into a *solid* state which entails a rapid deterioration of the expected converence rate in return for a modest further reduction in variance.

The investigation of the three parametric families of energy landscapes leads us to the following conclusions:

- ▷ A nonzero level of *randomization by design* is desirable in all cases.
- ▷ In all three cases, the optimal level of randomization is a monotonically increasing, convex function of the *steepness* of the basins of attraction (as captured by the parameters β, α and γ in the three families respectively).
- ▷ Similarly, in all three cases, the optimal convergence rate is a monotonically increasing, convex function of the *steepness* of the basins of attraction.
- ▷ The geometric characteristics of the optimal level of randomization as well as the resulting optimal convergence rate are largely insensitive to drastic variations in the geometry of the energy landscape.

This empirically observed robustness in the performance of appropriately randomized gradient descent algorithms is a desirable property for systems facing complex, largely unknown nonconvex energy landscapes.

It is worthwhile to provide a comparison of our conclusions to results from parallelization attempts for Simulated Annealing (SA). Specifically, if P_x is temporarily used to denote the measure in path space induced by sequential SA, then we know from [**Cat92**] that:

$$\lim_{N\to\infty} \inf_{\beta(\cdot)\nearrow} \sup_{x\in\mathcal{X}} \frac{1}{\log N} \log \mathrm{P}_x(f(X_N) > \epsilon) = -\frac{1}{D_f},$$

where

$$D_f = \max\left\{ \frac{H(x)}{f(x)} : x \in \mathcal{F}(\mathcal{D}f) \cap B(0) \right\}$$

and

$$H(x) = \max\{f(z) - f(y) : y \in \mathcal{W}(x), z \in \partial\mathcal{W}(x)\}.$$

In [**Aze92**], a variety of parallelization schemes are proposed for SA, all based on interacting multiple versions of the traditional, sequential SA. The convergence rate thus obtained becomes exponential with

$$\lim_{N\to\infty} \inf_{\beta(\cdot)\nearrow} \sup_{x\in\mathcal{X}} \frac{1}{N} \log \mathrm{P}_x(f(X_N) > \epsilon) = -\frac{1}{2eD_fK},$$

where K is the constant involved in the sequential SA convergence rate and P_x now refers to the path measure induced by the parallelized version of SA (see [**Aze92**]). Conceptually, the nonzero level of imposed randomness in the restart gradient descent algorithms discussed in this paper corresponds to non-monotonic annealing schedules in the context of SA. Furthermore, the Bernoulli restarts proposed here offer a generalization of the setup in [**Aze92**]. Finally, the optimal expected time between restarts is found to be independent of the overall time allowed, a property which is consistent with a constant Bernoulli success probability.

To recapitulate, the main findings of this research address the characterization of the desirabililty of a nonzero level of randomization. The optimized algorithm design is qualitatively invariant over a wide range of diverse energy landscapes. More work is required to develop a concrete understanding of the relationhip between the optimal level of imposed randomization and the range of p which strikes a balance between a competitive expected converegence rate and low variability for the convergece rate.

6. Appendix

This Appendix includes the notation used throughout the paper.

DEFINITION 6.1. Fix $\epsilon \in [0, \max_{x\in\mathcal{X}} f(x))$. Let

- Events up to time n

$$\mathcal{B}_n \triangleq \sigma(X_k, k \in [0, n]),$$

- Points at and below energy level ϵ

$$\mathcal{L}(\epsilon) \triangleq f^{-1}([0, \epsilon]),$$

- Points above energy level ϵ

$$B(\epsilon) \triangleq \mathcal{X} \setminus \mathcal{L}(\epsilon),$$

- Energy "well" (zone of attraction) of set A

$$\mathcal{W}(A) \triangleq \left\{ x \in \mathcal{X} : \lim_{k\to\infty} \left[\mathcal{D}^k f\right](x) \in A \right\},$$

- Points outside the "well" of $\mathcal{L}(\epsilon)$

$$\Gamma(\epsilon) \triangleq \mathcal{X} \setminus \mathcal{W}(\mathcal{L}(\epsilon)),$$

- Set of local minima of f

$$\mathcal{F}(\mathcal{D}f) \triangleq \{x \in \mathcal{X} : \mathcal{D}f(x) = x\},$$

- Number of gradient descent steps needed to go from x to its closest local minimum or $\mathcal{X} \setminus A$ (whichever is less)

$$d(x) \triangleq \min\left\{k \geq 0 : \left[\mathcal{D}^k f\right](x) \in (\mathcal{F}(\mathcal{D}f) \cup \mathcal{L}(\epsilon)) \cup (\mathcal{X} \setminus A)\right\},$$

- Mass assigned by μ to points in $d^{-1}(j)$ and inside $\mathcal{W}(\mathcal{L}(\epsilon))$

$$q(j) \triangleq \mu\left(\mathcal{W}(\mathcal{L}(\epsilon)) \cap d^{-1}(j)\right),$$

- Mass assigned by μ to points x in $d^{-1}(j)$, outside $\mathcal{W}(\mathcal{L}(\epsilon))$ and when gradient descent from x leads to a local minimum in A

$$p_1(j) \triangleq \mu\left(\Gamma(\epsilon) \cap d^{-1}(j) \cap \left[\mathcal{D}^j f\right]^{-1}(A)\right),$$

- Mass assigned by μ to points x in $d^{-1}(j)$, outside $\mathcal{W}(\mathcal{L}(\epsilon))$ and when gradient descent from x leads to $\mathcal{X} \setminus A$

$$p_2(j) \triangleq \mu\left(\Gamma(\epsilon) \cap d^{-1}(j) \cap \left[\mathcal{D}^j f\right]^{-1}(\mathcal{X} \setminus A)\right),$$

- Maximum $d(x)$ outside $\mathcal{W}(\mathcal{L}(\epsilon))$

$$a \triangleq \max\{j \geq 0 : p_1(j) \vee p_2(j) > 0\},$$

- Maximum $d(x)$ inside $\mathcal{W}(\mathcal{L}(\epsilon))$

$$b \triangleq \max\{j \geq 0 : q(j) > 0\}$$

From the above definitions, one notices a natural decomposition of $\mathcal{X}$ which we will use extensively:

$$\mathcal{X} = \mathcal{X}_1 \,\dot{\bigcup}\, \mathcal{X}_2 \,\dot{\bigcup}\, \mathcal{X}_3,$$

where

$$\begin{aligned}
\mathcal{X}_1 &\triangleq \left\{x \in \mathcal{X} : \left[\mathcal{D}^{d(x)} f\right](x) \in \mathcal{X} \setminus A\right\} \\
\mathcal{X}_2 &\triangleq \mathcal{W}(\mathcal{L}(\epsilon)) \\
\mathcal{X}_3 &\triangleq \left\{x \in \mathcal{X} : \left[\mathcal{D}^{d(x)} f\right](x) \in (A \cap B(\epsilon))\right\}.
\end{aligned}$$

Some clarification is due regarding the boundary and the closure of a discrete set. We use the following definitions:

DEFINITION 6.2. For any $A \subseteq f(\mathcal{X})$,

$$\partial f^{-1}(A) \triangleq (\mathcal{X} \setminus f^{-1}(A)) \bigcap_{x \in f^{-1}(A)} \mathcal{N}(x),$$

$$\overline{f^{-1}(A)} \triangleq f^{-1}(A) \cup \partial f^{-1}(A).$$

References

[Aze92] R. Azencott, ed., *Simulated annealing: Parallelization techniques*, John Wiley & Sons, Inc., 1992.

[Cat92] O. Catoni, *Rough large deviations estimates for simulated annealing. application to exponential schedules*, The Annals of Probability **20** (1992), 1109–1146.

[Dit96] Frank-Michael Dittes, *Optimization on rugged landscape: A new general purpose monte carlo approach*, Physical Review Letters **76** (1996), no. 25, 4651–4655.

[HLH97] Bernardo A. Huberman, Rajan M. Lukose, and Tad Hogg, *An economic approach to hard computational problems*, Science **275** (1997), 51–54.

[Kau95] Stuart Kauffman, *At home in the universe: The search for laws of self-organization and complexity*, Oxford University Press, 1995.

[MSK96] William G. Macready, Athanassios G. Siapas, and Stuart A. Kauffman, *Criticality and parallelism in combinatorial optimization*, Science **271** (1996), 56–59.

[Str93] D.W. Stroock, *Probability theory*, Cambridge University Press, 1993.

[The95] T.V. Theodosopoulos, *Stochastic models for global optimization*, Ph.D. thesis, MIT, May 1995.

Market Risk Management, BankBoston Corporation. Address: 01-11-08, 100 Federal Street, Boston, MA 02110

E-mail address: tvtheodosopoulos@bkb.com

Selected Titles in This Series

(Continued from the front of this publication)

14 **Robert Calderbank, G. David Forney, Jr., and Nader Moayeri, Editors,** Coding and Quantization: DIMACS/IEEE Workshop

13 **Jin-Yi Cai, Editor,** Advances in Computational Complexity Theory

12 **David S. Johnson and Catherine C. McGeoch, Editors,** Network Flows and Matching: First DIMACS Implementation Challenge

11 **Larry Finkelstein and William M. Kantor, Editors,** Groups and Computation

10 **Joel Friedman, Editor,** Expanding Graphs

9 **William T. Trotter, Editor,** Planar Graphs

8 **Simon Gindikin, Editor,** Mathematical Methods of Analysis of Biopolymer Sequences

7 **Lyle A. McGeoch and Daniel D. Sleator, Editors,** On-Line Algorithms

6 **Jacob E. Goodman, Richard Pollack, and William Steiger, Editors,** Discrete and Computational Geometry: Papers from the DIMACS Special Year

5 **Frank Hwang, Fred Roberts, and Clyde Monma, Editors,** Reliability of Computer and Communication Networks

4 **Peter Gritzmann and Bernd Sturmfels, Editors,** Applied Geometry and Discrete Mathematics, The Victor Klee Festschrift

3 **E. M. Clarke and R. P. Kurshan, Editors,** Computer-Aided Verification '90

2 **Joan Feigenbaum and Michael Merritt, Editors,** Distributed Computing and Cryptography

1 **William Cook and Paul D. Seymour, Editors,** Polyhedral Combinatorics